The Origin of Mountains

The Origin of Mountains is **ground breaking**. This highly illustrated book describes mountains from all over the world, emphasising their landforms, their rocks, and their structure and age. This leads to a deduction on the mechanism that formed them, causing the authors to reject the preconceived well-known hypothesis that plate tectonics and folding creates mountains.

The Origins of Mountains approaches mountains from facts about mountain landscape rather than from theory. It uses old and recent references, as well as field evidence. It shows that mountains are not made directly by folding, but result from vertical uplift of plains (planation surfaces) to form plateaus, which may later be eroded into rugged mountains. It also assembles the evidence that this uplift occurred in the last few million years, a time scale which does not fit the plate tectonic theory.

Another fascinating story is that the age of uplift correlates very well with climatic change. Mountain building could have been responsible for the onset of the ice age and the monsoon climate, and certainly created many new environments. Fossil plants and animals are used to work out the time of mountain formation, which in turn helps to explain biogeographic distributions.

Cliff Ollier is a Visiting Fellow, Centre for Resource and Environmental Studies, Australian National University and also Emeritus Professor, University of New England. **Colin Pain** is Assistant Director, Cooperative Research Centre for Landscape Evolution and Mineral Exploration, Australian Geological Survey Organisation.

The Origin of Mountains

Cliff Ollier and Colin Pain

London and New York

First published 2000
by Routledge
11 New Fetter Lane, London EC4P 4EE

Routledge is an imprint of the Taylor & Francis Group

Typeset in Garamond by Florence Production Ltd, Stoodleigh, Devon

Printed in the United Kingdom at the University Press, Cambridge

British Library Cataloguing in Publication Data
A catalogue record for this book is available from the British Library

Library of Congress Cataloging in Publication Data
Ollier Cliff.
   The origin of mountains / Cliff Ollier and Colin Pain.
    p.cm.
   Includes bibliographical references and index.
   1. Orogeny. 2. Mountains. I. Pain, C. F. II. Title.
QE621.O45 2000
551.43'2–dc21   99–058339

ISBN 0–415–19889–5 (hbk)
ISBN 0–415–19890–9 (pbk)

To Janeta and Josie

Contents

List of figures   viii
List of tables   xv
Preface   xvi
Acknowledgements   xviii

1 Introduction   1

2 Simple plateaus and erosional mountains   21

3 Fault block mountains   41

4 European mountains   60

5 Western North America   96

6 The Andes   112

7 Asian mountains   128

8 Mountains with gravity structures   150

9 Volcanoes and granite mountains   168

10 Mountains on passive margins   193

11 Plains and planation surfaces, drainage
   and climate   227

12 Problems of mountain tectonics   271

13 Science and the origin of mountains   311

      References   316
      Index   336

Figures

1.1	The Acholi Plain in northern Uganda	2
1.2	A typical cross-section in northern Uganda	2
1.3	Cross-section of the Western Carpathians	3
1.4.	Two naive illustrations of fold mountains	7
1.5	The relationship between mountains, plains and geological structure	9
1.6	Diagram of card pack	10
1.7	Tectonic plates and earthquakes	14
1.8	Some possible mechanisms for creation of mountains by plate tectonics	17
1.9	Geological time scale	19
2.1	Mt Connor, Central Australia	22
2.2	Table Mountain, Capetown, South Africa	22
2.3	Tepuis in Venezuela	23
2.4	Sketch map of the Tepuis of Venezuela and the Pakaraima Range	24
2.5	Successive cross-sections of the Gran Sabana, Venezuela	25
2.6	Aerial view of the Carr Boyd Ranges, Western Australia, showing bevelled cuestas	26
2.7	Diagram of planation cut across folded rocks, forming bevelled cuestas	26
2.8	Cuestas	27
2.9	Ayers Rock from the air	29
2.10	A distant view of Ayers Rock showing an apparently flat surface	29
2.11	Photo of ridges and grooves on top of Ayers Rock	30
2.12	Stages in the modification of an old landscape by glaciation	32
2.13	Snowdon, the highest mountain in Wales, showing remnants of an old pre-glacial surface after glaciation	33
2.14	Mount Elie de Beaumont, New Zealand, showing a cover of ice and snow	33

2.15   Mount Cook, the highest mountain in New Zealand   34
2.16   Mt Baker, one of the peaks of Ruwenzori   35
2.17   Diagram showing the formation of an inselberg   36
2.18   Inselbergs of Angola   37
2.19   The Sugarloaf at Rio de Janiero, Brazil   38
2.20   Inselbergs formed at the junction of the African
       Surface and the Acholi Surface in Uganda   39
2.21   Stone Mountain, near Atlanta, Georgia   40
3.1   Cross-section of the Sierra Nevada.   42
3.2   Tectonic evolution and extension of the Basin
       and Range Province over the past 40 Ma   44
3.3   Average trend of extension of all normal fault
       segments up to 10 miles (16 km) long within
       each square degree of the Basin and Range Province   45
3.4   Map of the Lake Albert Rift Valley   47
3.5   Three cross-sections of the Lake Albert Rift Valley   48
3.6   The rift valley wall on the east side of the Lake
       Albert Rift, near Butiaba, Uganda   49
3.7   Diagrammatic sections showing the evolution and
       extension of the east margin of the South Kenya
       Rift Valley   51
3.8   Diagrammatic section across the Rio Grande rift
       near Albuquerque, New Mexico   52
3.9   Mount Margherita, one of the summit mountains
       of the Ruwenzori massif   54
3.10   Antecedent drainage of Milne Bay, Papua New
       Guinea   55
3.11   The Shanxi Graben, China, with an en-echelon
       pattern   56
3.12   Long profile of the Fen River, showing the control
       of faults on terraces   57
3.13   The summit surface envelope of Fiordland,
       New Zealand   58
3.14   A plate tectonic model for the uplift of the
       Southern Alps of New Zealand   59
4.1   The main alpine mountains of Europe   61
4.2   Cross-section of the Alps   61
4.3   Proposed plate tectonic evolution of the Alpine
       system   65
4.4   The Apennines and surrounding seas   67
4.5   A SW to NE section of the Apennines, starting
       south of Naples   68
4.6   Graben and horst structure of eastern Italy   69
4.7   The planation surface near Ancona, Italy   70
4.8   The planation surface in Umbria, Italy   70

4.9	Cross-section of the Apennines from the Tiber to the Adriatic	71
4.10	Three cross-sections of the Apennines	71
4.11	Two major steps in the development of the Apennines	72
4.12	Drainage pattern of the central Apennines	74
4.13	Sketch of the Tyrrhenian Sea	76
4.14	Cross-section of the Po Valley	77
4.15	Some illustrations purporting to show the origin of the Apennines by subduction	78
4.16	The Apennine nappe front	80
4.17	Rotation of Italy, the Tyrrhenian Sea and Corsica–Sardinia forming the Apennines	81
4.18	Evolution of the Apennines by rotation	82
4.19	Mountain ranges of the Iberian Peninsula	83
4.20	Map of Pyrenees and diagrammatic cross section	84
4.21	Schematic sections showing the relations between the different units of the southern flank of the central Pyrenees	84
4.22	Central Cordillera of Spain	86
4.23	Map of the Carpathian Mountains showing the main divisions	87
4.24	Cross-section of the Western Carpathians	88
4.25	Caucasus location map	92
4.26	Cross-section of part of the Walbrzycg Uplands, Sudeten Mountains, Poland	94
4.27	Geographical divisions of the Atlas Mountains	95
5.1	Map of mountains and plateaus in western North America	97
5.2	The Precambrian cores of uplifts of the western United States	98
5.3	Diagrammatic section across the central Uinta Uplift, Utah	100
5.4	Suggested structure of the frontal zone of Owl Creek uplift, Wyoming	100
5.5	Overthrusts on the Front Range of the Rocky Mountains, Colorado	101
5.6	Sketch section of 'mushroom tectonics' as applied to the Rockies in Colorado	102
5.7	Some of the drainage anomalies in the Rocky Mountains region	104
5.8	Cross-section of the Canadian Rockies showing thrusting to the east over a decollement	106
5.9	A cross-section of the Ouachita Mountains, Arkansas	109
5.10	Mountains and plateaus of central America and Mexico	110
6.1	The main geographical features of the Andes	113

6.2	The Andean planation surface near Cuenca, Ecuador	114
6.3	The Andean planation surface north of Quito, Ecuador	115
6.4	Map and cross-section of Tierra del Fuego	117
6.5	Relation between Quaternary volcanicity and the Central Valley in Chile	120
6.6	Map and cross-section of Andes of Peru	122
6.7	Divergent faults in Andes	123
6.8	Distribution of Granite in the Andes	126
7.1	The Tibet Plateau, a vast, uplifted pediplain	129
7.2	Map of mountains and major faults in the Himalayas	130
7.3	Diagram of isostatic response to erosion at the edge of a plateau	133
7.4	Composite mean elevation profiles of the Himalayas and Tibet	134
7.5	Isostatic uplift of a single major valley	134
7.6	Anticlines along valleys	135
7.7	Section across Asia from the Verkhoyansk Mountains to the Arabian Sea	139
7.8	Simplified geography and tectonics of the Himalayan–Tibetan Plateau region	141
7.9	One-million-year-old basalt in horizontal gravels at the base of the Kunlun Mountains	143
7.10	The uplifted peneplain of the Bural-Bas-Tau in the Tien Shan	145
7.11	Lake Aiding on the floor of the Turfan Depression	145
7.12	Cross-section of the Turfan Basin	146
7.13	Diagrammatic cross-section across Lake Baikal	146
7.14	The peneplained surface of southern Japan	148
7.15	Morphotectonic profile from the Tibet Plateau to the Yunnan Plateau	149
8.1	The Agulhas submarine landslide, offshore southeast Africa	151
8.2	Cross-section of the Bengal Fan, showing folding at the distal end	151
8.3	The Niger Delta, showing many features of an orogen	152
8.4	Sections derived from seismic survey showing thrusts and decollement, in marine sediments	153
8.5	Cross-section of the Jura Mountains	154
8.6	Cross-section of the Muller Range, part of the Papuan Fold Belt, Papua New Guinea	155
8.7	Three versions of the Papuan Fold Belt	156
8.8	A cross-section of Taiwan	157
8.9	Cross-section of the Naukluft Mountains, Namibia, from NW to SE	158
8.10	Diagram of gravity collapse structures of Iran	159

8.11	Superficial gravity structures in Northamptonshire, England	161
8.12	Zagros valley bulges	163
8.13	Diagrammatic representation of the foundering circuit in the Dolomites	165
8.14	Cross-section of the mountains near Sasso Lungo, Dolomites	166
8.15	Lateral diapir on a valley side	166
8.16	Lateral Tectonic Extrusion. Uplift of the Tauern Window	167
9.1	Simple classification of igneous rocks	169
9.2	Chimborazo, a typical Andean stratovolcano in Ecuador	170
9.3	Tristan da Cunha, an island volcano in the middle of the Atlantic	173
9.4	Distribution of active volcanoes of the world	175
9.5	Hawaii hotspot trace	176
9.6	A plot of age against latitude of volcanoes of eastern Australia	177
9.7	Distribution of Quaternary volcanoes in Papua New Guinea	179
9.8	Volcanic activity on the Colorado Plateau	181
9.9	Stages in erosion of a volcano	182
9.10	Diversion of rivers by Mt Etna	183
9.11	Cross-section of the Mole Granite, New South Wales	186
9.12	The evolution of Mount Duval, New South Wales	187
9.13	Mount Kinabalu, Borneo	188
9.14	Diagram of the gneiss-mantled dome of Goodenough Island, Papua New Guinea	188
9.15	Generalised contour map of the Dayman Dome, Papua New Guinea	189
9.16	Contour map of part of the Dayman Dome, Papua New Guinea	190
9.17	Air photo of part of the edge of the Dayman Dome, Papua New Guinea	191
10.1	The basic geomorphology of passive margins with mountains	194
10.2	The Great Escarpment near Innisfail, south of Cairns, Queensland	195
10.3	The abrupt junction between the New England Tableland and Great Escarpment, New South Wales	196
10.4	The abrupt junction of the plateau and the Great Escarpment, Bakers Gorge, New South Wales	197
10.5	Wollomombi Falls, New South Wales	197
10.6	Contour map of the Carrai Tableland, New South Wales	198

10.7	Eastern Australia showing the Great Divide, the Great Escarpment, the continental margin, and areas of basalt	200
10.8	Major physiographic features of southwestern Western Australia	202
10.9	Evolution of the Cowan Drainage	203
10.10	The Western Ghats between Bombay and Poona	204
10.11	The Western Ghats in southern India, near Kodaikanal	205
10.12	Physiographic units of eastern North America	207
10.13	Diagram of the Appalachians, caused by simple compression due to continental collision	210
10.14	A typical example of the Great Escarpment in Scandinavia	212
10.15	The Palaeic surface lowered by glaciation. Southern Norway. This is a classic gipfelflur	213
10.16	A palaeoplain remnant in Dronning Maud Land	215
10.17	The possible morphotectonic similarities between different continents with passive margins	216
10.18	Great Escarpments of Australia, India and Southern Africa	218
10.19	From rift valley to seafloor spreading	219
10.20	The geography of eastern Gondwana before the break up	220
10.21	Contrasted styles of rifting	221
10.22	Retreat of the Great Escarpment where the rivers run roughly parallel to the escarpment, as in southern New South Wales	222
10.23	Retreat of the Great Escarpment where the divide is roughly perpendicular to the drainage	223
10.24	Two possible relationships of the regolith-covered palaeoplain to tectonics at a passive continental margin	225
11.1	A dissected planation surface may be detected by the accordant summit levels	230
11.2	A major pediment at the foot of the Flinder Ranges, South Australia	232
11.3	Valley widening by slope decline and by slope retreat	233
11.4	Double planation surfaces at Coober Pedy, central Australia	235
11.5	Minor pediment surface in Tibet	236
11.6	Erosion surfaces in relation to the Miocene volcano Mount Elgon, on the Uganda–Kenya border	238
11.7	Simple dendritic drainage pattern in which tributaries join the main stream at acute angles	242
11.8	Long profile of main stream and tributaries	243
11.9	The structural relationships of streams on dipping strata, and the preferred nomenclature of streams	244
11.10	Trellis drainage pattern. Montery, Virginia quadrangle	245

11.11	Outdated nomenclature for structurally controlled streams in regions of dipping strata	246
11.12	Drainage pattern of the Upper Lachlan and Upper Murrumbidgee (SE Australia)	248
11.13	Drainage in the Harrisburg region of the Appalachians	249
11.14	Diagrammatic representation of river capture	250
11.15	Successive river captures in Yorkshire, England	252
11.16	The 'leaping' of a drainage divide	253
11.17	The formation of barbed drainage	255
11.18	Reversal of drainage in the Lake Victoria–Lake Kyoga system	256
11.19	Barbed tributaries of the Clarence River, New South Wales	258
11.20	The dome of the Lake District, northern England	259
11.21	The original drainage pattern of Wales	260
11.22	Some effects of faulting on drainage systems	262
11.23	The Lake George Fault and associated features, near Canberra, Australia	263
12.1	Two interpretations of the same outcropping pattern of strata	274
12.2	Sawtooth Mountains, Montana	277
12.3	The 'wedge' explanation of thin skinned tectonics	277
12.4	Damaged cars and geological analogy	278
12.5	A generalised section across the Ucayali basin, eastern Peru	281
12.6	The Iranian Plateau, an example of a median plateau	287
12.7	Cross-section of the Andes showing symmetrical features despite one-side subduction	288
12.8	Cross-section of the Andes	289
12.9	Tectonics of the south-west Pacific	290
12.10	New Zealand interpreted as a symmetrical system	291
12.11	The Philippines interpreted as a symmetrical system	292
12.12	Papua New Guinea interpreted as a symmetrical system	293
12.13	Cross-section of Papua New Guinea to show the tectonic components of the symmetrical system	294
12.14	Structure of the upper mantle under the western Mediterranean, showing a possible asthenolith	295
12.15	Origin of the western Mediterranean, with a rising asthenolith and divergent spreading structures	296
12.16	Sketch of the Aleutian island arc	299

Tables

10.1   Some possible causes of formation of marginal swells
       on passive continental margins   226
12.1   Summary of plantation and uplift in mountain areas   304
12.2   Twenty causes of tectonic uplift   308

Preface

A few centuries ago mountains were viewed with fear, awe and suspicion. Nowadays they are almost universally regarded with admiration, aesthetic pleasure and affection. We too have enjoyed working in mountains in many parts of the world, and we believe that the joy of mountains is enhanced by trying to understand them. Some scientific understanding adds to the aesthetic experience, making for a more complete comprehension and pleasure. That does not mean we know all about mountains. Far from it, but the pleasure comes from learning, and always having interesting questions to ask.

This book is about the origin of mountains. We describe mountains in many different situations, and are led to believe that mountains result from vertical uplift of old plains, followed by erosion. We cannot claim it is a modern view of the subject, because at the time of writing that title must go to the plate tectonics theory of mountain building by continental collision. Neither is it old-fashioned, because old ideas of mountains resulting from a shrinking earth like the ripples on the skin of a shrinking apple would have that title. It is not even original, for the ideas that we present have been described by numerous writers, some almost a century ago.

This explains why we refer not only to many modern papers (if only to show we have read them) but also to many authors who saw and wrote about what we see ourselves. The swing to plate tectonics since about 1965 has caused much valuable earlier literature to be neglected or revised. This is nowhere more obvious than with Holmes' book, *Principles of Physical Geology*. The edition of 1965 was one of the finest geological books of the century. The editions that have been published since then have removed much of the great man's insight, and have replaced much of his factual information with speculation and models. Many other giants of the past have a place in our book, which finds their ideas still relevant. We have an advantage over them in that communication today is better than ever before. It is possible to travel widely, and literature from many countries is now readily accessible.

The present is an interesting time for Earth Science. Tectonics is no longer an isolated branch of geology related to major Earth features such

as the origin of continents or mountains. Its findings reflect on geomorphology (the evolution of landforms), biogeography (land bridges or
continental drift), climatology (to what extent do mountains control the
climate?), and other disciplines. Sadly, in our opinion, Earth Science has
become too concerned with theory, models, and dogma. We agree with
Claud Bernard (a physiologist) who wrote: 'Men who have excessive faith
in their theories or ideas are not only ill-prepared for making discoveries,
they also make poor observers.' We hope, in our small way, to encourage
people, from students to professionals, to have a new look at mountains,
without reference to pre-conceived theories, but with attention to what can
really be seen.

Cliff Ollier and Colin Pain
Canberra, May 1999

Acknowledgements

We wish to thank the colleagues who provided photographs, acknowledged in the captions to figures, and Bryan Ruxton and John Hepworth who read some sections of the manuscript.

1 Introduction

Mountains are a common feature of the Earth, easily recognised by ordinary people and scientists alike. Folded rocks are also observed in many places, and a dominant idea arose very early that the same forces that folded the rocks also formed the mountains. Over the past two centuries many theories of mountain formation have been proposed, mostly based on a mechanism that could also fold rocks. An early idea was the shrinking Earth, where fold mountains were created like folds in the skin of a shrivelled apple. Since about 1965 plate tectonic ideas have dominated geology. In this hypothesis mountains are formed where plates of the Earth's crust collide, and the movement of one plate below another (subduction) causes both folding and mountain building.

These grand theories go straight to the ultimate mechanism and driving force, and miss out on some of the vital landscape-forming processes, especially the former existence of plateaus where the mountains stand today.

Geomorphologists (scientists who study landscapes) have long studied plains, many of which are erosion surfaces cut across varied rock structures such as folds and faults. If an extensive plain is raised by regional uplift it becomes a plateau, such as the Tibetan Plateau, the Highveld of southern Africa, or the Tablelands and High Plains of Australia. Erosion of plateaus can create rugged topography – in fact it makes mountains. So to understand mountains we must first know more about their geomorphic history. In most mountain studies this is not done: instead the rock structures inside the mountains are described in detail, with the tacit assumption that whatever made the structures also made the mountains.

This assumption is not warranted. The plains of Western Australia and Africa are about as flat as any erosional land surface can get (Figure 1.1). Very complex structures including folds, faults and highly sheared metamorphic zones underlie the plains (Figure 1.2). Nobody suggests that these structures formed the planation surface. Yet when similar structures are found beneath mountains many geologists assume that the forces that made the structures also formed the mountains. Even when the rock structure includes great thrusts, as in the Carpathians, the reasoning should be the same: the major structures were planated long before the present mountains

Figure 1.1 The Acholi Plain in northern Uganda (photo C.D. Ollier).

came into existence (Figure 1.3). Thus structures in the Appalachian Mountains are commonly thought of as related to the Appalachian orogeny that folded the Palaeozoic rocks and also formed the Appalachian Mountains. In reality the Palaeozoic structures were planated, and it was later uplift of the planation surface followed by further erosion of valleys that created the Appalachians of today. Similarly in Scandinavia we are told that the Caledonian orogeny (which deformed the rocks) made the Caledonian Mountains. The reality is that the Caledonian structures were eroded to a plain, the planation surface was warped up much later to form a plateau, and later erosion made the mountains of Norway.

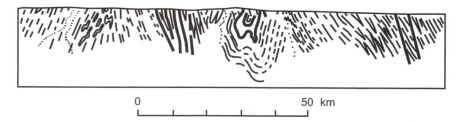

0   50 km

Figure 1.2 A typical cross-section in northern Uganda (J.V. Hepworth, pers. comm.).

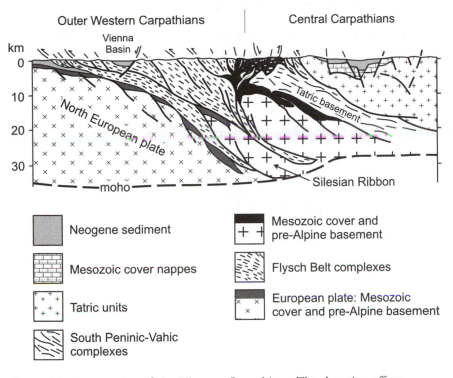

Figure 1.3 Cross-section of the Western Carpathians. The thrusting affects virtually all the crust, and in comparison all the structures are planated. The mountains of today are caused by much younger forces than those that caused the great structures. (after Plasienka *et al.* 1997).

Planation surfaces are formed by erosion to a base level, usually sea level. Uplifted planation surfaces (plateaus) indicate vertical uplift of a former low-lying plain. Widespread planation itself indicates a period of tectonic stability when erosion was not offset by uplift. Uplift is a vertical movement. Uplift does not affect the whole world, but a broad area. This is *epeirogeny*, and it is ironic that while the meaning of orogeny has changed from 'mountain building' to 'rock deformation' it now seems that epeirogeny is what makes mountains.

The study of the geomorphology of mountains all over the world leads to a remarkable conclusion. Most of the world's mountains are formed by erosion of plateaus, which are themselves uplifted erosional plains. Folding in these areas must be older than the formation of the planation surface, and of course older than the mountains. This means we must separate two processes, or sets of processes – one set causes folding and other structures, the other set causes plateau uplift. In fact we can distinguish four sets of processes:

1   processes that cause folds and other structures;
2   processes that make planation surfaces;
3   processes that cause uplift of a plain to form a plateau (gentle bends, monocline, fault block (horst), tilt block);
4   erosional processes that dissect a plateau into mountains (basically fluvial and glacial).

These four sections will be dealt with in the next chapters.

Furthermore, many of the world's mountain belts are on sites of geologically young uplift, which also leads to many interesting conclusions. Later in the book we shall present evidence for the age of mountains throughout the world, and examine the implications.

It turns out that uplift of land to make mountains results in tectonic, climatic and geomorphic changes. In other words there are feedback mechanisms between mountain building and other processes.

Volcanoes are different from other mountains in their mode of origin, and are treated separately in Chapter 9.

This book presents a detailed account of mountain building including both tectonic and geomorphic evidence so far as can be done in a single book. We also have to digress at times to discuss some basic ideas of geology and geomorphology. Finally we discuss some of the implications of this view of earth history.

Terminology

Uplift, orogeny and mountain building

This section is all about the confusing nomenclature associated with mountains. If you are not interested in details you can skip this part, but if you have a little learning, and especially if you think orogeny makes mountains, it will be best to read it before you read the rest of the book.

Nearly all modern books on mountain-building and orogeny are confused about the origin of mountains and the origin of structures inside them. Hsü's *Mountain Building Processes* (1982) is all about structures, and it is simply assumed by most contributors that 'orogeny' creates both internal structures and the present-day topographic mountains. In that book only Gansser, in his chapter on the 'morphogenetic phase' of mountain building, distinguishes the late, vertical mountain building from earlier compression. Schaer and Rogers' book *The Anatomy of Mountain Ranges* (1987) is likewise about internal structures, tacitly assumed to be related to present day mountains. Orogeny is still equated with mountain building by many geologists.

Orogeny

Orogeny is a word literally meaning the genesis of mountains, and when proposed it meant just that. Unfortunately in later years the idea of folding and mountain building being the same thing became entrenched, and with a further swing the term 'orogeny' came to mean the folding of rocks. *Orogeny is now used to refer to the folding of rocks in fold belts.* It does not mean mountain building, despite its etymology. We shall have to use the longer phrase *mountain building* to be clear.

If authority is needed for this practice we may note the following:

King (1969) wrote in his influential paper: 'In this account, and on the legend of the "Tectonic map of North America", "orogeny" is therefore used for the processes by which the rock structures within the mountain chains or fold belts are created.'

In *Orogeny Through Time* Burg and Ford (1997) claim that: 'To field geologists the term orogeny represents a penetrative deformation of the Earth's crust.' We are not convinced that all field geologists really appreciate this not-so-subtle change of meaning.

Jackson (1997), in what is virtually the bible for English-speaking geologists, wrote:

> *orogeny* literally, the process of formation of mountains. The term came into use in the middle of the 19th Century, when the process was thought to include both the deformation of rocks within the mountains, and the creation of the mountainous topography. Only much later was it realised that the two processes were mostly not closely related, either in origin or in time. Today, most geologists regard the formation of mountainous topography as postorogenic. By present geological usage, orogeny is the process by which structures within fold-belt mountainous areas were formed, including thrusting, folding, and faulting in the outer and higher layers, and plastic folding, metamorphism, and plutonism in the inner and deeper layers. Only in the very youngest, late Cenozoic mountains is there any evident casual relation between rock structure and surface landscape. Little such evidence is available for the early Cenozoic, still less for the Mesozoic and Paleozoic, and virtually none for the Precambrian – yet all the deformation structures are much alike, whatever their age, and are appropriately considered as products of orogeny.

However the modern usage has not filtered down to lower levels, and elementary books and dictionaries still commonly follow the old usage, and have figures like that in Figure 1.4. For example, the *Hutchin Pocket Dictionary of Geography* (1993) has the following entry:

> *orogeny* or *orogenesis* – the formation of mountains. It is brought about by the movements of the rigid plates making up the Earth's crust (described by plate tectonics). Where two plates collide at a destructive margin rocks become folded and lifted to form chains of fold mountains (such as the young fold mountains of the Himalayas).

The prestigious National Geographic Society in *Exploring Your World* (1993) states:

> Fold mountains are formed when two of the large plates that carry the earth's crust slowly collide and compress, or when one plate gradually folds and wrinkles as a result of this action.

They illustrate the idea with a diagram like Figure 1.4, leaving no doubt that they consider the folded rocks to be the direct origin of the fold mountain. Some more serious sources also maintain the old story, such as the definition in *The Cambridge Encyclopedia of Earth Sciences* (1981):

> *Orogeny* An episode of tectonic activity (folding, faulting, thrusting) and mountain building usually related to a destructive plate margin.

Some writers seem to adopt their own personal definitions of mountain building and orogeny. Miller and Gans (1997), for instance, wrote: 'Mountain building and orogeny (i.e. thickening of continental crust) are not necessarily the result of subduction.' We do not equate either mountain building or orogeny with crustal thickening, and suspect that few other workers do so.

Allmendinger and Jordan (1997) wrote: 'The term "Andean Orogeny" refers collectively to all tectonism that occurred between the Jurassic and Recent.' This would probably not be most people's idea of orogeny, but it certainly would not be synonymous with 'mountain building' for, as we shall see, the Andean Cordillera were created in only a twentieth of that time.

Epeirogeny

In contrast with orogeny, early geologists used the term 'epeirogeny' to mean the uplift of broad areas, as opposed to the narrow fold belts of mountain chains. Gilbert (1890, p. 40) coined the word 'epeirogeny' and was also one of the first to use the term 'orogeny', so it is useful to get his

---

Figure 1.4. Two naive illustrations of fold mountains
Top. Elementary diagram of fold mountains, in which the actual mountains have the same shape as the folds. This does not happen in the real world.
Bottom. A diagram showing the common misconception of 'fold mountains', with the folding of the rocks and formation of the actual mountains occurring simultaneously by compression of horizontal rocks (simplified from *National Geographic*, 1993). As explained in the text, real mountains are not formed in this way. This diagram, common in popular literature, shows that the mountains are supposed to be formed at the same time as the rocks are folded, and by the same force.

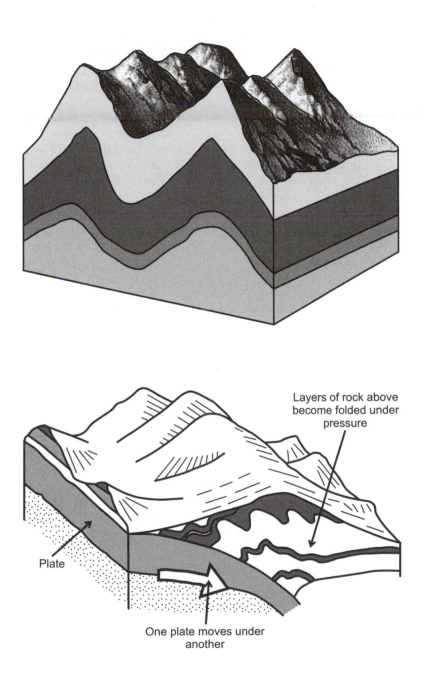

Layers of rock above
become folded under
pressure

Plate

One plate moves under
another

views: 'The process of mountain formation is orogeny, the process of continent formation is epeirogeny, and the two collectively are diastrophism.' Probably nobody follows this usage any more. 'Epeirogeny' is still a valid term for the uplift of broad areas, but it does not mean continent formation. We believe that mountains result from erosion of areas that have been uplifted epeirogenically.

The paradox was noted long ago by Stille (1936) who expressed it thus:

> As a matter of fact, orogeny in the tectonic sense generally fails as an explanation for the existence of the topographically great mountains of the earth, such as the Alps of Europe or the Cordilleras of North America. These mountains exist – or still exist – as a result of post-orogenic en bloc movements, for the most part still going on, and belonging to the category of epeirogenic processes. Thus arises the terminological contradiction, that the mountains as we see them today owe their origin not to what is called orogeny, but to an entirely different type of movement that is to be strongly contrasted with the orogenic process.

The distinction between orogeny and epeirogeny is acknowledged in the definition provided by the McGraw-Hill *Dictionary of Earth Sciences* (1984):

> *Orogeny* The process or processes of mountain formation, especially the intense deformation of rocks by folding and faulting which, in many mountainous regions, has been accompanied by metamorphism, invasion of molten rock, and volcanic eruption; in modern useage, orogeny produces the internal structures of mountains, and epeirogeny produces the mountainous topography.

This definition highlights another problem: there is no one-to-one correlation of folding and mountain building. Lowland plains may be underlain by intensely folded rock, as in much of Western Australia, the Canadian Shield or Finland. Mountainous areas may be underlain by horizontal strata, like the Drakensberg of South Africa or the Blue Mountains of New South Wales. Furthermore, granite or basaltic lava flows may be found under both plains and mountains.

The relationships of plains and mountains to areas of folding and non-folding

Plains occur on horizontal strata (Murray Basin).
Mountains occur on horizontal strata (Drakensberg).

Plains occur on folded rocks (Amazon Basin).
Mountains occur on folded rocks (Alps).

Plains occur on horizontal basalt (Western Victoria Plains).
Mountains occur on horizontal basalt (Snake River).

Plains may be cut across granite (Western Australia).
Mountains may be cut across granites (Sierra Nevada).

Plains may be cut across metamorphic rock (Finland).
Mountains may be cut across metamorphic rock (Scottish Highlands).

Clearly there is a need to divorce mountain building from folding. To emphasise this point the information is shown again in Figure 1.5.

Theories

Theories about the origin of mountains are based on certain assumptions, often unstated and sometimes dubious. Many hypotheses purporting to be about mountain building are in fact concerned with geosynclines, plate tectonics, or the origin of mobile belts – belts of deformed rocks which may or may not be coincident with mountain belts or former mountain belts.

Theories come in all varieties of sophistication, but many fail to distinguish clearly between the folding of the rocks and the formation of mountains. We must include here various ideas that are really concerned with fold belts but because they are called theories of orogenesis they imply a relationship with mountain chains.

Early in the twentieth century Geikie (1912) wrote *Mountains* in which he distinguished 'Original or Tectonic' mountains from 'Subsequent or Relict' mountains. The second group is discussed in only one chapter out of 12, despite the fact that much of Geikie's book is concerned with the eroded mountains of Scotland and Europe.

	FOLDED ROCKS	HORIZONTAL STRATA	HORIZONTAL BASALT	GRANITES	METAMORPHIC ROCK
MOUNTAINS	Alps	Drakensberg	Snake River	Sierra Nevada	Scottish Highlands
PLAINS	Amazon Basin	Murray Basin	W. Victoria Plains	Western Australia	Finland

Figure 1.5 The relationship between mountains, plains and geological structure. There is no simple relationship between mountains and folding, or any other structure.

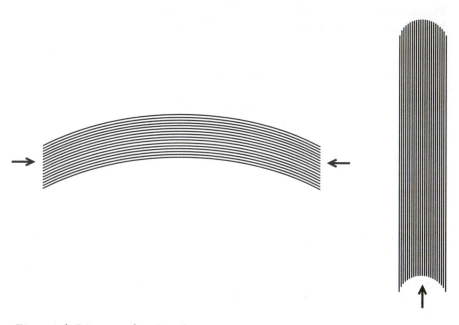

Figure 1.6 Diagram of card pack. The cards may be 'folded' with lateral
        compression (left) or with constant lateral dimension (right).

Two main concepts have been used to explain the creation of mobile
belts: the lateral (tangential) compression hypothesis, or vice concept, and
the vertical (radial) tectonic concept, with vertical uplift of broad areas.

There are then two possible related assumptions:

1   That the strata are shortened and that the two opposite sides of a sedi-
    mentary packet move towards each other during folding. As shown in
    Figure 1.6 it is possible to make folds without shortening,
2   That there is little shortening and there may even be extension
    (dilation).

We shall try to make these points clear in discussing real mountains in the
rest of the book.

Finally we may mention assumptions about earth volume, which include
the following:

1   Hypothesis of a contracting Earth. This was an old favourite, last
    presented seriously by Lees (1952), who wrote: 'Mountain building is
    the consequence of contraction of the interior of the earth and crustal
    compression from this cause has been dominant throughout revealed
    geological time.'

2   Hypothesis of an Earth of constant volume. This is by far the commonest assumption.
3   Hypothesis of an expanding Earth, most seriously proposed by Carey (1976, 1988).
4   Hypothesis of oscillating Earth volume. Nobody seems to have seriously proposed this, and it is included here merely to complete the list of possibilities.

The study of mountains

In the study of anything, scientists should work from observations to theory. Using any techniques available, they gather relevant facts, and then derive a hypothesis to explain them. This, we hope, is the method used in this book.

In the study of mountains we want to know:

1   the age and origin of the rocks;
2   the age and origin of geological structures;
3   the age and style of erosion;
4   the age of uplift;
5   the cause of uplift.

The last is the most difficult to know, but it is where most people start!

Outline of this book

The first chapter sets the scene for later discussions of mountains and presents essential definitions together with a brief outline of plate tectonics and geological time.

Next comes a series of chapters on mountains of particular regions, or arranged by a common theme or origin.

This is followed by a section where we discuss some generalities, processes and implications related to the study of mountains.

To put a thread through the story we present the following simple outline of mountain formation:

Some mountains are simple plateaus. Others are made by erosion of high masses of rock, and perhaps there is little trace of earlier landscape development.

The simplest kind of tectonic uplift is of fault blocks, so we deal with mountains related to simple block faulting, tilt blocks and multiple tilt blocks.

Next come those mountains that are eroded plateaus on folded rocks. Differential erosion occurs on a simple structure, but an earlier phase of planation is usually evident.

Beyond simple folds are the great thrust sheets of rocks called 'nappes'. We deal with several mountain ranges formed on nappes, but find that a planation surface was formed after the nappe movement, and then came vertical uplift.

Some uplifted blocks apparently spread as they rise, thrusting rocks over the surrounding lowlands. Numerous examples are described.

Erosion on the edges of uplifted blocks can make valleys so big that the land rises isostatically, inducing uplift of the edges of the plateaus.

Some continents have a marginal swell of high land which is eroded on the ocean-facing side into Great Escarpments, commonly known as mountain ranges. These are the passive margin mountains.

In very many mountainous regions the common sequence of events is:

1   erosion to a plain followed by
2   uplift to form a plateau, after which comes
3   erosion of the plateau to make isolated mountains.

There are, of course, many variations on this theme.

Two groups of mountains do not fit into this general picture but are treated for completeness, and for the light they throw on other mountains. Volcanoes are constructional landforms built of lava derived from deep in the Earth and erupted at the surface. Granite mountains are formed of rock formed deep in the Earth. Most granite mountains are simply the result of great erosion, but some result from the rise of granite above the general ground level.

A brief outline of plate tectonics

As soon as the voyages of discovery had revealed the similarity in shape of the opposite sides of the Atlantic, the idea that opposing continents had drifted apart was bound to arise. It was Wegener (1929) who brought scientific knowledge and scholarship together in a serious and concentrated exposition of continental drift.

In Wegener's day the concept of drift could be supported by evidence from the distribution of plants and animals, the distribution of fossils, ancient deserts and glaciated rocks, and by matching other geological evidence. Wegener also thought that the mountains along the Pacific coast of the Americas were pushed up at the front edge of drifting continents where the continental slab buckled against the Pacific.

Continental drift was not accepted readily by the geologists, especially in the Northern Hemisphere, and 'drifters' were lepers in the geological community. But in 1958 Carey showed that the 'fit' of the South Atlantic was remarkably good, within half a degree over 45 degrees of latitude, which could hardly be accidental, and by 1965 Bullard and others showed that the entire Atlantic could be closed to give an equally good fit.

However, the evidence that was to change our views on drift came not from the continents but from the study of the ocean floor. Ocean surveys in the nineteenth century showed that the Atlantic was shallowest along the centre. A mid-Atlantic ridge runs the full length of the Atlantic and remains very close to the middle, emerging above sea level in Iceland. The ridge is split by a great rift. Perhaps the greatest geological discovery of this century is that the ocean floors are spreading. Magnetic studies showed a symmetry about the mid-Atlantic ridge, and later studies showed that the sea floor gets consistently older with distance from the ridge. New seafloor is being created at the spreading sites, and old seafloor migrates away from them.

There is overwhelming evidence that the Atlantic Ocean has been formed by the drifting apart of the continents that bound it. The sea floor grows at the centre – the mid-Atlantic rift – and spreads sideways. It seems that the Atlantic Ocean only started to open in the Triassic, about 200 million years ago.

Seafloor spreading was later demonstrated in all the oceans, though the Pacific ridge is not in the centre of the ocean, but runs into the continents of North and South America.

But if all the oceans are growing, they must disappear somewhere, unless the Earth is increasing in size. The sites where ocean disappears are called 'subduction sites', and are located at island arcs and 'active' continental margins. The driving force is thought to be convection currents, rising at the spreading sites and sinking at subduction sites.

In simple continental drift the continents were thought to drift around on a 'sea' of sub-continental material, but the ocean floors were thought to be quite passive in the process. Plate tectonics differs in having the oceans as part of the system. Seafloor is created at spreading sites, and destroyed at subduction sites. If a spreading site were to originate beneath a continental area, the continent would be rifted and drifted apart and a new ocean basin formed in the middle.

Plate tectonics is a name for the concept that the earth's crust can be divided into several plates (Figure 1.7) and the world's main tectonic features are related to activity at the edges of the plates. New crust is created at spreading sites, and old crust is destroyed at subduction zones. Beyond this simple exposition many complications are possible, and variations from writer to writer, and from place to place are almost numberless. For the past 30 years plate tectonics has been the ruling theory in geology, and has been used to account for almost everything in Earth Science.

Many phenomena follow plate boundaries. Shallow earthquakes follow the line of the sub-oceanic ridges, the spreading sites. Deep earthquakes follow the line of subduction sites around the Pacific rim and lines of deep sea trenches. A few earthquakes follow the rift valleys which might be incipient spreading sites.

The distribution of volcanoes (Figure 9.4) shows a similar pattern. Volcanoes are distributed along lines of sea floor spreading, and on the areas bordering subduction zones.

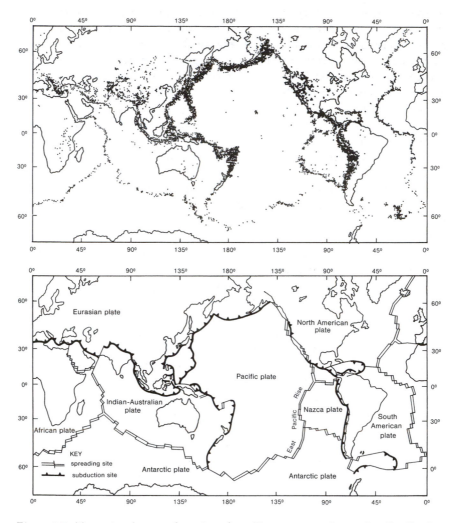

Figure 1.7 Tectonic plates and earthquakes. The top map shows the distribution
of earthquakes. The bottom map depicts the 'plates' of plate
tectonics, each bounded by lines of spreading or subduction.

Other evidence, including geochemical, gravity and heat flow measurements, all helps to confirm the same boundaries of the major plates of the
Earth's crust, and there is no doubt that the view of the Earth in terms of
plates has made a major advance in geological thinking, even though details
may be disputed.

For many Earth processes, including mountain building, the 'action' is
thought to occur on active margins, but as we shall see later there are also
mountains on passive margins, and some on spreading sites.

Spreading sites

These are mostly under the sea, and only affect the mountain story where the rift emerges as land, or where oceanic volcanoes achieve mountain proportions.

The geomorphology of the rift area can be seen in fair approximation in Iceland, which may be regarded as a piece of the mid-Atlantic ridge above sea level. There the rocks get older as one moves away from the rift, the rocks are dominated by fissure eruptions and the topography, where not primarily volcanic, is marked by many faults and fissures parallel to the main mid-Ocean (or mid-Iceland) rift. The volcanic intrusion is permissive. That is, dyke rocks come in to fill a space created by tension. We should not imagine a thin dyke forcing its way into a crack and pushing aside a plate many thousands of kilometres across and billions of tonnes in weight. Rather there is some force pulling the plates apart and the dykes come in to fill the space created when a crack appears.

Subduction sites

Subduction sites (known also as collision sites and active margins) provide more varied tectonic settings than spreading sites, including the following types. Some of the possible effects are shown in Figure 1.8.

Continent–continent collisions (Himalayan type)

The most spectacular example of this collision type is that presumed to have occurred between India and continental Asia. The idea is that India broke away from the supercontinent of Gondwana, drifted north, and was subducted under Asia, causing the rise of the Himalayas and the Tibet Plateau. This idea is discussed further in Chapters 7 and 12.

The collision of Africa with Europe is another example of the plate tectonic explanation of mountains, especially in Europe.

Continent–ocean collision

This is the commonest sort of collision envisaged in the plate tectonic theory, but can itself be divided into two.

Simple subduction (Andes type)

This is allegedly typified by the plate junction on the west side of South America.

The Pacific Ocean is said to be subducted under the South American continent. This is supposed to account for:

the deep trench along the edge of the continent;
the rise of the Andes;

the volcanic activity of the Andes;
the granite plutons of the Andes.

The course of the slab's progress is recorded by a series of earthquakes that show a fairly consistent increase in depth with distance from the plate edge making an inclined earthquake zone known as a 'Benioff zone'.

Sometimes sediments are scraped off the down-going slab and thrust on to the continent, perhaps to form 'fold mountains' (Figure 1.8d).

Alternatively, sediment may be carried under the continent (Figure 1.8e) where it is presumed to melt and form granite, or perhaps melt and provide the magma for volcanoes along the collision zone.

Island arc subduction

Around the western edge of the Pacific the map shows many islands arranged in festoons of arcs, either single lines or double rows of islands, making a quite clear pattern of curves. These lie some distance off the continent, and in front they have a deep trench very like that off the coast of South America. These deeps also bound a Benioff zone dipping towards the continent and marked by earthquakes reaching depths of several hundred kilometres.

Some arcs face east (Indonesian arc), some arcs have no continent behind them (South Sandwich Islands), and some lines of islands have all the attributes of arcs such as Benioff zones and trenches but lack the arcuate shape (Solomon Islands). Nevertheless the island arc complex makes one of the distinctive landform and tectonic assemblages of the Earth and, despite many complexities and variations, island arcs mark plate junctions.

Obduction

In a few collision zones a slab of ocean floor is thought to have over-ridden rather than under-ridden the continent. In Papua New Guinea a slab of basic rocks with petrology, layering and structures exactly like those thought

---

Figure 1.8 Some possible mechanisms for creation of mountains by plate tectonics.
   a. Continent–continent collision (Himalayan type)
   b. Continent–ocean collision with buckling up of the continent and subduction of the ocean floor (Andes type)
   c. Obduction of the sea floor over the continent with later isostatic rise to form mountains (Cyprus type)
   d. Thrusting of sediments on to the continental plate with foreland folding at the front (Appalachian type)
   e. Thrusting of sediment under the continent, where they might melt and form granite.
   f. Crustal thickening resulting from plate collision, possibly accompanied by gravity sliding of rocks near the surface (Pyrenees type).

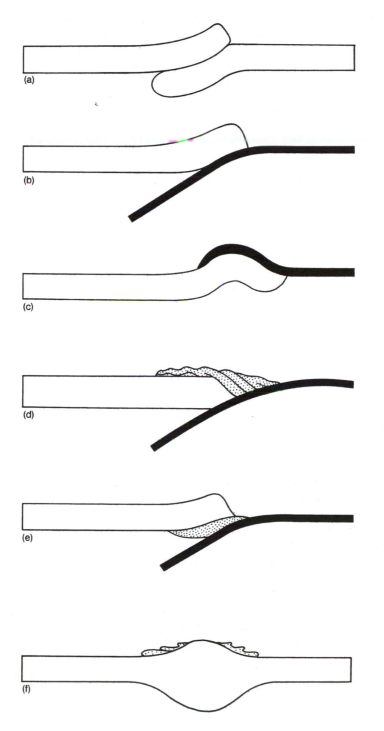

to be typical of ocean crust makes the mountainous terrain of the Papuan Ultramafic Belt. Another arc is the Troodos Mountains of Cyprus where dyke-ridden oceanic rocks override continental rocks. 'Obduction' is the term for this overriding, in contrast to subduction when the oceanic slab goes down below the continental mass.

The commonest mountains in island arcs are volcanoes, but some arcs have mountains made of bedrock, usually as fault block mountains.

Ocean–ocean collision

An arc may be formed, as in the South Sandwich Islands (Scotia arc) where a strip of spreading Pacific seems to be pushing a finger of oceanic crust into the oceanic crust of the Atlantic. The contact is marked by a trench, earthquake zone and active volcanicity. The Caribbean arc is somewhat similar, with the Pacific pushing into the Atlantic. The arcs may have sufficient elevation to cause mountains.

The curvature of the Indonesian arc may result partly from the Pacific Ocean spreading into the Indian Ocean.

Crustal thickening

In this hypothesis the collision of two plates causes a thickening of the crust, resulting in uplift, and perhaps sediments slide off the uplifted axis (Figure 1.8f).

Geological time

The Earth is about 4,500 million years old. The old rocks are known as Precambrian, and are normally devoid of fossils. For the past 560 million years life has been abundant and fossils are common in sedimentary rocks. This stretch of time is known as the Phanerozoic.

The Phanerozoic is divided into units of various types. A division into three Eras separates the Palaeozoic (old life), the Mesozoic (middle life) and the Cenozoic (recent life, also known as the Cainozoic). These three main groups are divided into smaller groups called Periods. The Cenozoic is frequently divided into the Quaternary – the uppermost Period – and the Tertiary. It is also sometimes found convenient to split the Cenozoic into the Palaeogene, which includes the Paleocene, Eocene and Oligocene, and the Neogene, which includes the Miocene, Pliocene and Pleistocene. The Quaternary is split into the Pleistocene, and the Recent (the last 10,000 years).

A geological time-scale is shown in Figure 1.9. In general we know more of recent events in Earth history than of remote times, in our landscapes recent events have more impact than remote ones, and the more recent tectonics are visible at the surface while ancient tectonic features may be

ERA		PERIOD	EPOCH		M. yrs ago
PHANEROZOIC	CAINOZOIC	QUATERNARY	Recent - about 10,000 yrs Pleistocene		
		NEOGENE (TERTIARY)	Pliocene		1.8
			Miocene		5
					24
		PALAEOGENE (TERTIARY)	Oligocene		37
			Eocene		58
			Paleocene		65
	MESOZOIC	CRETACEOUS			141
		JURASSIC			205
		TRIASSIC			251
	PALAEOZOIC	PERMIAN			298
		CARBONIFEROUS			354
		DEVONIAN			410
		SILURIAN			434
		ORDIVICIAN			490
		CAMBRIAN			545
PRECAMBRIAN		PROTEROZOIC			2500
		ARCHAEAN			4000
		HADEAN			
		AGE OF EARTH			4600

Figure 1.9 Geological time scale. Ages are as currently used by the Australian Geological Survey Organisation.

preserved in rocks but not in landscapes. We thus have a perspective of time, with a clearer view and larger scale for those features near us in time, and a more limited view of events and features in the distant, remote past.

Sometimes ages are given in years. This is usually in millions of years, which may be indicated by Myr or Ma. Ages determined by isotopic methods are often called absolute ages, but perhaps numerical ages is a better term.

Much of this book is concerned with the Neogene, and it is inconvenient that the time divisions are so irregular. The Quaternary or Pleistocene starts at 1.8 Ma and the Pliocene at 5.5 Ma, but the Miocene goes back to 22.5 Ma. We shall often be interested in events in the last ten million years, so will often refer to things as Upper Miocene or Mio-Pliocene or Plio-Pleistocene. The Cenozoic units are divided into smaller units, but they are numerous, hard to remember, and vary around the world, so we avoid using them.

2 Simple plateaus and erosional mountains

The simplest of mountain types are plateaus, but even here there is a wide range of variety. Several examples will be described which can be regarded as type specimens. Plateaus are eventually destroyed by erosion, and many stages can be traced from almost intact plateaus covering huge areas, through small remnants and eventually to mountains where no trace of original higher ground can be found. Erosional mountains, sometimes known as mountains of circumdenudation, are also treated in this chapter.

Caprock plateaus

Mountains with structural control by hard rocks are common, and the simplest type are in horizontal rocks.

Mount Conner in central Australia is a typical example. It has a caprock of quartzite, and this forms vertical cliffs above the concave slopes eroded across underlying shales (Figure 2.1). The rock is probably Palaeozoic. There is no way of knowing how much material has been stripped off the top, and there is no reason to suppose that the top ever corresponded to an erosion surface. Such mountains are common, and often have local names such as table mountain, including the famous Table Mountain of Capetown, South Africa (Figure 2.2), mesa, high plain, or simply plateau.

Mesas (Tepuis) of Venezuela and Guyana

Structurally controlled plateaus are very common, and because of vertical jointing in the rocks they often have near-vertical sides. It may be possible to find erosion surfaces in such a situation, but because horizontal rocks give horizontal ground surfaces anyway, it is difficult, to say the least.

In the northern part of South America, in Brazil, Venezuela and the Guyanas, lies a huge plateau known as the Pakaraima Mountains. It is underlain by horizontal Precambrian marine strata and results from simple vertical uplift in post-Cretaceous time. Since then it has been attacked by erosion. On the northern side is a Great Escarpment, deeply embayed and sometimes isolating individual plateau remnants. The name 'Pakaraima

Figure 2.1 Mt Connor, Central Australia (photo C.D. Ollier).

Figure 2.2 Table Mountain, Capetown, South Africa (photo C.D. Ollier).

Mountains' appears to be applied to both the plateaus and the bounding escarpment.

In the north west of Guyana the flat-lying Proterozoic sediments of the Roraima Formation form the Pakaraima Mountains. These are forested plateaus, generally at about 600–1000 m, with huge escarpments over which are spectacular waterfalls such as the Kaitur Falls and the world's highest waterfall, Angel Falls (979 m). Several mesas or 'tepui' rise above the general level (Figure 2.3) culminating in Mt Roraima (2730 m) at the point where Venezuela, Brazil and Guyana meet. The Tepuis provide wonderful isolated plateaus, and were used by Conan Doyle as a remote place in his book *The Lost World*.

In Venezuela there is a large area known as the Gran Sabana (Schubert and Huber, 1990; Briceño and Schubert, 1990), north of the Pakaraima

Figure 2.3 Tepuis in Venezuela (photo J. Dohrenwend).

Range. The Gran Sabana is lowland, but there are also several areas of plateaus of tepui, notably the Chimanta Massif (Figure 2.4). The mountains are plateaus, and clearly controlled by the rock structure, which is here horizontal Proterozoic sedimentary strata. For the past 545 million years life has been abundant and fossils are common in sedimentary rocks. This stretch of time is known as the Phanerozoic (Figure 1.9).

The tepuis are possibly the simplest of mountains. They are commonly regarded as being made by differential erosion of horizontal strata. However, Schubert and Huber (1990) propose a much more complicated scenario, with repeated planation and scarp retreat through Phanerozoic time. Their scheme is shown in Figure 2.5, and the caption explains their ideas.

The Kimberley and Carr Boyd Ranges

The Kimberley Plateau in north west Australia is underlain by gently folded sandstones and shales of Proterozoic age. The rocks were planated, and broad bevels are cut across the hard rocks. The plateau was uplifted, and softer rocks were eroded to form broad valleys. The process is more obvious in the nearby Carr Boyd Ranges where dips are steeper. The 'ranges' are strike ridges on the harder rocks (Figure 2.6). Strike ridges with sharp tops where two opposing slopes meet are called 'cuestas'. The strike ridges of the Carr Boyd Ranges have a flat bevel on their tops, and such ridges are known as 'bevelled' cuestas (Figure 2.7).

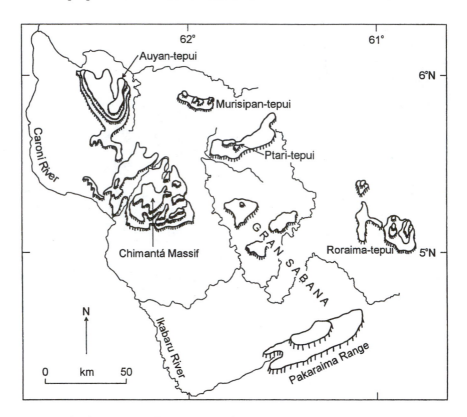

Figure 2.4 Sketch map of the Tepuis of Venezuela and the Pakaraima Range.

This is a suitable place to describe the origin and significance of bevelled cuestas. If a dipping bed of hard rock is simply eroded it develops a dip slope and a scarp slope (Figure 2.8a), with a rather sharp ridge (a cuesta) where these two slopes intersect. If the crest of this ridge is bevelled (Figure 2.8b), that is to say it has a level top, there is no doubt it is the remnant of a former erosion surface, for the rock structure would never develop a flat top unless a lateral erosion process was working at a particular base level. If there are a lot of bevelled cuestas, all at about the same elevation, then a certain confidence may be placed in the former existence of a broad planation surface.

The evidence for former planation surfaces must be considered in three dimensions. For reconstruction of old surfaces in folded sedimentary rocks, if the strike sections of ridges show accordant summits then there is good reason to think there was a planation surface not much higher, even if bevelled cuestas are absent.

The Carr Boyd bevelled cuestas shown in Figure 2.6 are clear indications of an old planation surface, for there is no mechanism to erode flat surfaces

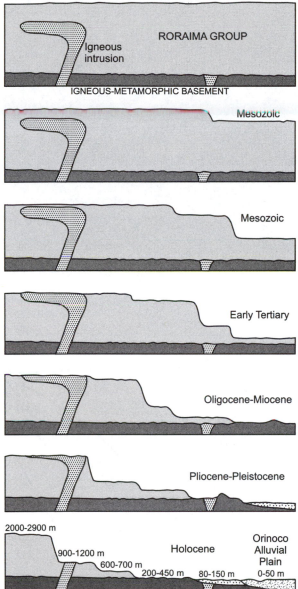

Figure 2.5 Successive cross-sections of the Gran Sabana, Venezuela (after
Schubert and Huber, 1990). A succession of planation surfaces are
formed, but according to these authors the upper surfaces are reduced
by general surface lowering. This is discussed further in Chapter 11.

Figure 2.6 Aerial view of the Carr Boyd Ranges, Western Australia, showing
bevelled cuestas. Note how the bevel can be traced around the nose
of the anticline (photo C.D. Ollier).

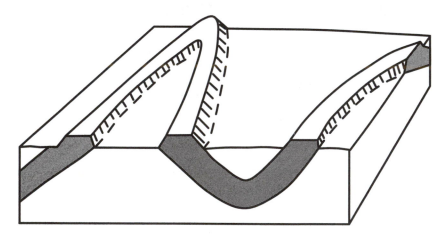

Figure 2.7 Diagram of planation cut across folded rocks, forming bevelled
cuestas.

across the hardest rocks *after* uplift and incision of valleys. So the folding
of the rocks is older than the erosion surface, and the folding did not make
the mountains. Uplift of a planation surface to form a plateau made the
area one of high relief, and subsequent erosion made the Carr Boyd Ranges.

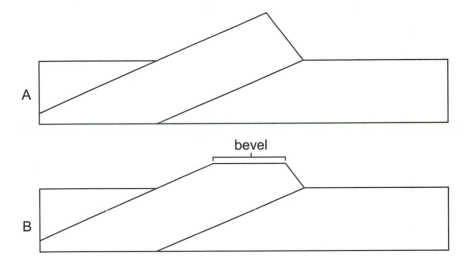

Figure 2.8 Cuestas. (a) A cuesta is a ridge formed by differential erosion of a dipping hard stratum. (b) A bevelled cuesta has a flat bevel that is a sure sign that an upper planation surface at about bevel level existed before differential erosion created the cuesta.

This simple story is repeated in a great many of the world's plateaus and ranges. The only unusual thing about the Kimberley–Carr Boyd region is the great age of the planation surface. The ancient Kimberley Plateau had already formed and been partly dissected by the time of a Proterozoic Glaciation, and has been a land area ever since (Ollier *et al.* 1988).

The Ozark Plateau

Like the Kimberley Plateau, the Ozark Plateau has an erosion surface cut across folded rocks, but it is more complex and has given rise to great arguments. It was described in detail by Bretz (1962), and it is worth repeating some of his observations here, together with his list of criteria that support a planation surface interpretation.

The Ozark Plateau is structurally and topographically an imperfect dome with a diameter of about 400 km, surrounded by lowlands. Drainage is essentially radial, with long radial interfluves that cut across the strata because structural dip is slightly greater than topographic slope. A typical summit flat near Raymondville is over 1.5 km wide for a length of about 16 km and ranges in altitude only 25 m. The flat on which the town of Licking stands has a slope of only 4 m in a kilometre, whereas the gradient of the radial valleys is about 200 m/km. Old river gravels on the summit surface appear to be pre-Pleistocene, and the age of the surface is Tertiary or older.

As support for a planation surface in the Ozarks, Bretz lists the following:

1   relict shallow valleys on the summit flats;
2   dissection of the main divide to leave wide, isolated summit flats inter-
     spersed with much dissected stretches carrying the same summit
     accordance and composed of the same rocks;
3   nonparallelism of divide and stream valley longitudinal profiles;
4   relict hills that rise above the even crest lines of the divides;
5   continuity of profile of benchlands in valley heads with interfluve
     summits downstream;
6   failure of rock and structure contrasts to appear in the upland topog-
     raphy;
7   failure of escarpments to influence river course;
8   the bevelling of escarpment crests to conform to neighbouring radial
     summit flats;
9   entrenched meanders that necessitate changes of base level;
10   clay fill in many Ozark caves;
11   bedded stream gravels on uplands.

The planation surface is covered in part by gravels from ancient streams,
which are probably of late Tertiary age, and the uplift is also presumed to
be of late Tertiary age. This area is of particular significance in the history
of geomorphology, because it is the type area for Dynamic Equilibrium, a
hypothesis of landscape evolution advocated by Hack (1960).

Ayers Rock

Ayers Rock (Figure 2.9), known to Aboriginals as Uluru, is a mountain in
central Australia created by erosion of all the surrounding rocks. The strata
of Ayers Rock consist of arkose (feldspathic sandstone) and are very steeply
dipping. Since it is 70 km to the next mountain group, the Olgas, the
amount of erosion is impressive. Most tourists wonder 'how did that rock
get up there?' but the real question is 'where did all that rock go?' The
impressive feature in the photograph is the planation surface, not on top
of a plateau, but at ground level. This is a planation surface under construction.

Figure 2.10 gives a different view of the rock, and suggests that it might
indeed be a plateau, with the remains of a former planation surface just
gone. The top of the rock has no trace of ancient sediments or landforms
but is grooved by ridges due to modern weathering (Figure 2.11).

Between Ayers Rock and the Olgas there is an ancient lake lying on the
lower planation surface, with Pliocene sediments. This suggests that Ayers
Rock has looked very much as it does since Pliocene time.

Figure 2.9 Ayers Rock from the air. Note that the lower surface is a very well
developed planation surface, much flatter than the top of the rock.

Figure 2.10 A distant view of Ayers Rock showing an apparently flat surface. This
surface is modified by weathering but there is no regolith, so the
surface is lower than any original planation surface. Note also the
great flatness of the lower surface. This has a veneer of sediment in
part, but is essentially an erosion surface cut across steeply dipping
rocks, such as those of Ayers Rock itself (photo C.D. Ollier).

Figure 2.11 Photo of ridges and grooves on top of Ayers Rock. The surface is
entirely modelled by weathering, making strike ridges and many
weathering pits. In these circumstances, with no correlative sediments or regolith, the age of the upper surface is very hard to
determine. Here the best hope is to find datable sediments on the
lower surface, to provide a minimum age for the mountain.

Ancient plateaus

While Ayers Rock has a thoroughly eroded top with no accumulation of soil, some old plateaus have preserved weathered rock and soil from long ago, testifying to a remarkable lack of erosion. Many of the plateaus of eastern Australia are deeply weathered, and indeed weathered shale is usually the common brick-making material of the area. Even in glaciated Scandinavia, pockets of ancient weathering profiles of probable Cretaceous age are preserved (Lidmar-Bergström, 1995).

Rugged mountain ranges

As erosion attacks a plateau, the amount of remaining top surface land decreases, and the amount of land with steep valley sides increases. When valley sides meet at the top, a sharp ridge is formed and the plateau is destroyed. This makes a common mountain landscape of sharp ridges and V-shaped valleys, known in geomorphology as a feral landscape.

When the rivers have cut down to base level they very soon start valley widening, with the creation of flat valley floors and flood plains.

The true feral relief, with angular ridges and valleys, can only occur while the valleys are actively cutting down, which usually means that the region is undergoing active uplift. In fact wherever feral relief is found, it is an indication of active uplift. Hack (1960) provided a theory of equilibrium slope development based on regions with feral relief that was rapidly embraced by geomorphologists and applied to many and diverse areas. We believe that Hack's equilibrium ideas have an application strictly limited to areas of advanced fluvial erosion in regions of active tectonic uplift.

Glaciated mountains

In most of the world where glaciation is a major landscape factor, the glaciation remodels older landscapes that were created by fluvial erosion. Figure 2.12 shows several stages in landscape modification by ice. In some situations the old landscape is almost intact with just a few glacial cirques. In others glacial erosion has removed all but a few remnants of the old surface. Snowdon, the highest mountain in Wales (1085 m), is such a mountain (Figure 2.13). Eventually all trace of former pre-glacial landscape may be removed.

Great ice sheets have variable erosion styles but in general they produce glaciated plains rather than mountains. Nunataks are mountains that project above an ice sheet. When the ice finally disappears, old nunataks may be distinguished by their angular appearance compared to the rounded country occupied by the ice sheet.

The quickly flowing, wet-based ice of valley glaciers is the dominant erosive agent in glaciation. At the head of a glacier over-deepened hollows

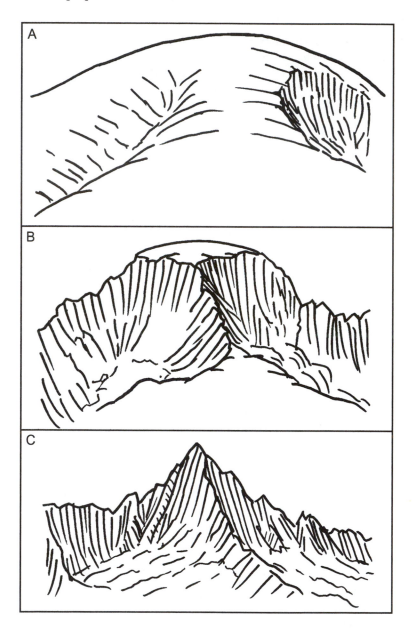

Figure 2.12 Stages in the modification of an old landscape by glaciation.

called 'cirques' may form. The main valley glacier carves the well-known U-shaped valleys (which contrast with the V-shaped valleys of fluvial erosion). Eventually the steep glaciated slopes recede, meet, and form a landscape of ridges and horns such as the Matterhorn of Switzerland.

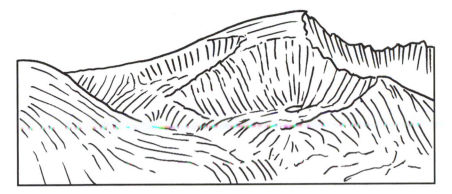

Figure 2.13 Snowdon, the highest mountain in Wales, showing remnants of an old pre-glacial surface after glaciation.

Contrasting landscapes can form depending on whether the mountains retain a cover of ice and snow, like Mount Elie de Beaumont in New Zealand (Figure 2.14), or if the snow is shed in avalanches from steep sided peaks like Mount Cook, New Zealand (Figure 2.15). Mountains in glaciated areas that do not collect snow are much attacked by frost, giving very jagged mountains, and steep rock upper slopes over scree-covered lower slopes (Figure 2.16).

Some mountains have suffered numerous glaciations, others few. In Australia for instance, the island of Tasmania had at least four glaciations

Figure 2.14 Mount Elie de Beaumont, New Zealand, showing a cover of ice and snow (photo C.D. Ollier).

Figure 2.15 Mount Cook, the highest mountain in New Zealand. The snow
          cannot accumulate to any great extent on the higher slopes, but
          avalanches to feed glaciers lower down (photo C.D. Ollier).

in the past 3 million years, but on mainland Australia only the last glaciation (20 000 m years ago) is represented, and it covered only 25 square
kilometres. Mt Kosciusko, Australia's highest mountain, is essentially an
old plateau with quite trivial glacial modelling, while some of Tasmania's
mountains are of classical alpine glacial form.

   Once all trace of a pre-glacial landscape has been removed it is difficult
to determine the early geomorphic history. There is no evidence of former
planation surfaces, except for accordant height of many mountains over a
wide area, so a Gipfelflur may be recognised, as in the Swiss Alps. Similar
accordant levels may be recognised in other places such as the Himalayas
or southern New Zealand, but the evidence is not accepted by all observers.

Inselbergs and rock domes

Inselbergs are steep sided hills or mountains of essentially bare rock that
rise above their surrounding plains like islands in the sea. There are two
main types.

   The first kind is associated with deep weathering, which penetrates over
100 metres beneath the ground surface. There is often a knife-sharp surface
between unaltered rock and completely altered rock which nevertheless
retained all rock structures intact (saprolite); this results from isovolu-
metric weathering. The contact between fresh and altered rock is called the

Figure 2.16 Mt Baker, one of the peaks of Ruwenzori. This is too sharp to be a snow collector, and it has a frost riven cliff and scree slope below (photo C.D. Ollier).

Figure 2.17 Diagram showing the formation of an inselberg by deep weathering followed by stripping of the saprolite so that the pre-formed inselberg protrudes from the surrounding regolith. Isolated core-stones are lowered as the saprolite is removed. Inselbergs frequently break up by unloading cracks parallel to the dome surface.

Figure 2.18 Inselbergs of Angola (photo Ilio Do Amaral).

Figure 2.19 The Sugarloaf at Rio de Janiero, Brazil. The inselberg is made of granite but the surrounding weathered and eroded rock is gneiss (photo C.D. Ollier).

weathering front, and it is very irregular. When the saprolite is removed by later erosion the higher parts of the fresh rock start to emerge, and they stick up as inselbergs (Figure 2.17). Details of how such inselbergs form are still much debated, and the arguments are summarised in Thomas (1994). Such inselbergs are very widespread in Africa (Figure 2.18), but also occur in many other localities.

The second kind of inselbergs, in Africa and elsewhere, have the same silhouette but the flat surface is cut across fresh rock, not saprolite. Particularly fine examples occur in Namibia. Some authors use the term 'bornhardt' for a particular kind of inselberg, but the exact differences between bornhardts and inselberg seem to be a matter of opinion. Perhaps bornhardt most often refers to inselbergs in deserts, but they are found in a wide range of climates.

Inselbergs are particularly common on granite, but also occur on gneiss, and similar landforms occur on sandstone (Ayers Rock), conglomerate (the Olgas), and on other hard rocks when joints are few and bedding plane partings absent.

In many instances the boundary of the inselberg does not correspond with any structural or lithological discontinuity in the bedrock. In a few cases there does seem to be some structural control. The Sugarloaf at Rio de Janiero (Figure 2.19) is a good example. The Sugarloaf itself is granitic, but the weathered rock all around which has been lowered by erosion is gneiss. Although the Sugarloaf gets most attention from tourists and photographers there are many other such inselbergs in the Rio vicinity.

It is difficult to date inselbergs or their surroundings. Sometimes the surrounding plain can be dated by overlying volcanoes or sediments. Some inselbergs have been related to particular planation surfaces, such as the African Surface, which may be roughly dated themselves. Sometimes the 'surfaces' may be misleading. For instance in Uganda the landscape was often described in terms of the African Surface and the Acholi Surface. But soil surveys showed the African Surface to be underlain by saprolite, and

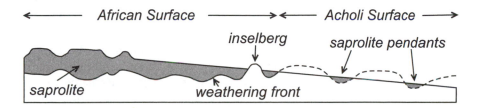

Figure 2.20 Inselbergs formed at the junction of the African Surface and the Acholi Surface in Uganda. The Acholi Surface is a younger surface cutting back and removing the saprolite that covers the African Surface. Since the weathering front is very irregular there are many inselbergs at the edge of regolith stripping, and some pendants of saprolite on the Acholi Surface (after Ollier, 1992).

Figure 2.21 Stone Mountain, near Atlanta, Georgia. A granite dome (sketched
        by C.D. Ollier from a photograph).

the Acholi Surface to be cut across bedrock. Ollier (1992) suggests that a
young planation surface cut across an old deeply weathered surface obliquely
(Figure 2.20). Inselbergs are particularly common around the edge of the
saprolite.

   The sharp contrast between weathered and unweathered rock at the weathering front is not confined to the tropics, or to plains. In many other areas,
bare rock domes, formed by deep weathering followed by stripping, protrude
through the landscape. In Australia, Bald Rock is a typical granite dome
that actually lies on the Great Divide of eastern Australia. If it rose from
a surrounding plain it would be an inselberg, but being surrounded by
steep slopes it is simply a granite dome. Stone Mountain in the United
States is a fine example (Figure 2.21), and from some accounts it seems
that Mt Monadnock (New Hampshire, US), which gave its name to the
last remaining remnant on a late-stage peneplain, is in fact a granite dome.

3 Fault block mountains

Some mountains are the dissected remains of blocks caused by faulting. The blocks may be tilted, and may bear the remains of an earlier planation surface.

Oregon-type fault block landscapes

This is the simplest type. It consists of originally flat-lying rocks, which in the type area are lava flows. There the faulting is so recent, the erosional and depositional modification so slight, that the region displays almost block–diagram simplicity.

Steens Mountain, Oregon, is a single fault block of Pliocene lavas with gentle tilt, bounded by fault scarps to east and west, and represents the simplest and most obvious fault block mountain of the horst type. The Grand Teton mountains are an example of a deeply eroded horst.

The Afar triangle in Ethiopia consists of fault blocks of young lavas which could be classified as Oregon type.

Sierra Nevada

The Sierra Nevada, California, is the remains of a huge, single tilt block, about 100 km by 600 km high on the eastern fault scarp edge, sloping gently to the west, with a massive fault separating it from Death Valley, 85 m below sea level (Figure 3.1). Sedimentary strata were folded in the Late Palaeozoic and intruded by huge batholiths. This had been eroded to a plain during a standstill in Eocene and Oligocene times when it was fairly low. Uplift and tilting to the west happened in Late Miocene to Pleistocene times. The single colossal block of the Sierra Nevada was partly covered by volcanic flows, especially to the north.

A precursor of the Tuolumne River in the central Sierra Nevada cut a large valley that was filled by volcanic rocks 10 million years ago. This allows various calculations and reconstructions. The uplift of the crest of the range over the past 10 million years has been calculated at 1830 m, which may be compared with 2150 m estimated from the San Joaquin River 30 km to the south.

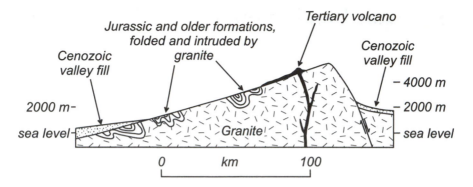

Figure 3.1 Cross-section of the Sierra Nevada.

Ten million years ago an ancestral range of hills occupied the present site of the Sierra crest. Although it had only moderate relief it was a barrier to westward drainage, even before the later uplift. The modern channel of the Tuolumne River is 915 m lower than the abandoned channel, and as much as 1525 m of channel incision has taken place in the last 10 million years (an average rate of about 150 metres per million years).

Axelrod (1962) used biogeographical evidence to determine the time of uplift of the Sierra Nevada. At present the Owens Valley, east of the Sierra Nevada, has a very different climate from the lowlands to the west of the range, a climate modified by the existence of the high mountains between. In the early Pleistocene a similar vegetation, a pine–fir ecotone, was established right across the whole region, a situation that could only exist if the present climatic barriers were absent, so the major uplift, on this basis, is well into the Pleistocene. Axelrod favoured a post-Pliocene uplift.

Unruhe (1991) used progressive tilting of strata to determine the age of uplift and tilting in the Sierra Madre. He found that major western tilting began 5 Ma ago, and approximately uniform tilting continued at a rate of 0.28° per million years. He points out that this does not correspond to the onset of the major Basin and Range extension. Nor does it support plate tectonic models that link uplift of the Sierra Nevada to migration of the Mendocino triple junction. He further notes the coincidence of timing with uplift of the Cascades, Colorado Plateau, northern Basin and Range and the southern Rockies and suggests the uplift 'may have occurred as part of a Cordillera-wide uplift event'. We agree, and merely extend it to a world-wide event.

Inevitably plate tectonic explanations have been applied to the Sierra Nevada. Crough and Thompson (1977) suggest that uplift is related to the passage of a subducted slab under the ranges. Others have gone into more detail, suggesting that subduction of the Mendocino triple junction migrated below the Sierra Nevada.

Basin and Range fault block landscape

The Basin and Range Province in the western United States is a classic example of an area of numerous tilt blocks, and can be well dated by associated volcanoes (Dohrenwend *et al.*, 1987).

In this type the bedrock has been folded and metamorphosed, though planated and perhaps capped by lava before faulting produced the present tilt blocks. This is associated with considerable shearing, but more importantly with crustal extension, a feature common in areas of block faulting and rift valley swells. Very good evidence of extension is provided by several elliptical to circular fault-bounded depressions in Oregon such as Summer Lake and the Upper Alvord playa, described by Fuller and Waters (1929). If they were due to compressional faulting then the forces must have acted centripetally like the closing of a camera shutter. Seemingly a dome would have been a more logical structure under these conditions. Upper Alvord playa is less then 5 km across, yet the bounding fault scarps have been thrust up over 300 m; the exact mechanism is not clear, but some sort of downfaulting of the depression during general extension seems the most plausible. The geomorphology of the classical Basin and Range province of the United States is complicated by aridity and the deep accumulation of sediments in the basins.

During Mesozoic and early Tertiary time the region was uplifted and eroded providing vast amounts of sediment to the Rocky Mountains and Coast Range geosynclines. In late Tertiary times and Quaternary times the crust was distended, with collapse into fault blocks, strike slip movement, and volcanic activity. The crust broke into blocks, forming the basins and ranges, with the fault block mountains being bounded by high angle faults dipping from 45° to 70°, and the block surfaces are tilted. The faults are curved – steep near the surface and flattening out at depth (listric faults).

The Basin and Range province has been the site of extensional orogeny for the past 40 million years according to Elston (1978). He believed it results from a three-part history: first an Andes-type volcanic arc, followed by a spreading ensialic backarc basin, which was in turn succeeded by intraplate block faulting and rifting (Figure 3.2). The total extension may have exceeded 100 per cent.

The first half of the Basin and Range evolution reached its climax in a great 'ignimbrite flareup' between 35 and 25 million years ago, when the province was inundated by 10 million km^3 of silicic volcanic rocks. The second half was dominated by extension, faulting and rifting which give rise to the present topography. The province overlies planed off earlier fold belts and the boundaries between cratons, eugeosynclines and miogeosynclines, but seems to be quite unrelated to them. There was a quiet period between the Laramide orogeny (Late Cretaceous to Early Tertiary) and the Middle Tertiary events.

Elston's estimate of extension over 100 per cent is much greater than that of most other workers who have extension of about 5 or 10 per cent.

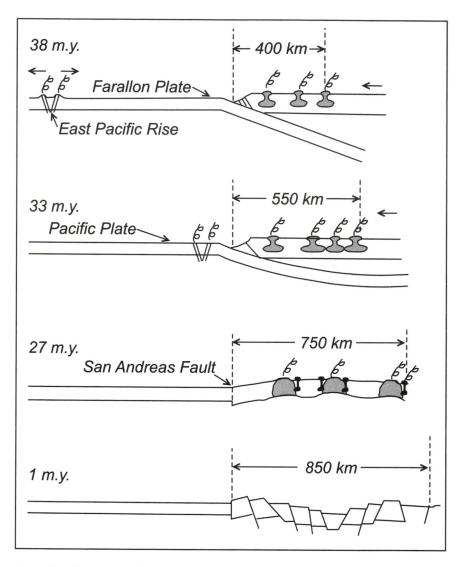

Figure 3.2 Tectonic evolution and extension of the Basin and Range Province over the past 40 Ma. The sections run from about the northern end of Baja California to north-central New Mexico (after Elston, 1978).

The low estimates are based only on the geometry of Basin and Range faults, but Elston also considers the volume occupied by plutons and thinning of the lithosphere during extension. Plutons are thought to underlie at least one-third of the Basin and Range province, and the crust has thinned from the average 40 km or more under the Sierra Nevada, Colorado Plateau and Great Plains, to 25–35 km under the Great Basin.

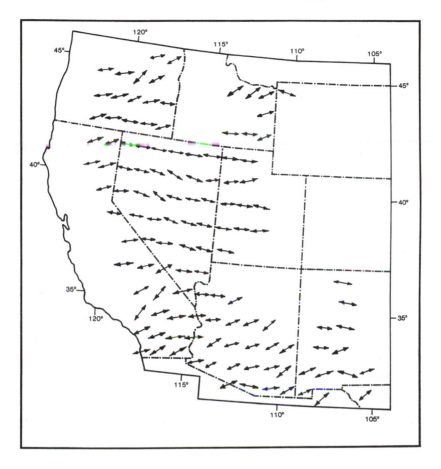

Figure 3.3 Average trend of extension of all normal fault segments up to 10 miles (16 km) long within each square degree of the Basin and Range province (after Gilluly, 1970).

The extension of the western United States is also brought out by Figure 3.3 which shows the trend of extension, based on all normal faults up to 16 km (10 miles) long. The arrows are drawn at right angles to the faults and show the direction of extension but not the amount, which must be determined from other geological considerations (Gilluly 1970).

The highest mountains in the Basin and Range Province are great volcanic plateaus, such as the San Juan Mountains and the Mogollon Plateau. They may have been raised to their present height (up to 4500 m) by the buoyancy of underlying batholiths during regional extension.

In the Basin and Range Province of southwestern Arizona external drainage systems replaced earlier internal drainage some time between 10.5

and 6 million years ago, and since then progressive erosion has created the present landscape.

Cenozoic extension has doubled the width of the Basin and Range Province. Faults reach depths of 20 km, and range in age from Oligocene to Holocene (Hamilton 1987). The traditional view of the Basin and Range Province is that it is formed by high angle normal faults. More modern work shows there is a great deal of low angle normal faults that have accommodated profound crustal stretching (Wernicke, 1981; Davis, 1984). Many of these detachment faults are of Oligocene–Miocene age. They are cut and offset by the high-angle normal faults.

Axelrod (1956) used fossil flora as a measure of palaeoaltitude, and argued that the northern Basin and Range Province has been uplifted 470–630 m since Miocene–Pliocene times.

Nitchman *et al.* (1990) examined the planation surface in the western Basin and Range and determined that it formed between 7 and 4 Ma. The disruption by the Basin and Range style of normal faulting began about 4 Ma.

The Basin and Range Province lies between the Sierra Nevada to the west and the Colorado Plateau and Rockies to the east. Plate tectonic compression has been applied to both the Sierra Nevada and the Rockies, but it is difficult to account at the same time for a broad zone of continued extension.

Basin and Range extension continues into Mexico east of the Sierra Madre Occidental both north and south of the Volcanic Belt (Figure 5.10). There, activity took place from middle Tertiary to present.

Rift valleys

Rift valleys are elongate grabens of large size, often with mountains along their flanks, and frequently associated with volcanic mountains. The East African rift valley system has a total length of over 3000 kilometres between the Red Sea and the lower Zambesi in southern Africa. Over this great length the system has remarkably uniform widths, as shown below:

	km
Western Rift (L. Albert)	35–45
Tanzania	40–50
Nyasa (Malawi)	40–60
Dead Sea	35
Turkana (Rudolf)	55

They are bounded by normal faults with huge throw, so that the base of the graben may be below sea level. The basement in the Dead Sea graben is from −2600 to −1229 m; Tanzania varies from 2150 to 2560 m; Nyasa varies from 1565 m to −1005. Opposite walls of the rift may be at different height, so clearly the outer blocks have risen by different amounts, as well

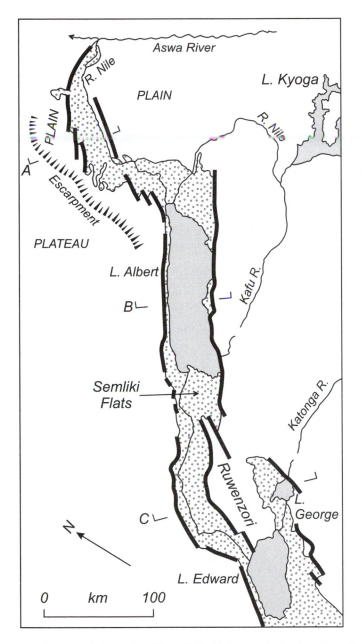

Figure 3.4 Map of the Lake Albert Rift Valley, and the location of cross sections shown in Figure 3.5.

as the centre block having sunk. The continental rift valleys have marked
negative gravity anomalies associated with their thick sedimentary fills.

In rift valleys the emphasis is on the grabens that commonly run along the
crests of continental arches, though in some instances the rise to the rift is
some tens of kilometres from the faults. Rift valleys are commonly bounded
by high land, as the Vosges and Black Forest border the Rhine graben, and

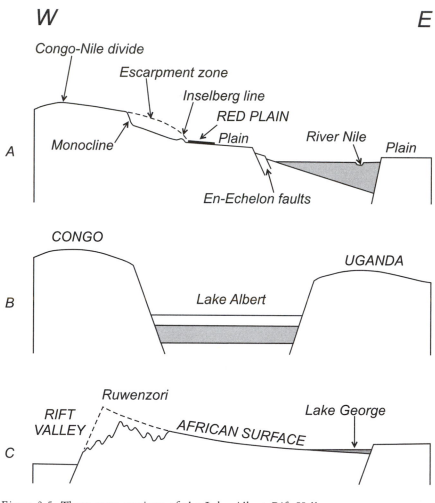

Figure 3.5 Three cross-sections of the Lake Albert Rift Valley.
　　　　　Top. Half graben, with hinge zone on west of the Nile
　　　　　Middle. Simple graben of central Lake Albert
　　　　　Bottom. The upwarp of the African surface to form Ruwenzori
　　　　　massif. It is assumed that all the fault movements are of about the
　　　　　same age, so the uplift of Ruwenzori is the same age as the forma-
　　　　　tion of the Rift Valley.

in places there are great topographic contrasts, such as the Albert Rift Valley which has a base (below lake sediments) about 1000 m below sea level and is bordered by Ruwenzori reaching 5110 m. The Lake Albert Rift also illustrates the wide variety in rift valley form (Figures 3.4 and 3.5). There are half graben, simple graben, and highly asymmetrical graben, and simple faults, en echelon faults, and scissor faults. The flatness of the African Surface is very striking in contrast to the steepness of the rift valley walls (Figure 3.6).

The rift system may be explained by vertical tectonics with uplift forming a broad arch, with the rift giving more localised compensating depression; simple tension, which accounts for the rifts but not the bordering arches or high fault blocks; compression (based on an improbable explanation of gravity data); and strike–slip faulting, which undoubtedly occurs in some rifts such as the Dead Sea but does not itself explain the rift form.

The pattern and timing of rift evolution and uplift has been reviewed by Partridge (1997).

In the Afar Triangle in the north faults bounding the Ethiopian Rift began to develop in late Oligocene–early Miocene times, and by mid-Miocene the rift was well established. Elevation of at least the northern part by 2000 m occurred since the late Miocene, and probably most of the movement (1000–1500 m) is referred to the late Pliocene. A major phase of tectonic movement of the western scarp of the Ethiopian Rift occurred

Figure 3.6 Photo of the rift valley wall on the east side of the Lake Albert Rift, near Butiaba, Uganda. Note the evenness of the planation surface (the African Surface) above the rift, with occasional erosional mountains rising above it. The rivers have not yet cut their valleys down to lake level, indicating relatively recent uplift (photo C.D. Ollier).

between 2.9 and 2.4 million years ago. Partridge summarises the evidence as showing subsidence of the rift and at least 1000 m of uplift beginning later than 3.5 million years ago It most probably started about 2.9 million years ago, and finished by 2.4 million years ago.

Further south the rift valley system splits into the Kenya and Western rift systems. There are broad swells bordering the rifts, 100–200 km wide and uplifted about 1000 m above the surrounding plateau.

Along the Kenya Rift volcanism and uplift started in the early Miocene. Doming started about 20 million years ago, but is no more than about 300–500 m. A large part of the topography around the Kenya Rift is associated with major volcanic eruptions (140,000 km^3) since the Miocene. Major uplift occurred later than 3 million years ago and reached a maximum during the Plio-Pleistocene. In these movements the rift shoulders were raised by 1200–1500 m in places.

In the Western Rift Miocene doming was smaller. Uplift of the rift shoulder averaged 1500 m but in Ruwenzori reached 4,300 m. Most of the rift movement occurred between 3 and 2 million years ago. Adjacent to the graben floor are steep fault scarps that rise up thousands of metres. The highest occur on the Ruwenzori Block, where Margherita Peak, at 5110 m, is more than 4000 m above the Semliki Plain below. The Livingstone Mountains rise over 2000 m above the northern end of Lake Malawi. In the Kenya Rift the Aberdares reach over 2000 m above the rift floor.

Pickford *et al.* 1993 have made a significant contribution to knowledge of the Western Rift in Uganda, summarised as follows.

Prior to the development of the rift the area was at an elevation of about 500 m. A shallow downwarp developed in the middle Miocene (15 Ma ago) which accummulated evaporites and fluviatile sediments. Lacustrine conditions began about 10 Ma ago and a permanent lake was present about 7 Ma ago. There are about 4 km of sediments in the Lake Albert Basin.

In summary, most of the significant mountain building associated with the rift valleys occurred in the last 3 million years. A modern review of the East African rifts is provided by Hampton (1997).

Many writers believe that rift valleys, especially the African rifts, make use of older structures – they are suggesting resurgent tectonics, and reactivation of old faults. In some places the grain of the basement rocks is parallel to the rifts, but in others it is oblique. In places undisturbed Precambrian rocks, virtually horizontal, overlie complex older rocks and are not folded, though cut by the later rift valleys. There is a danger that some older structures such as thrust faults might be attributed to the present rift activity when in fact they have no connection.

Vulcanicity is associated with many rift valleys, but not all. Some parts of the rift valley consist wholly of volcanic products, but here the chronology is relatively easy to work out, and the tectonic evolution can be followed in some detail. Part of the South Kenya rift valley is of this type (Baker and Mitchell, 1976). The tectonic evolution of the eastern margin followed

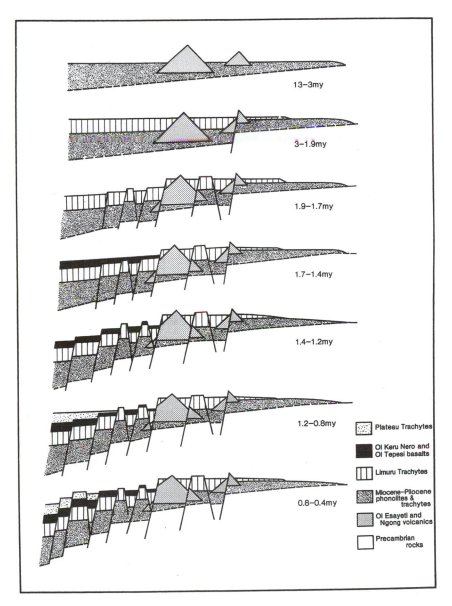

13–3my

3–1.9my

1.9–1.7my

1.7–1.4my

1.4–1.2my

1.2–0.8my

0.8–0.4my

Plateau Trachytes

Ol Keru Nero and
Ol Tepesi basalts

Limuru Trachytes

Miocene–Pliocene
phonolites &
trachytes

Ol Esayeti and
Ngong volcanics

Precambrian
rocks

Figure 3.7 Diagrammatic sections showing the evolution and extension of the east margin of the South Kenya Rift Valley (after Baker and Mitchell, 1976).

pre-rift eruption of lavas from shield volcanoes starting about 13 Ma ago. An eastern boundary fault formed 3.3 Ma ago and was followed by formation of an inner graben from 1.9 Ma ago. This was followed by an almost continuous succession of block fault movements (Figure 3.7).

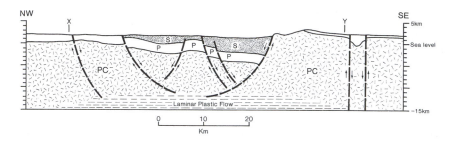

Figure 3.8 Diagrammatic section across the Rio Grande rift near Albuquerque, New Mexico. Extension of 8 km is calculated between points X and Y. S = sedimentary fill; P = Palaeozoic to Oligocene rocks; PC = Precambrian rocks (after Woodward, 1977).

Several rifts are known to be extending at measurable rates. The Rio Grande rift is about 850 km long, running roughly north–south from Colorado to New Mexico. There has probably been about 8 km of crustal extension in a north west–south east direction near Albuquerque since about 26 Ma ago (Woodward, 1977), an estimate based on the assumption that the normal faults bounding the graben decrease in dip downwards and extend to a depth of about 10 to 15 km, below which the crust undergoes plastic deformation (Figure 3.8). The rate is 300 B (Bubnoff units), which compares with a rate of 400 to 1000 B for the eastern rift system of Africa (Baker *et al.* 1972) and 100 B for the Rhine graben (Illies, 1972).

As noted in a previous section, the formation of swells seems to initiate faulting, rifting and extension, and it is interesting that the rift valley system of Africa can be traced continuously to the Red Sea, and thence to the Carlsberg sub-oceanic ridge. The Red Sea has major differences from rift valleys, as it has a positive gravity anomaly and a simatic sea floor. It is in fact an example of the earliest stage of sea-floor spreading, and it is easy to imagine the spreading site working along the rift system and isolating East Africa from the rest of the continent, just as in the past a similar break-up presumably separated Madagascar.

The Baikal rift

The Baikal system of rift valleys is remarkably similar to that of East Africa in many ways, including scale. It extends about 2500 km and there are about 12 major depressions ranging from 100 to 700 km in length and 15 to 18 km in width. Rift structures are graben along arched uplifts. They are usually asymmetric, with a steep northern flank and the other margin more gently sloping, either downwarped or with minor faults. The depressions are mainly dry, but there are lakes in places, such as Lake Baikal itself. Lake Baikal is huge, with an area of about 32,000 km², a length of

630 km, width 80 km, and a greatest depth of 1700 m. It holds a large portion of the world's fresh water.

The Baikal system is essentially Cenozoic, but tends to follow earlier structural trends. Before rifting, in Cretaceous and early Cenozoic times, the area was one of tectonic stability with subaerial weathering and erosion. Sediment thicknesses are around 1500 to 2500 m, but the exceptional Baikal depression probably has up to 6000 m. The graben fill of continental sediments and volcanics can be divided into two groups. A lower group is mainly lake, swamp and river deposits of Oligocene to Pliocene age, with thicknesses from 1000 to 3000 m: an upper group consists of 500 to 1200 m of very diverse sediments accumulated under cold conditions from later Pliocene to Holocene times. They were produced from mountain relief which increased through time. Intense gravity lows are found over Lake Baikal, resulting mainly from the great thickness of light sediment.

Volcanism, mainly basaltic, is largely independent of the rifts and is mostly on the uplifted blocks and arches. At the southern end of the lake there are some small Quaternary volcanoes. Some north-flowing rivers are antecedent to the Lake Baikal fault.

The Rhine Graben

In Germany the Rhine Valley is a broad rift valley on the Rhenish Plateau. The plateau was mainly above the sea after the Jurassic, but with marine incursions in parts in the upper Cretaceous and the Oligocene. The main uplift started at the end of the Oligocene and continues to the present, but uplift accelerated in the Pliocene. The Rhine Valley has been in existence since the Middle Miocene.

The Ruwenzori Mountains

The Ruwenzoi Mountains are the fabled Mountains of the Moon in Central Africa and were the reputed source of the Nile in ancient times. They are located in the middle of the African plate, bordering a rift valley, and could not be affected by subduction.

Ruwenzori is a dissected massif of Precambrian rocks that has been uplifted to a great height, and makes by far the highest non-volcanic mountain mass in Africa. To the west lies the Western Rift Valley. The eastern side of Ruwenzori is in some places simply warped and merges into the African erosion surface without a break (Figure 3.5c). Elsewhere the mass is faulted, and the northern part is a horst sloping down to the north. The age of the erosion surface in this part of Africa is poorly controlled, but is probably Mesozoic. Of course there is no trace of an erosion surface on the actual summits of Ruwenzori. These peaks are purely erosional mountains, with glaciers still surviving despite the equatorial location, and evidence of much greater glaciation in the past (Figure 3.9). On the reasonable assumption that the

Figure 3.9 Mount Margherita, one of the summit mountains of the Ruwenzori massif (photo J.F. Harrop).

main Ruwenzori uplift occurred at the same time as other tectonic activity in the Western Rift, the uplift probably occurred within the last 3 million years.

Owen Stanley Mountains at Milne Bay

On a simple tilt block rivers might be expected to flow in the direction of the regional ground surface, but an antecedent river will maintain its earlier course, even if it is in the opposite direction. A splendid example comes from the area south of Milne Bay, Papua New Guinea (Figure 3.10). Here the regional ground surface slopes gently up from south to north, and then descends quickly: it is an obvious tectonic tilt block. The main rivers rise at an elevation of only 50 m near the south coast, but flow north in gorges cut through the Owen Stanley Range 1000 m high. The rivers evidently originated as north-flowing rivers when the general ground surface also sloped to the north. The rivers continued to flow north while the tilt block was uplifted, and erosion kept pace with uplift so the rivers now flow through gorges cut through the tilt block (Pain and Ollier 1984). The high part of the ridge includes some remnants of a Plio-Pleistocene erosion surface, so uplift and antecedence are Quaternary events, though the original drainage may date back to the Upper Tertiary.

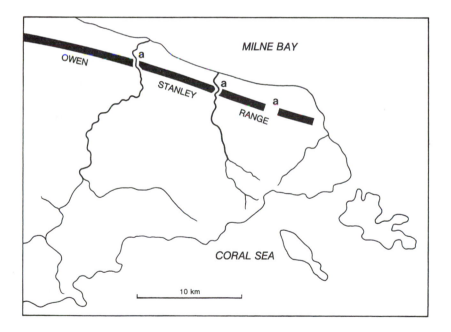

Figure 3.10 Antecedent drainage of Milne Bay, Papua New Guinea. The rivers rise near the south coast below 50 m and cross the Owen Stanley Range 1000 m high to enter Milne Bay in the north.

Shanxi Mountains and Shanxi Graben system

In China the western part of Yunnan Province resembles the Basin and Range Province of Arizona. It is formed by listric faults parallel to the eastern edge of the basin, which have produced steep fault scarps, dissected into triangular facets. Faulting was initiated in the Neogene, and there are

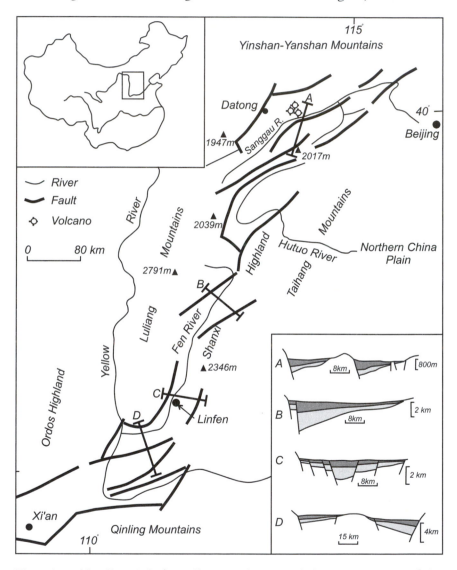

Figure 3.11 The Shanxi Graben, China, with an en-echelon pattern. Most of the basins are half-grabens. The dark sediment shown in the graben is Quaternary, the lighter sediment is Neogene. Although the Fen River follows roughly the line of basins it repeatedly cuts across horsts and is an antecedent stream.

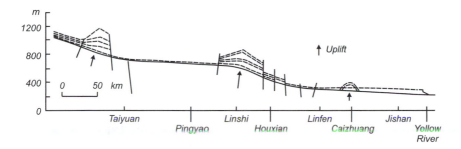

Figure 3.12 Long profile of the Fen River, showing the control of faults on terraces.

up to 1900 m of Neogene sediments over Paleogene sandstones and shales (Wu and Wang, 1988).

East of the Ordos Plateau the Yellow River flows north to south. East of this is the Fen River, which flows in the same direction. This river maintains its course across a series of horst and graben structures that lie obliquely across its course (Figure 3.11). The whole system is about 1200 km long, and the Luliang and Taihang Mountains that bound it have average altitude of 1400–2000 m. The fault blocks move differentially, as indicated by deformed river terraces (Figure 3.12). In the Miocene the area was planated, fault movement began in the late Pliocene, and uplift was at a maximum in the middle Pleistocene (Li *et al.*, 1998).

The Fen River has a strange course, almost parallel to the Yellow River (Huang Ho), but it flows across a whole series of horsts and grabens, repeatedly leaving the lowland which would provide an easy passage to the Yellow River and plunging into gorges that cross the uplifted blocks.

New Zealand mountains

The Southern Alps are located on the west side of the South Island, New Zealand, and more mountains are found in North Island.

In the Oligocene New Zealand was largely under the sea. In late Miocene–Pliocene times earth movements began, and raised the planated surface in the Kaikoura Orogeny. The mountains are essentially fault block mountains, and the uplift is essentially Pliocene: 'the history of tectonic mobility of most of the New Zealand region was such that in many areas the major vertical movements crucial to the development of the main ranges and lowlands did not take place until the Pliocene' (Suggate, 1982). In many parts of New Zealand the old erosion surface is very obvious, and even where erosion is quite intense, as in glaciated Fiordland, a summit envelope surface depicts the old planation surface or Gipfelflur quite well (Figure 3.13).

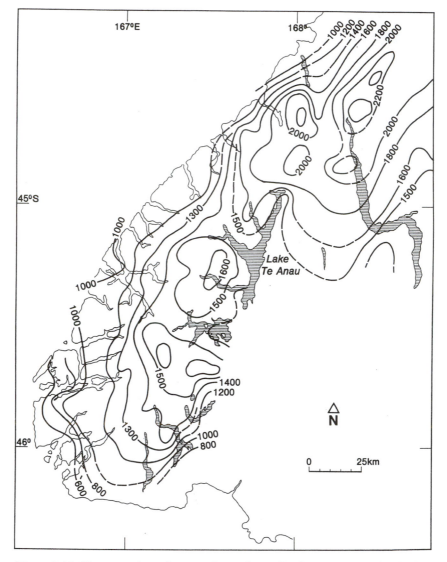

Figure 3.13 The summit surface envelope of Fiordland, New Zealand, which
approximates to the old erosion surface (after Augustinus, 1992).

As the main ranges of the North Island were uplifted to their present
height during the Pleistocene, they carried Pliocene marine sediments to
elevations up to 1200 m (Stevens, 1980). A great example is located near
the Manuatu Gorge, where the Manuatu River follows an antecedent course
through the main range. Pliocene marine sediments lie on top of the range
on either side of the gorge.

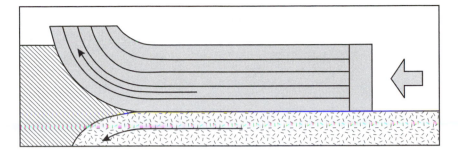

Figure 3.14 A plate tectonic model for the uplift of the Southern Alps of New Zealand, with subduction, obduction and rotation. The driving force is mysterious, but presumably seafloor spreading. The origin of the 'planation surface' on the Alps is not explained (after Wellman, 1975).

There are, of course, many plate tectonic explanations of New Zealand structure and landscapes, including subduction of the Pacific Plate under North Island, subduction of the Indo-Australian Plate under Fiordland (south-western New Zealand), or the formation of a compensating root under Mount Cook, the highest mountain. These are illustrated in Selby (1985, p. 82). Wellman (1975) provides a rather extreme example shown in Figure 3.14. The most significant drawback of plate tectonic explanations is the difference in time scale between seafloor spreading and the young uplift of the mountains, and the lack of plate tectonic explanation for the pause that allowed extensive planation of New Zealand.

The Sierra de Famatina

This range in north-western Argentina appears to be a tilt block uplifted along a major thrust fault. It became a positive feature for the first time about 6 million years ago (Tabbutt, 1990).

4 European mountains

The mountain ranges of Europe (Figure 4.1) fall into several groups. The mountains of Scandinavia are treated in Chapter 10 with passive margin mountains. The others include the Apennines of Italy, various ranges in Iberia, the European Alps and the Carpathians. In this chapter we also deal with some related mountains of Asia and Africa.

The European Alps

The European Alps have become a standard and type area for the study of the geology and geomorphology of mountains. This is the area where great battles were first fought over nappe theory, those vast sheets of rock that were thrust for great distances. It seems to be generally forgotten that these nappes have very little to do with the present Alpine topography. The whole region was planated in the Pliocene, and then broadly uplifted and eroded to the present spectacular topography (Figure 4.2). The old erosion surface has virtually gone, but is preserved as the Gipfelflur or summit level (Heim, 1927; Rutten, 1969).

This idea is not new, and was described, for example, by Heritsch (1929) who wrote:

> The morphological studies in the Eastern Alps have further proved, from the summit-level (*Gipfelflur*) of the peaks . . . that these erosion-horizons have no sort of relation at all to the geological structure. A further result of research on East-Alpine morphology is the recognition of the fact that the upheaval of the Alps is not connected with the production of the leading features of the internal structures, but that it is related to a later process of elevation, which was of vigorous character.

The Alps are divided into the Western Alps or 'Swiss Alps', and the Eastern or 'Austrian Alps' and the Southern Alps. The Western Alps are further divided into many geographical units including the High Calcareous or Helvetian Alps, and many others. The European Alps have a complex geological structure.

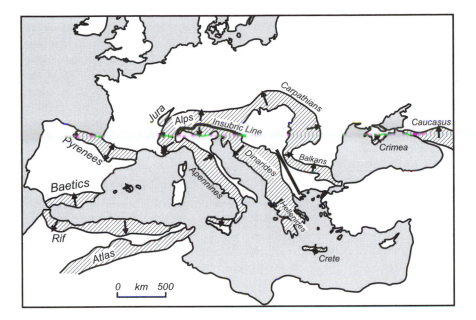

Figure 4.1 The main alpine mountains of Europe.

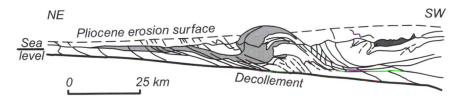

Figure 4.2 Cross-section of the Alps. The complicated nappe structures were planated before the uplift of the present mountain mass (simplified after Spencer, 1965).

The Alpine story in southern Europe starts with deposition of sediments in geosynclines that were part of the Tethys Sea. The main Alpine folding was in the Early and Middle Tertiary. Terrestrial sediments (molasse) are common from the Oligocene or Miocene onwards. There are variations in time in the main folding between different parts of the Alpine system, the Pyrenees and the mountains of Provence having been folded earlier than most of the other parts of the chain. Despite many variations all the mountain chains belong to the same Alpine 'orogeny', that is the same general period of rock deformation.

The rocks are thrust as a series of nappes. In the north, thrusting was to the north, and the southern Alps have thrusting to the south. Between the two is a complex zone called the 'root zone'. Prior to about 1880 the rocks of the Alps were thought to be more or less in the place in which they had originally been deposited (autochthonous) although it was evident that they had been folded. By about 1900 it was clear that many of the rocks had travelled long distances from their site of deposition (allochthonous) and the term 'nappe' came to be applied to any large, more or less horizontal sheet of rock that had been displaced a considerable distance.

At this stage the classical theory of alpine orogeny was evolved, based on nappes and the 'root zone' roughly along the Swiss–Italian border. The general concept is of vice-like compression between the forelands of Europe and Africa, with squeezing out of the deposits of sediments in the middle, forming a root zone of vertical and metamorphosed rocks, together with the far travelled nappes.

The root zone is nowhere connected geographically to the nappes, and probably has little direct geological connection. It is a zone, 400 km by 20 km or less, of vertical gneisses, highly metamorphosed and obscured by later plutons. It is a distinct structure separating the northern Alps from the southern. It is a major crustal cicatrise, but how and when it was formed, and what is its relationship to nappes is not clear. The root zone is not to be confused with the root of the Alps postulated to explain negative Bouguer anomalies. The root zone is immediately to the south of the northern nappes, so if they are far-travelled they have been followed closely by their own root zone – the root has travelled with them.

According to Trumpy (1980) the climax of 'orogeny' occurred in the Eocene but by the Pliocene the Alpine region was worn down to a chain of low hills. Later irregular erosion reduced the Pliocene plateau to a Gipfelflur, roughly accordant summit heights suggesting a former continuous surface (Rutten, 1969).

In the Eastern Alps northward movement of nappes across the basin of molasse deposition ceased before the Upper Miocene. After the Miocene the development of the topography continued with differential vertical movements on steeply dipping faults. Fault-bounded basins of marine or lacustrine deposition were formed in places, most importantly the fault-bounded Vienna Basin which cuts across the Alps and buries their continuation to the Carpathians under several thousand metres of Miocene to Quaternary marine and later lacustrine sediments.

Frisch *et al.* (1998) divide the Eastern Alps into two parts, the western and eastern. They say the western Eastern Alps were already mountainous in Late Oligocene and Early Miocene times, whereas the Eastern Alps formed lowlands. This area became a mountainous area not before Late Miocene.

According to the same authors the Tauern window has grabens and pull-apart basins filled with Miocene sediment. The Tauern window was opened about 13 Ma.

They also say the topographic evolution of the Eastern Alps show that elevations that would allow a differentiation into climatic provinces as they exist today were not attained before Late Miocene/Pliocene times. They support a mechanism of tectonic extrusion, described further in the section on Gravity in Chapter 12.

The Central (mainly Swiss) Alps display high relief as well as high mean and maximum elevations, while in the Eastern (mainly Austrian) Alps these parameters decrease from west to east. According to Frisch *et al.* (1998), in the Eastern Alps some pre-Miocene landsurfaces are preserved, but the region became a mountainous area not before Late Miocene times. The topographic elevations that caused climatic provinces of today (Atlantic, Mediterranean and Pannonian) were not attained until Late Miocene/Pliocene times. This topic is discussed further in Chapter 11.

The Planation surface

The Alpine summits levels form a broad arch between the Molasse Basin in the north west and the basin of the Po Plain in the south east, with minor undulations along the arch and across it. Towards the close of the Pliocene the Alps had been reduced to a region of low relief, the complex underlying structures being truncated by the erosion surface.

This old erosion surface was described many times in the past but it seems to have been forgotten in the many plate tectonic explanations of recent years.

Plate tectonic explanations

Plate tectonic models applied to the Alps are meant to explain the topography as well as the structures. Dewey *et al.* (1973), for example, claim 'The contemporary Alpine system develops a spectrum of stages in the building of mountain belts . . . the Alpine system is the result of activity of accreting, transfer and subducting plate boundaries between Europe and Africa.' They totally ignore the planation, so do not explain why uplift occurred so long after collision, or why tectonic movement apparently ceased during erosion of the summit surface.

In plate tectonic terms the Alpine belt is a continent–continent collision belt formed by interaction of the African and Arabian plates with several European plates. Numerous plate tectonic scenarios have been presented including that by Dewey *et al.* (1973), and Windley (1977) presents a summary. The evolution is complicated even by plate tectonics standards because the evolution took about 200 Ma and involved a large number of plates and microplates (Figure 4.3). For example, according to Dewey *et al.* (1973) there were six periods of basalt formation, seven of ophiolite formation, three of ophiolite subduction, 11 of deformation and seven of high tectonic plate metamorphism taking place diachronously in different areas.

A synopsis is as follows:

1   fracturing of pre-Alpine continental Pangaea and extrusion of flood basalts in Triassic times;
2   formation of new ocean crust between platelets;
3   deposition of Triassic evaporites, carbonates and redbeds;
4   collapse of carbonate shelves and deposition of deep water shales in the Jurassic;
5   cretaceous development of island arcs with ophiolitic melanges, acid volcanism, blueschist metamorphism, deposition of flysch in trenches, thrusting and obduction;
6   Eocene–Oligocene continent–continent collision, with thrust sheets (nappes), further flysch deposition and opening of marginal basins in the western Mediterranean;
7   later orogenic uplift, and deposition of Upper Tertiary molasse in foredeeps. Late Miocene (Messinian) evaporite deposits in Mediterranean and other basins.

There is a great deal of variation in plate tectonic models, as indicated by interpretations of the Insubric Line, a suture dividing the Southern Alps from the northern, Austro-Alpine nappes. Ernst (1973) considers the Southern Alps overrode and metamorphosed the northern block; Oxburgh (1972) suggested the Austro-Alpine nappes were flaked off a northward descending southern plate, and Laubscher (1971) regards the Insubric Line as a strike–slip fault with up to 300 km of dextral movement.

Platt (1997) provides another view. He wrote:

> Convergence between Africa and Europe from mid-Cretaceous time onwards, was dominantly northwards, and this is traditionally regarded as the direction of Alpine shortening. Structural data from thrust faults in the Alps, however, suggest that motion was everywhere roughly normal to the local trend of the chain right around the Alpine Arc . . . this radial pattern of motion would require a large amount of extension along the trend of the Alps, for which there is little evidence.

This view contrasts with that of Frisch *et al.* described earlier. Platt describes the Alps as due to the collision of two passive margins, but he is not referring to the type of passive margin described in Chapter 10, with marginal swell and Great Escarpment. He means that in the Jurassic Pangaea split up, forming the Tethys Sea in the Alps area, which had passive margins that subsided steadily through the Mesozoic, and then the two 'passive margins' moved together again.

A modern account of the evolution of the Eastern Alps by Frisch *et al.* (1998) is basically a plate tectonic explanation of the evolution of the area during the Neogene, but differs from many such accounts in attempting

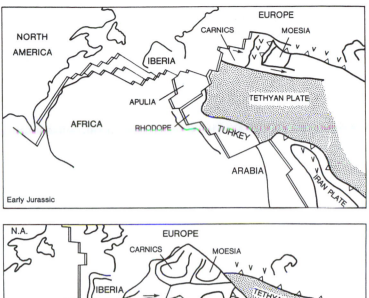

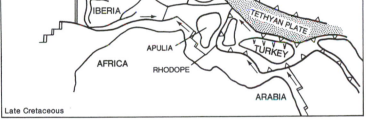

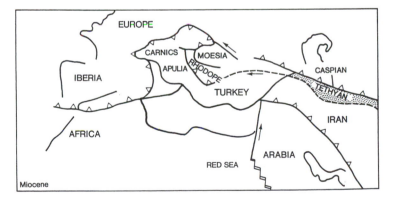

Figure 4.3 Proposed plate tectonic evolution of the Alpine system (after Dewey *et al.* 1973).

to reconstruct the Neogene palaeotopography and palaeohydrology as well as structural evolution.

The Jura Mountains lie in north-west Switzerland, separated from the Alps by the central Swiss plain. They rise about 1000 m above the Swiss

Plain and maintain a fairly constant altitude, suggesting a dissected plateau. The rocks are mainly Jurassic limestones, and they appear to have been folded during the Late Miocene and Pliocene. They provide the classical example of decollement, with the upper rocks folded as they detached from and moved along the sole of Triassic evaporite horizons (Figure 8.5). Many regard this as classical gravity tectonics. Platt (1997) on the other hand, wrote that the Jura are 'the youngest expression of Alpine convergence resulting from shortening' and 'This motion reflects the final stage of the convergence of Adria with Europe.' The shortening direction in the Jura, indicated by stylolites, varies from north to west around the arc, mirroring the pattern of the Alps as a whole.

The Apennines

The Apennines of Italy run in a broad curve or arc, convex to the east (Figure 4.4). They are bounded to the east and north by the Adriatic and its onshore continuation, the Po valley, to the west by the Tyrrhenian Sea, and to the south by the Ionian Sea. The Adriatic–Apulia Block (in southern Italy) is a foreland against which the Apennine rocks have been thrust. The Tyrrhenian Sea and Ionian Sea are spreading sites. The present Apennines reach about 3000 m, with mean altitude of about 1500 m.

Like the Alps, the Apennines are built of stacks of rocks (nappes) that moved horizontally for over 100 kilometres. But neither the nappe movement nor the intense folding and faulting at their front made the mountains. After the nappe movement came a period of erosion which created an essentially flat planation surface cutting across earlier rock structures. Regardless of arguments about the age and perfection of the planation surfaces, the old structures associated with the nappes are eroded, and vertical uplift made the present Apennines, mainly in the last 2 million years.

Outline geology of the Apennines

The Apennines consist dominantly of sedimentary rocks, mostly Mesozoic and Tertiary. A small area of metamorphic rocks is present in the west, and Quaternary volcanoes are also restricted to the west.

Structurally the Apennines consist of a series of nappes with marine and terrestrial components, including many layers that made suitable slip planes. Some nappes several kilometres thick show repetition of a Triassic–Jurassic–Cretaceous sequence, with nappe movement of over 100 km. Thrusting and deformation increases to the east (front) of the thrusted mass, but further to the west (back) the layers of sediment are parallel and unfolded. The unthrust foreland also consists of a nearly horizontal stack of Triassic, Jurassic and Cretaceous strata.

The stack of nappes has a total thickness of over 10 km, very much greater than the height of the Apennines, so much of the faulting

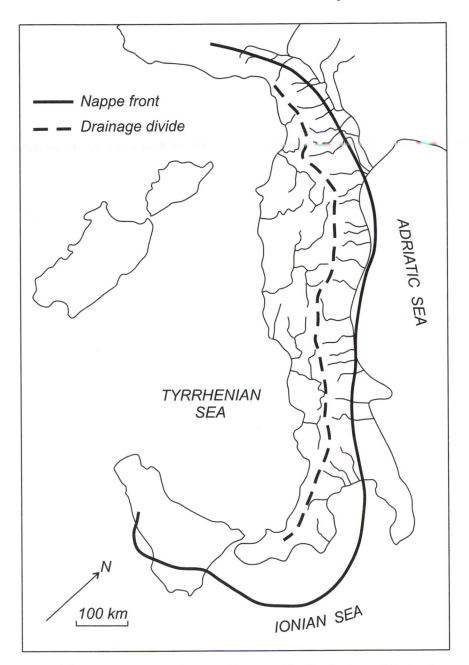

Figure 4.4 The Apennines and surrounding seas. The Tyrrhenian and Ionian Seas are spreading sites, but the Adriatic has continental seafloor that continues on land in the Po Plains. The major drainage divide is parallel to the nappe front, and even at this scale the difference in drainage style on opposite sides is evident.

occurred when the sediments were beneath the sea. The main thrusts and nappes were therefore formed before the Apennine topography had been created. Nappes extend across the entire country south of the Po Plain, and the source must have been to the west, where the Tyrrhenian Sea now lies.

The geology can be described most simply with reference to a typical cross section, Figure 4.5. In the east there is a simple conformable sequence of Triassic, Jurassic and Cretaceous sediments. To the west this conformable sequence is affected by fairly high angle thrust plains, but is then overlain by the first great low angle thrust or nappe plain and the Lagonegro–Molise sequence. This in turn is overlain by the Apennine unit, with conformable Triassic, Jurassic and Cretaceous strata. In the west this is overlain unconformably by Late Miocene rocks. The highest nappe unit of all, hardly shown on this section, is the Ligurian nappe, which is the oldest of all. The Ligurian rocks are highly deformed and includes shales, olistostromes, slices of oceanic basement, serpentine, gabbro, and peridotite. The Ligurian nappe forms the highest parts of the mountain chain in the northern Apennines. The western end of the section shows normal, mainly listric faults that are younger than the Late Miocene sequence, and older than the Middle Pliocene to Quaternary sediments that fill the basins created by downfaulting. It is evident on the section that the deformation of the nappes increases to the east. Since the normal faults are younger than Late Miocene, the formation of the large nappes must be older than Late Miocene.

Other sections are generally similar, but there are some variations. Quaternary volcanics are present in some areas, as are other nappes. Some show more intense thrusting against a foreland. The Ligurian nappe is present over a large area, and is at the easternmost end of the nappes, illustrating the idea that the oldest nappes travel furthest. Some nappes extend into Middle Pliocene–Quaternary deposits, showing they were still being thrust to some extent at the time these sediments were deposited. The nappes break against a foreland known as the Apulian foreland. A north–south section across Sicily shows similar features, with thrusts breaking against the Saccense foreland.

SW   NE

```
                                                                    ┌ 0
                                                                    ├ 5 km
                                                                    └ 10
```

Figure 4.5 A SW to NE section of the Apennines, starting south of Naples. Several nappes are thrust against the foreland in the NE, but the whole has been eroded and affected by normal faults which creates much of the present topography. The three shaded layers on the foreland (right) are Cretaceous, Jurassic, Triassic. The same sequence is repeated in the upper nappe in the SE. The unshaded layer beneath it is another nappe with rocks of the same age (simplified after a section in Cassano *et al.*, 1986).

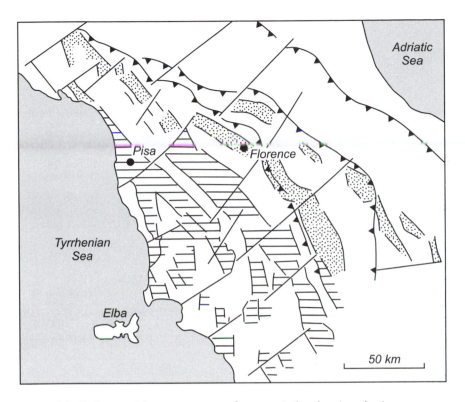

Figure 4.6 Graben and horst structure of eastern Italy, showing the importance of extension in the morphotectonics of the area. The basins with line shading have mainly marine fill, dating from Messinian (uppermost Pliocene) to Quaternary. The stippled basins have terrestrial fill dating from lower middle Pliocene to Recent.

The normal faults create a staircase of blocks, mainly downfaulted to the west to make a series of grabens and half-grabens (Figure 4.6). These make topographic basins which collect younger sediment, either marine if offshore, or lacustrine if onshore. The sedimentary fill is usually cut through by later erosion, though a few sedimentary basins are still intact. Grabens with lake sediments cover a significant area of the Apennines today. Pleistocene sediments are found at elevations over 500 m.

Some, perhaps most, normal faults tend to flatten at depth (listric faults). Several different types occur in the Larderello area of Tuscany. During the Early and Middle Miocene faults flattened in Triassic evaporites level. In the Messinian (Late Miocene) the faults flattened in Palaeozoic phyllite, and the Pliocene to Recent faults flatten deep in the brittle–ductile boundary.

A good and clear account of the geology of the Apennine nappes is provided by Rodgers (1997) but although his paper is about orogenic belts he does not deal at all with the mountain building phase of the Apennines.

Figure 4.7 The planation surface near Ancona, Italy (photo C. D.Ollier).

Figure 4.8 The planation surface in Umbria, Italy (photo C. D. Ollier).

The planation surface

The Apennines have an erosion surface that cuts across the basic geological structure. It is obvious in the field (Figures 4.7, 4.8) and has been described by many authors, including Demangeot (1965), whose interpretation is shown in Figure 4.9, Bartolini (1980), and Coltorti and Pieruccini (2000) who provided the basis for Figure 4.10. From their investigations they concluded:

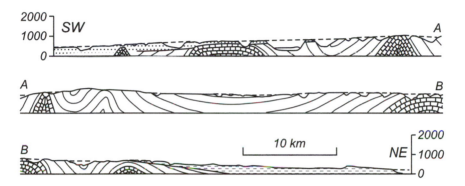

Figure 4.9 Cross-section of the Apennines from the Tiber to the Adriatic showing the planation surface cutting varied rocks and structures (simplified after Demangeot, 1965).

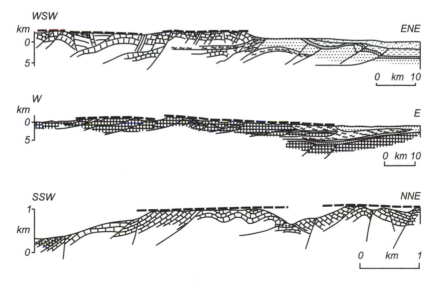

Figure 4.10 Three cross-sections of the Apennines showing the planation cutting earlier structures (simplified after Coltorti and Pieruccini, 2000).

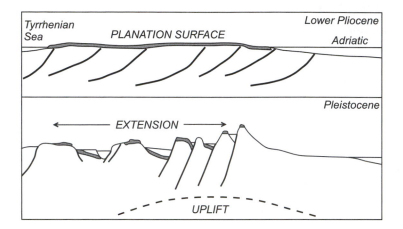

Figure 4.11 Two major steps in the development of the Apennines in central Italy.
Top. Pliocene planation, cutting across older thrust faults.
Bottom. Extension and block faulting.
(simplified after Coltorti and Pieruccini, 2000).

1   there was a single planation surface across the Italian Peninsula;
2   it was very flat;
3   it is better preserved on harder rocks and in the higher parts of the
       local relief;
4   it cuts terrain ranging in age from Palaeozoic to early Lower Pliocene;
5   it smoothed tectonic structures developed on the older terrain;
6   it is buried in places under continental and marine deposits younger
       than late Lower Pliocene;
7   after the end of the Lower Pliocene it was displaced and deformed by
       limited thrust re-activation and by high angle normal faults;
8   the planation surface was created in a relatively short time, much less
       than is usually assumed.

The planation surface is not preserved on the higher mountains because the
high elevation led to dissection or glacial erosion.

   The history is shown in very simplified form in Figure 4.11. It illustrates the main theme of the present book, that most mountains result from
vertical uplift after planation, and in this case accompanied by extension.

   Their most surprising conclusion is that the surface is a marine abrasion
platform, not a terrestrial planation surface such as a peneplain.

Emergence and uplift of the Apennines

The landmass of Italy today is bigger than it was earlier in the Quaternary.
The Northern Apennine completely emerged as a narrow belt of low relief

in the Messinian (latest Pliocene) and became a mountain range in Early–Middle Pleistocene (Bartolini, 1980). He suggests that 600 m on average was eroded in the Quaternary, and 850 m in the Pliocene, giving a total average erosion of 1.5 km in the Plio-Pleistocene. The Apennines seem to have a higher erosion rate than the Alps. In the south, the island of Sicily was uplifted in the Plio-Pleistocene.

Eustatic changes of sea level have been significant, especially in the Quaternary, and interact with the tectonic uplift of the Apennines. Pleistocene marine sediments extend well up the valley of the Tiber and other major rivers. Middle Pleistocene marine deposits are at 600 m on the Adriatic side. A 600 Ka (i.e. thousands of years old) marine terrace in Calabria (on the instep of Italy) is at a height of about 420 m. Near Rome Middle Pleistocene deposits are at 300 m.

Coltorti and Pieruccini (2000) point out that there are fluvial and marine terraces at progressive elevations revealing Lower–Middle Pleistocene uplift. From this, and their dating of the planation surface, they conclude that the mountain building (epeirogeny) in the Italian peninsula started during the Middle Pliocene and is still going on.

Using their age of the planation surface it is possible to calculate rates of uplift. In the Gran Sasso area uplift is about 3000 m in 3.5 Ma, so the average uplift is 0.85 mm per year (850 B, or Bubnoff units). In the lower mountain belt the average would be 0.42 mm per year (420 B). There are also some basins where tectonic lowering occurred. In the East Tiber basin the planation surface is 80 m below sea level, giving a negative value of −0.02 mm/yr.

It is an undulating composite surface and by no means a flat plane, but compared with the deep incision and high relief in other parts of the Apennines this is equivalent to a palaeoplain. In many places it is evident as a relatively flat interfluve between steep sided valleys, especially on the Adriatic side. In the east it is often the pre-fault surface bounding grabens. Residual hills rise above the planation surface. The palaeoplain was arched by broad tectonic movement and makes an envelope surface of the Apennines. The palaeoplain was eroded long after the creation of the nappes, but is itself faulted by normal faults, and in detail there has been some renewed movement on some old faults.

Drainage patterns

On the broad palaeoplain surface, drainage was initiated and rapidly evolved into a pattern that persists to the present (Figures 4.4, 4.12). The asymmetrical drainage pattern reveals an interesting feature of the tectonic geomorphology. On the Adriatic–Po side there is a simple parallel drainage, initiated on a simple planation surface. In marked contrast, most of western Italy has an angular drainage pattern related to post-nappe normal faults, so it is post-nappe tectonics that controls the topography by dividing it

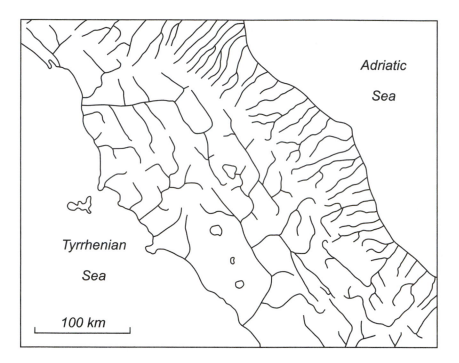

Figure 4.12 Drainage pattern of the central Apennines. Note the angular
         drainage on the western side, associated with normal faulting after
         nappe thrusting (simplified after Mazzanti and Trevisan, 1978).

into a series of fault blocks. Volcanicity is entirely Quaternary, and confined
to the eastern side near the coast.

To the Po–Adriatic there is a parallel to dendritic drainage with high
drainage density. The major rivers flow directly downslope and there is
little stream abstraction. The main rivers are also parallel to strike slip
faults oriented in the direction of movement of the nappes.

On the eastern, Tyrrhenian side, the drainage is less dense and has an
angulate pattern. This is related to block faulting, rather than the simple
slope to the sea. It is dominated by the faults related to extensional spreading
and tectonic lake formation.

The main divide between the Tyrrhenian and Adriatic drainage follows
the curve of the Apennine thrust fault. The highest mountains are not on
the major divide, but consistently east of it. This suggests headward erosion
by the rivers draining to the Po–Adriatic area. Even on the small-scale maps
shown in this chapter some of the fault-controlled drainage of western Italy
appears to be captured by the headwaters of the rivers that flow to the
Adriatic. Earlier, Mazzanti and Trevisan (1978) had suggested that the main
Apennine divide has migrated to the north east between the Upper Miocene
and Present as a 'tectonic wave' crossed Italy, but this is unlikely.

Some complications may relate to specific structures. For instance there appears to be a large pincer drainage pattern of the Arno and Tiber that could mark the scar of a great Tuscan mega-landslide, a structure verging on gravity tectonics. The course of the Tiber has been pushed to the east by volcanic deposits. Younger faults and landslides have affected many river courses in detail.

Volcanoes

Volcanoes in the Apennines are Pleistocene to modern, and are located in the west. They probably line up along a major extensional lineament on a fault that reaches to the Moho.

Volcanicity is a late event in the history of the Apennines, and is a minor associated feature, not a major contributing factor as in many volcanic island arcs. The most significant feature of the volcanicity is that it emphasises the extensional regime in the Apennines in modern times.

The surrounding seas

The Tyrrhenian Basin

The Tyrrhenian Basin is a topographic basin, bounded by many normal faults that cut through basement and some Tertiary strata (Figure 4.13). It has high values of Bouguer anomaly and heat flow, and is regarded as a spreading site with oceanic crust in the centre which extends roughly within the 3000 m isobath. West of the Tyrrhenian Sea is the Corsica–Sardinia massif which has not undergone any appreciable compressional deformation since the Upper Miocene (Mantovani *et al.* 1992). Below the Tyrrhenian Sea is a dome-shaped asthenolith with high heat flow and a crustal thickness of only 11 km.

The Ionian Basin

The Ionian Sea to the south of Sicily is an abyssal basin and a seafloor spreading site, bounded to the north by the Calabrian arc. The zone between Sicily and Italy around the Straits of Messina is an area of high uplift of Quaternary sea level markers, and this zone may be a link between the two spreading sites of the Tyrrhenian and Ionian.

The Adriatic–Po Basin

The Adriatic is not a spreading site but is floored by continental rock. It has long been an area of subsidence, and it continues into the Po Valley, which has also been an area of subsidence through much of the Tertiary. The northern Adriatic is partly filled by the prograding sediments of the Po Valley.

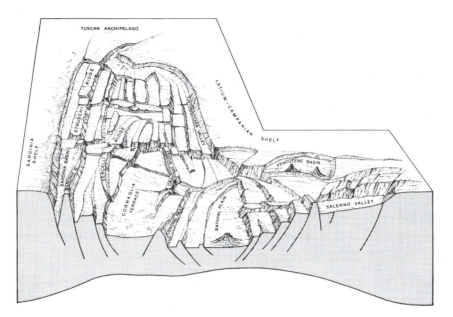

Figure 4.13 Sketch of the Tyrrhenian Sea. Note the normal faults and extension, and the thinned crust beneath the deep sea plain as revealed by seismic survey (after Wezel, 1982).

Quaternary sedimentation in the Adriatic is extensive (Ciabatti *et al.*, 1986). The pre-glacial deposition was large and fed from both northern and southern sources, including the Apennines, which already existed after Late Pliocene uplift. The youngest deposits represent the outbuilding of the Po delta. Room was available for progradation in the outer Adriatic shelf even during low stands of sea level owing to subsidence in the order of 1–4 mm/year that more than compensated for it.

The Po Valley and subduction

Evidence from the borders of the Po Plain provides a very interesting illustration of the impossibility of subduction, at least in this area. The northern Apennines are thrust to the north; the southern Alps are thrust to the south (Figure 4.14). The two blocks seem to be on a collision course, but the apparent collision by no means gives rise to compression and mountain building. The Po Basin has been subsiding since the early Tertiary, and has level sediments many kilometres thick. If Italy and the Southern Alps are converging, it is hard to see how the convergence has failed to fold the sediments of the Po Valley.

If subduction is responsible for the formation of the Southern Alps, the Po Plain has to move north and be subducted under the Alps. If the

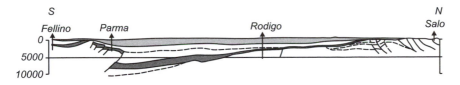

Figure 4.14 Cross-section of the Po Valley. Apennine thrusts move in from the south, and Southern Alps thrust in from the north, but instead of collisional compression there is subsidence and horizontal sedimentation. Upper shaded layer is Quaternary; lower shaded layer is Upper Miocene sediment. Bedrock is Mesozoic. Simplified after a section in Cassano *et al.*, 1986.

Apennines are made by subduction, the Po Plain has to be subducted and move south. It is hardly possible for the Po Plain to move in two opposite directions and be subducted under both mountain fronts. The cross section, derived from much factual data, also fails to suggest subduction at either boundary. Nor is there any of the usual evidence for subduction zones in the form of trenches or Benioff zones. In fact the thick sediments under the Po Plain clearly indicate stability with slow subsidence for the past 25 million years – a time span much longer than the Apennines have been in existence.

Both the Apennines and the Alps have also been attributed to collision of Africa and Europe, but the undisturbed Po Plain does not seem to be a good contender for the site of the subduction suture.

Plate tectonics explanations

The mountain building of the Apennines is commonly considered to be a direct consequence of compressional events that thrust the nappes. Most geologists nowadays use some sort of plate tectonics hypothesis as their working hypothesis, usually focusing on the nappes and using some combination of compression or subduction to create the structures, and by implication to create the Apennine mountains. Some alternative ideas are summarised below, and a few illustrations are shown in Figure 4.15.

In plate tectonics explanations of the Apennines, both the mountains and the nappes are explained as the result of subduction. The nappe front makes a huge arc and, to account for it, there are proponents for subduction from almost all quarters. Directions range from north, through east to south and even south west. It is hard to conceive of a mechanism to produce pressure from such a range of directions, and there is a great space problem where all these underthrust rocks collide at depth. Tectonic spreading from an uplifted centre might be a more feasible explanation. Many hypotheses relate uplift to collision of Europe with Africa, but how could such a collision make an east-facing arc and a north-facing arc, and why did the landscape-

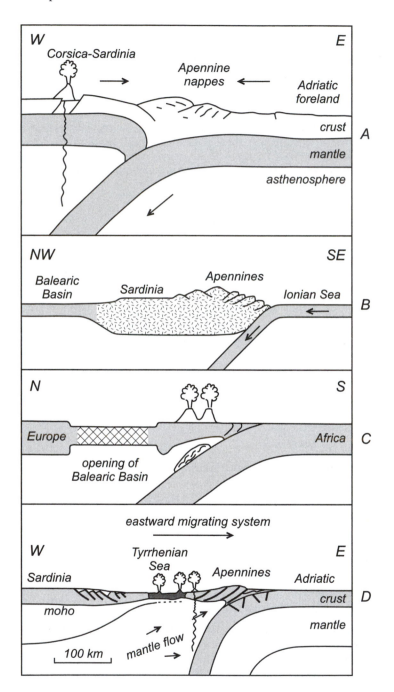

forming uplift occur so much later than any possible collision? The collision hypothesis fails to explain, and usually even to mention, the extensional tectonics in western Italy.

At the simplest, Italian tectonics is seen to result from collision of Africa and Europe. However, since it is clear that the Italian mountains are not parallel to the borders of either Africa or Europe, and as there are spreading sites in the Mediterranean such as the Tyrrhenian Sea separating Africa from Europe, more complex scenarios are usually presented. Mantovani *et al.* (1992) wrote: 'The most intriguing aspect seems to be the opening of a relatively wide oceanic basin, such as the Tyrrhenian one, in a zone where two large continents, Africa and Eurasia, have been converging.' They propose a model that interprets 'the opening of the Tyrrhenian basin and the other major tectonic events in the adjacent regions as consequences of a peculiar deformation pattern of the African/Adriatic promontory and surrounding belts driven by the Africa–Eurasia convergence.'

An INQUA Guide Book to the Apennines says:

> The Northern Apennines orogenic belt has been developing since Late Cretaceous times within the frame of Africa–Europe convergent movements, with subduction of the Adria continental margin below Europe [and] Later on, after the Africa–Europe collision, the sedimentary sequences lying on the Adria continental edge were progressively involved in NE-shifting thrusting.

The alleged subducting plate is located in different places by different authors. Some place it on the Tyrrhenean side, others on the Adriatic side, with either the Tyrrhenian Plate or the Adriatic Plate being subducted, and some have subduction south of Sicily.

The Apennine arc is a significant shape in relation to alternative hypotheses of formation, especially true if the submarine recurved southern part of the Apennine arc is included (Figure 4.16). There are proponents for subduction from almost all quarters. Directions range from north to south, which gives a space problem where all these underthrust rocks converge at depth.

Figure 4.15 (opposite) Some illustrations purporting to show the origin of the Apennines by subduction (after various authors).
    A.   Subduction of the Adriatic foreland beneath the Corsica–Sardinia block.
    B.   Subduction of the Ionian Sea under the Italian block.
    C.   Subduction of the African Plate under the edge of the European Plate, with contemporary opening of the Balearic Basin.
    D.   Thrusting in Sardinia and the Apennines with subduction of the Adriatic Plate and extension of the Tyrrhenian Sea.
    This is just a small sample of the many possibilities that have been proposed, for there are very few constraints on plate tectonic speculation.

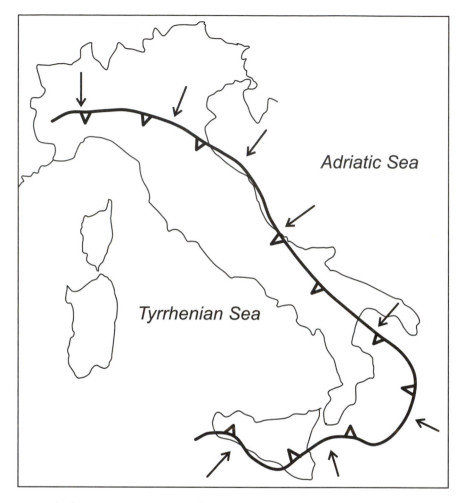

Figure 4.16 The Apennine nappe front. If subduction created the Apennines, as
commonly assumed, it came from many different directions,
converging on the Tyrrhenian Sea which is a spreading site.

Many hypotheses relate uplift to collision of Europe with Africa, but how
could such a collision make an east-facing arc, and why did the landscape-
forming uplift occur so much later than any possible collision?

Locardi (1988) suggests that the 'arc' of the Apennines results from
a rotation of Italy around a fulcrum in the north: 'a unique arc existed
from Oligocene to Middle Miocene times as a result of the rotation of the
Sardo-Corsico continental block'. The concept is illustrated in Figure 4.17.
This arc was disrupted into segments 'which advanced and shortened' because
of resistance from the rigid Adriatic foreland. Locardi has a map showing

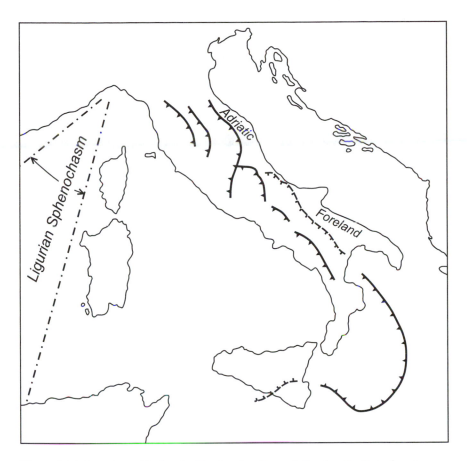

Figure 4.17 Rotation of Italy, the Tyrrhenian Sea and Corsica–Sardinia forming the Apennines by collision with the Adriatic foreland as a result of opening of the Ligurian sphenochasm (simplified after Locardi, 1988).

a Ligurian sphenochasm, equivalent to the Balearic Basin, but does not have a sphenochasm equivalent to the Tyrrhenean Basin. He says that 'the Apennine belt is bordered on the Tyrrhenian side by volcanic arcs, mostly Pleistocene in age. The Aeolian Islands are regarded as an andesitic volcanic arc.'

Another version of rotation is provided by Castellarin *et al.* (1992) as shown in Figure 4.18, with a different centre of rotation. In either rotational model it is odd that the Po Plain–Adriatic is not compressed, and neither model explains the spreading site of the Tyrrhenian Sea.

Luongo (1988) presented a model that 'concerns the spreading of the Tyrrhenean basin due to the upheaval of the mantle and the bending of the Italian peninsula caused by the convergent motion of the African and

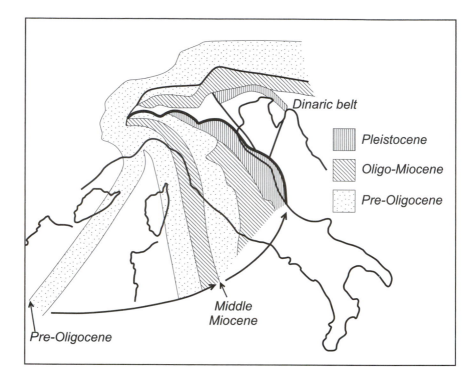

Figure 4.18 Evolution of the Apennines by rotation (simplified after Castellarin *et al.*, 1992).

European plates'. This model involves 'subduction of the Ionian lithosphere plate under the Tyrrhanean plate'. In contrast Di Girolamo and Morra (1988) present a model for subduction of the Adriatic plate causing back arc spreading and the opening of the Tyrrhenian Abyssal Plain.

There is no direct evidence for subduction, there are no convincing Benioff zones, and there are no trenches. The time scale does not seem right for attributing the tectonics of the Apennines to collision between Africa and Europe. The plates have been moving since the Jurassic (say 200 million years), but thrusting in the Apennine region occurred in the Neogene, and uplift of the Apennine Mountains in the Plio-Pleistocene.

Mountains of the Iberian Peninsula

The Iberian Peninsula is largely a plateau, but there are several distinct ranges (Figure 4.19).

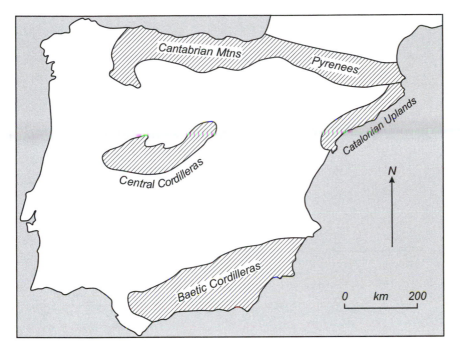

Figure 4.19 Mountain ranges of the Iberian Peninsula.

The Pyrenees

The east–west trending Pyrenees are a 600 km-long range between the Mediterranean and the Atlantic. The area is a typical symmetrical tectonic belt, with south-dipping faults in the north, and north-dipping faults in the south (Figure 4.20). From incised erosion surfaces at 1000–2000 m on the north flank of the Pyrenees and an extrapolation to 3000 m on the crest of the range, de Sitter (1952) inferred 2000 m of uplift since the beginning of the Pliocene. Uplift of the Iberian Plateau could, of course, be older.

The usual story of collision tectonics, as Africa collided with Europe, has been described by many authors, with various differences in detail. For example Duff (1992) says it is important to note that: 'Deformation related to the convergence between Iberia and Europe ceased in the Miocene.' Deformation appears to be symmetrical about an axial zone, with thrusting or sliding away from the axis (Figure 4.20).

After deformation came planation and uplift.

'The planation surfaces include relief forms of Tertiary or early Quaternary age, representing phases when denudation rates have kept pace with or even exceeded the rates of tectonic uplift or deformation' (Sala, 1984c).

In the axial zone numerous planation surfaces cut across the structure, between 2000 and 2900 m. Examples include the Carlin and Campcardos

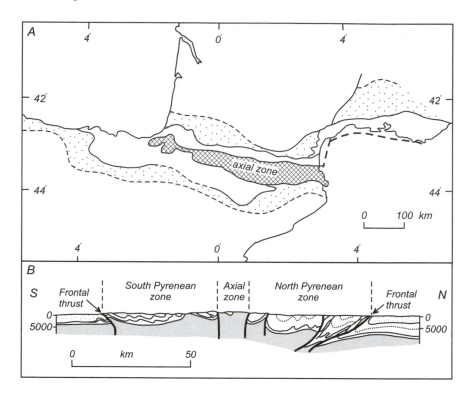

Figure 4.20 Map of Pyrenees and diagrammatic cross-section. Note the symmetry of the folds and thrusts or slides away from the axial zone.

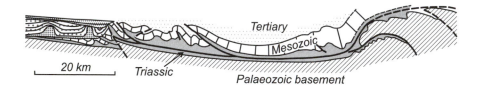

Figure 4.21 Schematic sections showing the relations between the different units of the southern flank of the central Pyrenees. The section appears to be a typical gravity slide of Mesozoic rocks over a planated Palaeozoic basement, utilising Triassic beds as the glide plane or decollement (simplified after Choukroune and Séguret, 1973).

plains. Their surfaces are uneven in height, not because of different levels of erosion but as a result of later tectonic deformation on old fault lines. Authorities disagree on the age of the older surfaces, but they are approximately Plio-Pleistocene, or uppermost Miocene.

In other places in the cordillera pre-Quaternary levels are more difficult to find, because of intense later fluvial erosion, but remnants have been mapped, and are thought to be of upper Pliocene age.

The role of gravity sliding in the Pyrenees has long been recognised. Choukroune and Séguret (1973) believed that the nappes were caused by gravity sliding, but assumed compression to give the uplift that gave rise to the slides. 'A result of this fundamental compression was the gliding of the nappes and detached masses of the south slope' (Figure 4.21). This compression is purely speculative, and other modern suggestions might have been thermal uplift or the rise of an asthenolith beneath the Pyrenees.

Cantabrica Range (Cordillera Cantabrica)

This range runs along the northern coast of Spain, as a topographic continuation of the Pyrenees, but compression between the Iberian peninsula and Europe is here impossible. Subduction also seems to be impossible, so perhaps these mountains are related more to those of passive margins described in Chapter 10. The continuity of high land from the Cantabrica Range to the Pyrenees presents significant problems to simple theories of mountain building.

Gale and Hoare (1997) report at least five episodes of glaciation in these mountains, and the most extensive is no younger than early Quaternary in age, and may represent one of the early Cenozoic glaciations. These findings contradict the conventional idea of the mountains as essentially young and dynamic.

The Central Cordilleras of Spain

Much of Spain is a plateau, and in the central area the Central Cordillera separates the two high plateaus of Castille (Old and New), and the drainage of the Duero from the drainage of the Tajo (Figure 4.22). Its highest peaks are 2430 and 2959 m. The range is about 50 km wide. It is made of granite and Palaeozoic metamorphics.

Overall the Central Cordillera is a great horst, but it is broken up into smaller blocks of horsts and grabens. Some graben, like the Lozoya Rift, contain Neogene sediments. Planation surfaces have been well studied (Sala, 1984a), and integrated into the story of landscape evolution. There were two major periods of planation in the Oligocene and Middle Miocene, but the major uplift is Plio-Pleistocene.

This mountain area is of particular significance, because being in the middle of a huge plateau it cannot be explained at all by plate tectonics. It has been totally ignored not only by plate tectonics workers, but also by the many workers who have tried to draw maps linking the various Alpine chains. We have no idea why the area was uplifted, but the mountains fit very well into the theme of this book, with Pliocene planation and Pleistocene uplift.

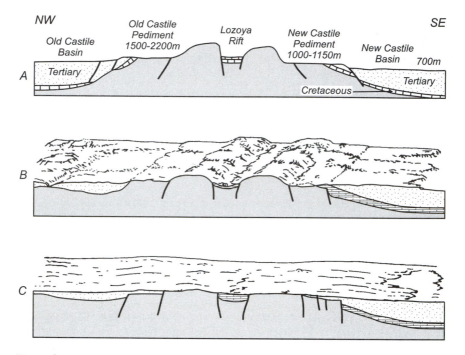

Figure 4.22 Central Cordillera of Spain. (a) Simplified view of the system. (b) The present day situation. (c) The Pontian (Uppermost Pliocene) peneplain.

Baetic Cordillera

This alpine range in southern Spain, known in part as the Sierra Nevada, runs west from Cadiz, and the structure continues into the Mediterranean to the Balearic Islands.

The core of the mountains is of Palaeozoic sedimentary rocks intruded by granite. Mesozoic strata were deposited on top, but have been thrust north. The northern part of the cordillera is characterised by masses of Jurassic and Cretaceous strata that have slid by gravity tectonics on a layer of Triassic gypsum. The symmetrical southward thrust nappes are found in the Rif mountains of North Africa.

Remnants of summit planation surfaces have been preserved in parts of the Baetic cordillera, and it is generally agreed that their age is Pontian (uppermost Miocene). The Baetic Mountains were uplifted in the early Pliocene (Choubert and Faure-Muret, 1974; Rondeel and Simon, 1974).

The Carpathians

The Carpathian mountains run in a large arc around the Hungarian Plain (Pannonian Basin) from north-eastern Austria to Romania (Figure 4.23).

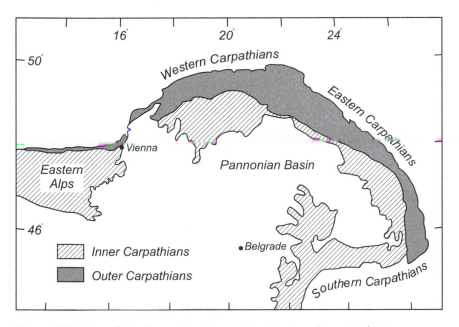

Figure 4.23 Map of the Carpathian Mountains showing the main divisions.

Geographically they are divided into Western, Eastern and Southern Carpathians. Geologically they are divided into outer and inner Carpathians. In the inner Carpathians planation surfaces of Paleogene or even Mesozoic age have survived. In the outer Carpathians relief developed mainly during the Neogene.

The outer Carpathians have sedimentary rocks, thrust into nappes, but block faulting has more to do with the present topography.

The inner Carpathians consist of several separate structures including crystalline and metamorphic cores, and volcanic areas. In the Inner Carpathians later block movements have dislocated the Tertiary planation surfaces, leaving a legacy of fragmentary flat-topped relief forms situated at different altitudes and deeply incised valleys on the sides of the block ranges. Some rivers, such as the Vah and Hornad in the Czech Republic and Slovakia and the Olt in Romania, pass through the mountain ranges in deep gorges, and are presumably antecedent.

The Pannonian Basin is a roughly circular depression surrounded by mountains, formed by subsidence of the basement in the late Tertiary together with uplift of the surrounding mountains. It became land during the Upper Pliocene, at the same time as uplift of the surrounding mountains and eruption of the intra-Carpathian volcanic belt.

Western Carpathians

The Western Carpathians are bordered in part by the sub-Carpathian structural depression (Figure 4.24). This appears to be an extensional region, of post-Miocene origin because Miocene sediments fill some of the grabens. The section does not show the compressive structures that might be expected of subduction.

The Western Carpathians are situated between the North European Plain and the Pannonian Plain. The highest part, the High Tatra, reaches 2655 m south of Krakow, and is formed of Hercynian granite and metamorphosed Palaeozoic rocks. Folding and thrusting of nappes occurred in mid-Cretaceous and in Miocene times.

The Western Carpathians (west of Bratislava) form a big arc about 400 km long. It is divided into mountains and basins by horst and graben structure. The inner zone is more dissected, and the whole is asymmetrical with the inner Carpathians more subdued and more dissected than the outer Carpathians. The axis of the upwarp, marked by the Tatra Mountains in the north east, occupies an eccentric position, only about 90 km from the fringe of the Western Carpathians.

The development of the Western Carpathians relief passed through four stages:

1   Alpine folding;
2   relief became subdued in the Upper Miocene, making the summit level of the present day;
3   uparching of the Carpathians *en bloc*. The summit level was deformed during the Pannonian (Pliocene) and a 'middle mountain planation surface' formed;
4   another phase of uparching in the Rhodanian (also Pliocene) phase. Towards the end of the Pliocene extensive valley pediments developed along the river courses.

The inner Western Carpathians consist of fault blocks, including the Slovakian Ore Mountains and the central Slovakian volcanic fault blocks.

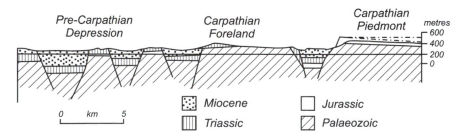

Figure 4.24 Cross-section of the Western Carpathians showing a graben, the sub-Carpathian depression in front with Miocene sediments.

The outer Western Carpathians are also block faulted, but not to the extent of the inner Carpathians.

Evidence of at least three planation surfaces can be seen in the Western Carpathians. The upper planation surface shows all the features of a regionally developed planation surface probably cutting the whole region of the outer Western Carpathians. Neotectonic movements dislocated this surface and raised it to various altitudes.

Eastern Carpathians

The Eastern Carpathians form an arc from western Ukraine southwards into Romania at about 2000 m. The eastern flank is strongly overthrust. The Eastern Carpathians are separated from the Transylvanian mountains by some Tertiary volcanoes.

In the Polish and Czech sectors of the Eastern Carpathians remnants of three planation surfaces can be observed, but in the Russian sector planation surfaces are more doubtful. The so-called mountain planation surface is preserved on sandstones.

In the Romanian Carpathians the oldest planation surface is the so-called Carpathian pediplain, thought to have developed between the Cretaceous and Oligocene under a warm and humid climate. The planation is said to have involved almost all of Romania. The different altitudes at which the remnants of the Carpathian pediplain now lie are the result of neotectonic movements that took place after the Tortonian (Upper Miocene). Younger planation surfaces are of Middle Pliocene age.

A volcanic chain of cones, craters, calderas and volcanic plateaus was erupted in the Eastern Carpathians from Hungary through Slovakia and Ukraine to Romania. The belt is 700 km long and 10 to 100 km wide, silicic to intermediate in composition, and consists of small to medium strato volcanoes, with domes, calderas and lava flows. Volcanicity lasted about 20 million years from early Miocene to late Pleistocene (Karatson, 1999).

Southern Carpathians

The Southern Carpathians (or Transylvanian Alps) extend from the Danube in the west to the Dimbovita River in the east. This area is divided into a wider northern and narrower depression southern part by a longitudinal depression, and the highest point is 2543 m. The Danube breaks through the ranges at the Iron Gate Gorge and is possibly an antecedent river.

The region consists of Hercynian crystalline blocks with a folded sedimentary cover. Overfolding and thrusting occur to the south east and south with a major uplift occurring in the Pliocene.

Despite the complex structure of the Carpathians, geomorphological studies suggest there is evidence for a simple upland surface formed from an uplifted peneplain with monadnocks. Planation surfaces in the Southern Carpathians include the Carpathian pediplain formed between the Cretaceous and the Oligocene. It is now preserved as bevelled ridge tops and small plateaus. The highest and most uniform altitudes attained by the pediplain are at 1900 to 2000 m. Younger surface developed in the Pliocene.

An overview of the Carpathians is provided by Földvary (1988), who summarised the later history as follows: 'The Pliocene Epoch saw the gradual drying up of the Pannonian Sea, forming an inland lake for a while, then a marshland. . . . This was the age of the basalt volcanicity and the overall uplifting of the Carpathian area.'

Plate tectonic explanations

Plate tectonic explanations of the Carpathians are concerned essentially with the Mesozoic and early Tertiary evolution of rock structures. Many, though not all, writers realise that the later Tertiary and Quaternary evolution was different. As with the Apennines and some other areas there is the problem of getting subduction from north west, north, north east, east, south east and south, which might reasonably be expected to cause accumulation of subducted crust under the Pannonian Plain, causing uplift rather than subsidence.

Some remarkably complex examples have been proposed, with subduction causing not only compression and underthrusting, but also 'subduction roll' and 'subduction pull'. The following examples come from Grecula *et al.* (1997):

> Subduction at the front of the Western Carpathians ceased by the Late Miocene. The back-arc extension was induced by subduction pull in the Eastern Carpathians, and/or by the post-rift thermal subsidence above upwelled hot mantle material. Intense sedimentation occurred only in the Danube Basin. In the Quaternary a positive tectonic inversion is observed in the Western Carpathians, and only some depocentres in the Vienna and Danube Basin are still subsiding.
>
> (p. 24)

The latitude available to plate tectonic hypotheses is illustrated by the following:

> The most discussed question remains the distance and dip of the European Plate underthrusting that probably changes in various segments due to the fact that the collision had an oblique orientation, mainly in the western segment, and was combined with extension in the background and with shifting of individual blocks of the Western Carpathians to the NE. The original configuration of the European plate boundary had probably considerable influence as well.
>
> (p. 32)

The Neogene development of the Western Carpathian intermontane basin was influenced by the subduction and collision process between the orogene and platform, by the back-arc rifting and consequent thermal subsidence in the Pannonian domain.

(p. 43)

The third type of underthrust crustal segments is the downbended part of the foreland crust typically occurring in collisional mountain belts. Both the slab-pull of the later detached oceanic lithosphere and loading by progressing frontal accretionary wedge of the outer Flysch Carpathians participated at the peripheral flexure of the rigid crust of the North European Platform.

(p. 40)

The authors of this work clearly believe the plate tectonics mechanism includes the mountain forming events.

Trunkó (1996) has provided an account of the geology of Hungary which is almost devoid of geomorphology and rich in plate tectonics speculation. He wrote: 'Doubtless the driving force of the Alpine tectonics was a relative movement of Africa against stable Europe.' He believes that the driving force was either a push from the rear, or subduction roll-back. The latter, as in the Apennines, has spatial problems on the great curve of the Carpathians. Nevertheless he does acknowledge that the Pannonian Basin is due to extension of Neogene age.

The Caucasus

The Caucasus Mountains lie between the Black Sea and the Caspian Sea, and mark the southern border between Europe and Asia. The two main masses are the Great Caucasus and the Lesser Caucasus with a lowland trough between (Figure 4.25). The volcanic Mt Elbrus is the highest mountain in Europe at 5633 m. The axis of the Great Caucasus shows the greatest neotectonic activity, with Neogene and Quaternary uplift of about 4500 m.

Planation surfaces

Subaerial planation surfaces are well developed on the watershed ridges of both the Great and Lesser Caucasus. They often consist of accordant summit bevels, in the highlands, and well-preserved erosion surfaces are found at lower levels in the foothills of the Great Caucasus. These surfaces have gravels that can be correlated and dated, and are Plio-Pleistocene.

Around Mt Elbrus, remnants of planation surfaces have weathered mantles, covered in some places by Upper Pliocene volcanic rocks. The surface and weathering must therefore be not younger than Upper Pliocene, but could be as old as Miocene. In the south east Caucasus the high surface at a level

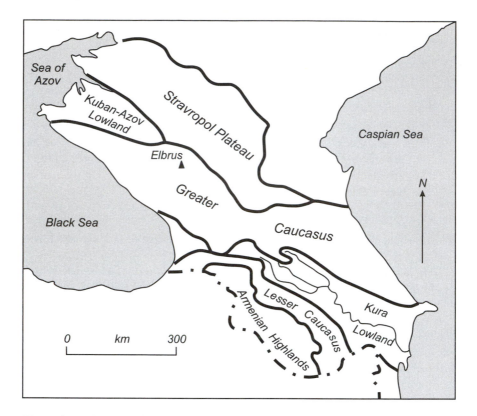

Figure 4.25 Caucasus location map.

of 4000–4200 m has been dated as Miocene by correlative sediments. In the eastern end of the Lower Caucasus weathered mantles and marine sediments of Miocene age have been described. The older surfaces exhibit deep weathering and stripping (etchplanation), which are thought to relate to warm climates in the early Tertiary. Upper Pliocene pedimentation is evident in many places.

Denudation chronology

In some mountain areas only a simple scenario of planation followed by uplift can be presented, but, of course, the real evolution is not of two instant events. In the Caucasus a more detailed account can be given. Uplift of the Caucasus region started in the Palaeogene when a narrow ridge emerged from the sea – the precursor of the Great Caucasus. By the Oligocene there was a low mountain system, but most of the Great Caucasus and all the Lower Caucasus consisted of erosional lowlands with some flat areas of deposition. The mountains continued to grow in the Upper Miocene and

Lower Pliocene. Volcanic activity gave rise to lava plateaus and volcanic mountains in central Armenia and south Georgia. Dissection began in the Middle Pliocene, but rapid uplift and intense erosion was in the Upper Pliocene when the relief of the region took the form that we know today. A complication in the area that drains to the Caspian Sea is that river incision occurrred during the Middle Pliocene because the water level of the Caspian fell 500 m, from a combination of tectonic and climatic causes. The first glaciation was in the Upper Pliocene. Uplift continued through the Pleistocene.

The Ural Mountains

The boundary of Europe and Asia is conventionally taken to be along the north–south range of the Ural Mountains, which runs for 3000 km from the Arctic Ocean to the Aral Sea in central Asia. The range has an average altitude of 1000–1300 m, with a highest point of 1894 m. The northern Urals run from Novaya Zemlya to the city of Perm.

The Urals have folds with Precambrian rocks exposed in some cores. This basement is broken up into blocks by faults, and fault scarps bound the Urals on both sides. The main structural forms are elongated and divided by parallel deep-seated faults. According to Bashenina (1984) there are also transverse faults that divide the Urals into seven units, each with its own typical pattern of relief.

The Mesozoic and Palaeogene were characterised by prolonged tectonic stability and extensive areas were worn down to surfaces of low relief covered by weathering products. The northern Urals were planed down in the Upper Oligocene, when relief was low and elevations up to 500 m. They are now up to 1000 m, reflecting younger vertical movements.

In the extreme south of the Urals there is an important sub-Cretaceous unconformity beneath the marine Cretaceous and Palaeogene deposits. These are being stripped off to reveal the old surface. This surface was bevelled by subaerial processes before the Cretaceous marine transgression.

Planation surfaces are evident in the Central Urals, but correlative sediments are not known.

After planation, neotectonic movements then built the topography, the nature and intensity varying from place to place. The differences in altitudes of river terrace sequences in different blocks show that differential block faulting has continued until recent geological time. The age of the uppermost terrace is always Oligocene–Miocene.

In the Polyarnry and Zapolyarny Urals, neotectonic movements can be shown to have commenced close to the Pliocene–Pleistocene boundary. Again, the basic reference is the planation surface evolving up to the late Tertiary, which became modified by differential tectonic movements and denudation into a slightly dissected upland, not unlike the present day Pay–Khoy upland. Further intense movement took place in the Middle

Pleistocene, indicated by the appearance of coarse sediments of this age in the foothills.

On the western slopes of the northern Urals five river terraces can be distinguished. The ages are probably Lower Pliocene, Middle Pliocene, Upper Pliocene, Middle Pleistocene, Upper Pleistocene.

The Sudeten

The Sudeten Mountains in south west Poland and the north east Czech Republic are part of the Bohemian Massif, a planated craton. Part of the Sudeten, the Walbrzych Upland, has been studied in detail, and provides a history that possibly applies more widely.

The Sudeten consist of a series of horst and grabens, formed since the Late Oligocene with the main uplift in the Pliocene. According to Krzyszkowski and Stachura (1998), 'Recently the Sudeten Mts have remained as a passive basin-and-range region, i.e. have no or very little tectonic movement.'

Valley shape is useful in this area to work out some of the landscape history. The whole area was covered by an ice sheet in the early Saalian (the third classic glacial stage of Europe). Valleys pre-dating this glaciation show no evidence of tectonic activity and are wide and shallow. Valleys post-dating the glaciation are narrow and deep (Figure 4.26).

Migon and Lach (1999) describe the Kaczawa sector of the Sudeten marginal fault, and again say the uplift was in the Pliocene–early Quaternary. They write, 'The tectonic uplift of the Sudetes relative to the foreland took place, in principle, during the Pliocene and Quaternary, and postdates the origin of Late Miocene denudation surface.' The actual scarp of the Sudetic Marginal Fault was not formed until the Early Pleistocene, when uplift was about 60–70 m (Krzyszkowski and Biernat, 1999).

The Atlas Mountains of North Africa

Though not European mountains, the Atlas mountains of North Africa are treated here because of their relation to the alpine mountain chains of Europe (Figure 4.1).

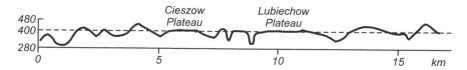

Figure 4.26 Cross-section of part of the Walbrzycg Uplands, Sudeten Mountains, Poland. There are distinct plateau remnants, with hills rising above. Preglacial valleys are wide and shallow, post early-Saalian glaciation valleys are narrow and deep, indicating uplift between the two periods of valley formation (after Krzyszkowski and Stachura, 1998).

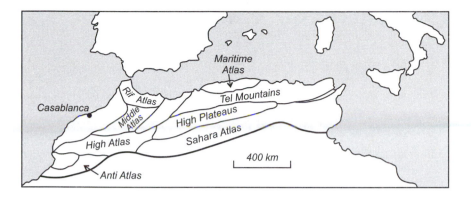

Figure 4.27 Geographical divisions of the Atlas Mountains.

The mountainous area north of the Sahara desert forms part of Africa known as the Maghreb or the Atlas lands. The Atlas Mountains are divided into several geographical regions as shown in Figure 4.27.

The Rif Atlas is a mirror image of the Baetic Cordillera of southern Spain, with thrusting to the south. The mountains are rugged, and glaciated in parts. The Rif mountains were uplifted in the early Pliocene (Choubert and Faure-Muret, 1974; Rondeel and Simon, 1974).

The Tel Mountains consist of a double line of mountains, formed in the Miocene.

The Saharan Atlas consist mainly of Cretaceous limestones and marls, faulted and folded.

The High Plateau lies between the Tel Mountains and the Saharan Atlas. It has an elevation of 750–1000 m, lower than the high peaks of the mountains that bound it. The plateau surface is mantled with Neogene marls and sands (Bridges, 1990, p. 35) The high plains are often covered in salt lakes (chotts) and external drainage is often lacking. The region seems to be a median plateau with bounding outwardly thrust mountain belts.

5 Western North America

Many mountain regions in western North America are plateaus, and some are known by name as such, while others are called 'ranges' or simply 'mountains'. The numerous mountain ranges, cordilleras and plateaus in western North America present a confusing picture at first sight. We will try to simplify it.

The Pacific border region is fairly simple, consisting of two main ranges with a depression between them. The western range is known as the Coast Range. The inner line consists of the Cascades to the north and the Sierra Nevada to the south. The situation is complicated by two barriers across the central depression. The Klamath Mountains connect the Coast Range to the Cascades, and divide the Californian Great Valley from the great valley of Oregon and Washington. The second barrier is the Transverse Ranges, around the latitude of Los Angeles, which separates the Californian Valley from a series of basins and valleys that lead to the Gulf of California.

The coastal ranges can be regarded as continuing into Canada, but the vast system of fjords breaks up the topography very much.

East of the Cascade–Coast Range, or Pacific Border mountains, is a broad belt of plateaus, basins and ranges, broken in the middle by the volcanic Columbia Plateau.

East of this central plateau belt is the line of the Rocky Mountains, which continues north into the Mackenzie Mountains and the Brooks Range, and south into the Sierra Madre Occidental of Mexico.

A broad distribution map is shown in Figure 5.1.

The Laramide orogeny

Trimble (1980) summarised the history of the region as follows: The onset of the Laramide orogeny was about 70 Ma ago. The last sea was displaced. Uplift and erosion followed but the plains remained stable. Old marine strata were depressed below sea level by the weight of overlying sediment. At about 50 Ma the Laramide orogeny ended. The whole region was planated and an erosion surface was cut across all. There was deep weathering, and then about 35 Ma a huge ignimbrite was deposited. Sometime after 29 Ma

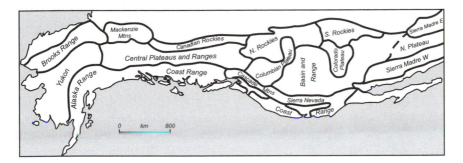

Figure 5.1 Map of mountains and plateaus in western North America.

there was some uplift, with erosion on the west and deposition to the east, but major uplift occurred within the past 5 million years, There was differential movement of mountain blocks, but basically the Colorado Plateau was raised 2100 m, the Western Plain uplifted 1500 m, and the Eastern Plains 300 m.

Many of the mountains are related to uplift of blocks or diapirs of Precambrian rocks (Figure 5.2) that pushed up overlying strata, which often moved away in gravity slides.

Trimble (1980) compared the tectonic history of the Rockies–Colorado Plateau area with that of the adjacent Great Plains. The interpretation of the tectonic history of the plains:

> indicates that during uplift of the mountain blocks the plains were tectonically unaffected and only slightly depressed by isostatic adjustment. Had compressional stresses been responsible for the rise of the mountains, the adjacent basin sites should have been tectonically downfolded . . . Obviously they were not, and I conclude from this that the vertical uplift of the mountains . . . was unrelated to horizontally directed compressional stress.

The Colorado and other plateaus

The Laramide orogeny refers to tectonic movements between Cretaceous and Tertiary that extended in a long belt from the Yukon Plateau in Alaska to the Sierra Madre in Mexico. The area was broken into many uplifts and basins. The Precambrian basement shows very little horizontal displacement, but often displays vertical uplift of many kilometres.

At the close of the Cretaceous the Colorado region was nearly at sea level. It was then raised about 2000 m, with a gentle tilt from south to north. The uplift was in two phases: first a small uplift in the Late Miocene, accompanied by much volcanic activity around the borders and intrusions in the interior,

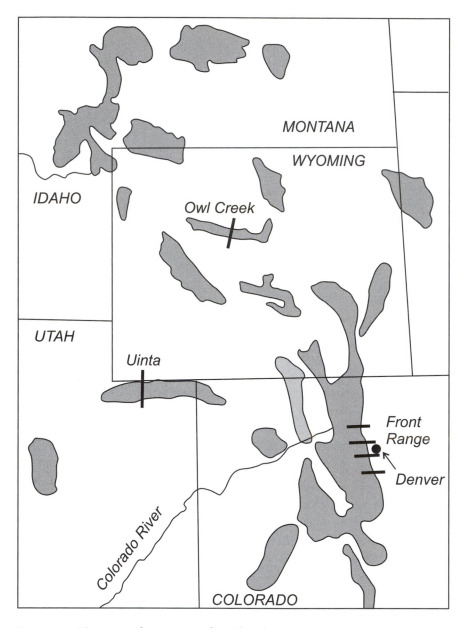

Figure 5.2 The Precambrian cores of uplifts of the western United States. Lines of section for Figures 5.3, 5.4 and 5.5 are indicated.

and second a much bigger uplift in the Late Pliocene to Recent, to form the Colorado Plateau of today, with only minor volcanic activity.

Lucchita (1979) interpreted the uplift of rocks deposited near sea level in Arizona during late Neogene times as evidence that the Colorado Plateau has been uplifted at least 880 m since 5.5 Ma. This can be compared with the estimate of Trimble (1980) that the Colorado River has incised its canyon up to 915 m in the past 5 Ma.

The southern edge of the Colorado Plateau, often called the Mogollon Rim, marks the southern termination of Permian cliff–making strata. According to Peirce *et al.* (1979) a plateau edge escarpment had evolved by mid–Miocene times, prior to the onset of Basin and Range faulting and the development of the modern Grand Canyon. The preservation of Cretaceous marine strata along portions of the present plateau physiographic edge at elevations over 2 km indicates uplift relative to sea level, but not the time of the uplift.

Graf *et al.* (1987) reviewed the geomorphology of the Colorado Plateau. They see the sequence of landscape-forming events as:

1   the uplift of the plateau margin took place during the late Cretaceous (Laramide) earth movement, creating a broad erosion surface 'much like the surface presently observed over much of northern Arizona';
2   the Laramide Surface was buried by up to 90 m of Rim gravel;
3   the Laramide was followed by Late Eocene–Early Oligocene tectonic quiescence and regional weathering;
4   the Colorado River began to emerge as an integrated drainage system in Late Miocene to Pliocene times.

In brief, the broad geomorphic framework of the Colorado Plateau appears to have been shaped by Laramide cliff recession under climatic and base level controls very different from those associated with Pliocene canyon incision.

Valley anticlines of the Colorado Plateau are described in Chapter 8. These seem to be an isostatic response to major valley incision, like that found in the Himalayas. Volcanic activity associated with the Colorado Plateau is described in Chapter 9, and illustrated in Figure 9.8.

In Wyoming Evanoff (1990) showed that the Eocene surface was quite rugged, and valleys were filled in with the little-deformed Oligocene White River Formation. The regionally widespread low-relief summit surface was developed in the Miocene, and is not part of the late Eocene unconformity.

North of the Colorado Plateau the Uinta Mountains comprise a plateau-like uplift about 70 km across, bounded by steep reversed faults and deep sedimentary troughs (Figure 5.3). The total vertical movement is about 10 000 m, accompanied by very considerable spreading of the upper part of the plateau, which must be distinguished from regional extension. It is a horst uplift bounded by reverse faults, showing that the rocks of the Precambrian core 'bulged outward as they ascended'. (Holmes, 1965).

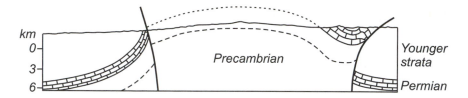

Figure 5.3 Diagrammatic section across the central Uinta Uplift, Utah, showing the outward spread of the uplifted Precambrian rocks along outwardly curving faults.

The gravity sliding associated with the Owl Creek uplift (Figure 5.4) was described by Wise (1963).

The Bighorn Mountains have a granite core, flanked by steeply dipping Palaeozoic and Mesozoic rocks. 'Sizeable expanses in the Bighorn Mountains exhibit surprising low relief and have been considered remnants of once more extensive erosion surfaces' (Thornbury 1965, and see his Figure 20.16). The stillstand that permitted formation of the surface is presumed to be Middle Tertiary, and the uplift at the close of the Tertiary and in the Pleistocene.

The Rocky Mountains

The name Rocky Mountains might be misleading, because they consist largely of a dissected plateau, with 'ranges' at the edges, and show much in common with the Colorado Plateau and other plateaus of the central

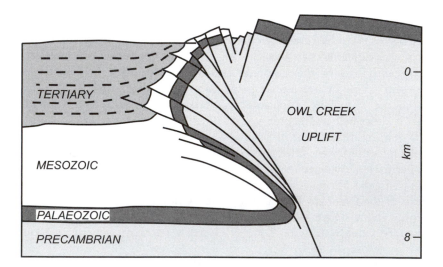

Figure 5.4 Suggested structure of the frontal zone of Owl Creek uplift, Wyoming. No vertical exaggeration (after Wise, 1963).

plateau belt. The Southern Rocky Mountains plateau is bounded by the Front Range on the east and the Park Range on the west.

The Rocky Mountains massif is just one of many examples of Precambrian rocks that were uplifted, and the others, like the Rocky Mountains, are surrounded by upturned and overturned younger sedimentary strata.

The Front Range on the western edge of the Rocky Mountains is over-thrust (Figure 5.5). The mountain front is straight on small–scale maps, but detailed work shows thrusting (Bieber, 1983). The situation suggests original vertical faults which spread under gravity, as explained in Chapter 8. The eastern side of the Rocky Mountain plateau is also thrust. Jacob (1983) used the phrase 'mushroom tectonics' to describe this phenomenon, and illustrated it by an example from the Front Range between South Park and the Denver Basin (Figure 5.6).

It seems that the Rocky Mountain Plateau was formed by vertical uplift, but the rock strength was insufficient to hold vertical fault scarps at the edges, and the plateau spread out, forming thrusts at the edges of the plateau. Such spreading occurs in many other regions, and the mechanisms and field indications of spreading will be described later.

The Southern Rocky Mountains and their continuation into Mexico, according to Eaton (1987), are the crest of a symmetrical continental feature of large dimensions, which he called the Alvardo Ridge. It is characterised

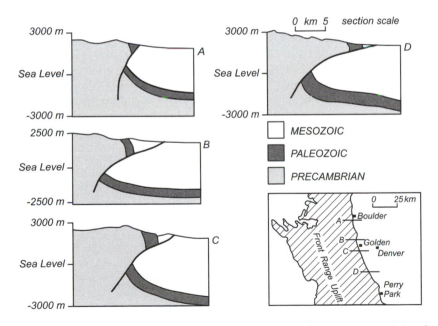

Figure 5.5 Overthrusts on the Front Range of the Rocky Mountains, Colorado, eastern margin of Rocky Mountains (after Bieber, 1983). Section lines at bottom right. Note that despite the low thrust angle in places, the fault in plan is fairly straight.

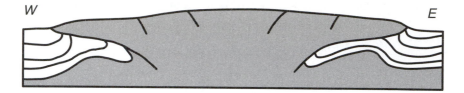

Figure 5.6   Sketch section of 'mushroom tectonics' as applied to the Rockies in
Colorado. Front Range to the east, Park Range to the west. Precambrian
rocks (shaded) spread over younger rocks on both sides (after Jacob,
1983).

by gentle rises which were originally covered by continental sediments of
Miocene and Pliocene age, tens to hundreds of metres thick. The Southern
Rockies have Precambrian nuclei described earlier that were upwarped
during the Palaeozoic and Mesozoic, and then affected by further vertical
movement in the Laramide diastrophism. Eardley (1963) pointed out that
the largest uplifts droop over surrounding basins and so are gravitational,
not compressional.

Foose (1973) gave an excellent summary of the bi–causal pattern of
tectonism in the Rockies.

> Throughout the Middle Rocky Mountains a tectonic style may be
> observed that emphasises the role of two major forces that acted on the
> crust during the Laramide orogeny . . . The primary and earliest force
> was that of vertical tectonism . . . The secondary force . . . was that of
> gravity, which extensively remodelled the basically simple geometry
> of the initial blocks in the Middle Rocky Mountains by creating a
> variety of structural features that provided for the release of stress within
> the blocks. Release of stress was accompanied by movements of parts
> of the block along the newly created structures in directions outward
> and downward toward the adjacent basins.

Loomis (1937) provided an early account of the physiography of the
Rockies. He wrote: 'The Northern Rocky Mountain Province is a complex
of many ranges that have been revealed by dissection of a peneplain which
can be traced over the whole area. This peneplain was later elevated to
above 9,000 feet [3000 m], and folded and faulted in places.' He further
wrote of the Rocky Mountain Peneplain: 'In climbing Pikes Peak, if one
stops just above the 9,000 – foot level, the sea of ridges all rising to approximately the same height makes the observer feel conscious of the existence
of the peneplain.' This is an American description of what the Europeans
call the gipfelflur.

Atwood (1940) wrote 'Many persons who have studied the Front Range
of Colorado have called the remnants of this old-age erosion surface in that

section the Rocky Mountains Peneplain. A remnant in the San Juan Mountains in southwestern Colorado has been called the San Juan Peneplain.' He lists many other plateaus in the area such as the Green Ridge Peneplain and the Medicine Bow Peneplain. It is interesting to note that the Rocky Mountain Peneplain was one of the first peneplains to be described.

Bradley (in Madole *et al.* 1987) has provided an extensive review of the erosion surfaces of the Front Range. Most workers now believe that a single, widespread erosion surface exists in the Front Range, commonly named the Rocky Mountain, Sherwood or Late Eocene Surface. Bradley believes this surface has regional significance far beyond the Front Range, that it may have had much greater regional extent, and it may transgress time (in which case Late Eocene is an unsuitable name). Questions remain about how the surface was formed, how much it has been lowered in post-Eocene time, and about the significance of small surfaces that have been called the Flattop Surface. According to Evanoff (1990) the age of the Rocky Mountain sub-summit surface varies from place to place. In central Colorado it is demonstrably Late Eocene, but in much of the Laramide and Medicine Bow mountains it is Miocene.

To the east of the Northern Rocky Mountains lies an area known as the 'Disturbed Zone', where Mesozoic and Early Tertiary rocks of the Great Plains are overturned and thrust. Low angle thrust sheets such as the Lewis Thrust have moved eastward several tens of kilometres. Precambrian rocks come to lie over Cretaceous and younger rock. Heart Mountain is an outlier of the Lewis Thrust and is a classic example of an erosional klippe. The thrusts are demonstrably gravitational slides (Reeves, 1946; Pierce, 1957). They have evidently slid over the existing relief at the ground surface. The margin of the Middle Rockies also has strongly folded Palaeozoic and Mesozoic rocks thrust from the west over Tertiary formations.

Sato and Denson (1967) noted an increase in grain size of sediment, and an increase in basement derived heavy minerals in Late Neogene sediments as evidence of uplift and erosion of the northern and middle Rocky Mountains during Late Miocene and Pliocene time. Eaton (1987) presented evidence for rapid epeirogenic uplift of the southern Rocky Mountains starting between 4 and 7 Ma.

Drainage anomalies in the Rockies–plateau belt

As Hunt (1967) expressed it, many of the rivers in the Rocky Mountains have developed anomalous courses 'through the mountains, not around them'. A number of the anomalous places are shown in Figure 5.7.

The Colorado River enters the plateau after crossing the mountainous uplift of the White River Plateau, and the Green River enters after crossing the Uinta Mountain Uplift. The Gunnison River enters after crossing the uplift of the Black Canyon.

The Arkansas River flows south between the Sawatch Range and the Mosquito Range, and then turns east through Royal Gorge, a steep sided

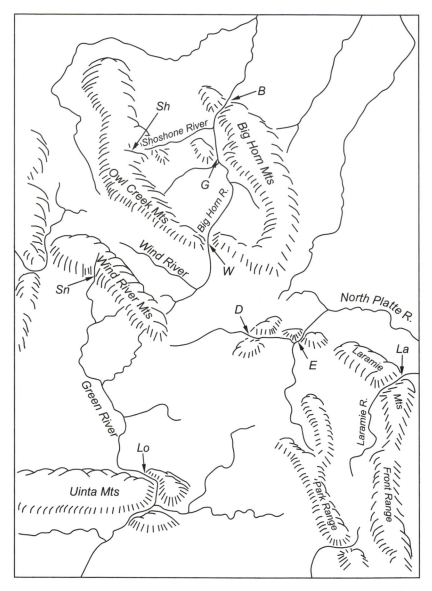

Figure 5.7 Some of the drainage anomalies in the Rocky Mountains region.
  B   Bighorn Canyon through Bighorn Mountains
  Sh   Shoshone Canyon
  G   Greybill Canyon, on Bighorn River
  W   Wind River Canyon through Owl Creek Mountains
  Sn   Snake River Canyon
  D   Devils Gate on Sweetwater River
  E   North Platte River near Alcova
  La   Laramie River through Laramie Mountains
  Lo   Lodore Canyon on Green River across Uinta Mountains.

canyon 600 m deep which is utilised by the Rio Grande Railroad to cross the Rocky Mountains.

The Laramie River rises on the high Laramie Basin and flows through gorges to the Great Plains 600 m below, crossing the anticlinal raised ridge of Precambrian rocks forming the Laramie Range.

The North Platte River has several complications, including the place where it flows towards an isolated mountain (the Seminoe Range), and crosses it through a gorge 300 m deep instead of taking the easy route round the mountain.

The Colorado River rises on the flank of the Front Range, only about 30 km from the Great Plains, but flows westward, crossing the Park Range and the Gore Range through Gore Canyon, over 600 m deep. After being joined by the Eagle River, the Colorado River crosses the White River Plateau through Glenwood Springs Canyon, almost 1000 m deep.

The Bighorn River first cuts a gorge through the Owl Creek Mountains, and then instead of taking the easy route to the Great Plains it cuts another gorge across the Bighorn Mountains.

Other examples are the Lodore Canyon through which the Green River crosses the Uinta Mountains, and the Black Canyon of the Gunnison River.

All these anomalies, on rivers flowing in many different directions, are almost certainly antecedent courses, and were commonly regarded as such by most early workers. Thornbury (1965) discussed arguments for super-position and antecedence and came down, surprisingly, in favour of the former. But the cover that has to be removed to allow superimposition has to be over half a kilometre deep on a regional scale, burying whole moun-tain ranges under debris, and it is hard to find a source for such debris that is higher than the ranges buried. Antecedence seems a better option to us. For further explanation of antecedent rivers see Chapter 11.

A good account of the history of ideas on the drainage and tectonics of the Colorado Plateau was provided by Baars (1972). He notes that the present valleys often contain meandering stream patterns reminiscent of the classic meanders of streams flowing across very flat country. Could such courses be maintained in the face of rising landscapes? In opposition to the antecedent theory, some eminent geologists, including Davis, felt that the rivers had been let down on pre-existing structures that had been buried under later sediments to mask the structure and produce fairly uniform plains across which the meandering streams could flow without reflecting the buried structure. This was the superposed theory. A still later group produced a very elaborate story of river capture to explain the course of the Grand Canyon, but Baars asks 'what of the canyons that truncate other major uplifts? And if river capture is the explanation, why are the entrenched meanders of the Goosenecks of the San Juan River perched on top of the Monument upwarp, where headward erosion and piracy should have been most active?'

He notes that 'the deepest and most spectacular canyons are today seen to dissect the major uplifts'. This seems to indicate that the uplifts followed the valley downcutting. Such a process is described later in regard to the Himalayan rivers, and it seems to be an obvious candidate for the Colorado Plateau uplifts. It has been specifically utilised for some valleys in Canyonland National Park by Potter and McGill (1978), who wrote 'all these anticlines appear to post-date the formation of the valleys'. The amplitude of folding decreases up section but the anticline is still visible 500–600 m above the river bed in the rim of Cataract Canyon. The anticlines are narrow and steep limbed, like those along the Himalayan rivers.

If a valley anticline is formed by pressure release, the actual rock must deform by expansion, shearing, or other details. In some accounts of such valley anticlines the change in a particular bed is sometimes interpreted as the fundamental cause of the anticline, instead of part of the response to gravity unloading. In the case of the Colorado Plateau valley anticlines the story might be complicated by flow of evaporites.

Further details and explanations of river capture, superimposed and antecedent drainage are provided in Chapter 11.

The Canadian Rockies

The Rocky Mountains continue into Canada, but the form is somewhat different, and a major lowland lineament, the Rocky Mountain Trench (Tintinna Trench in the Yukon), lies west of the Rocky Mountains, separating them from the Interior Plateau Province. There are imbricate faults and decollement, suggesting either nappe thrusting to the east, subduction from the east, or gravity tectonics (Figure 5.8). The highest peak is Mt Robson (3964 m).

The Brooks Range, with peaks of over 3000 m, consists of folded Palaeozoic strata in the middle and Mesozoic rocks on the flanks. The area was planated in the Tertiary and then uplifted.

The Coast Ranges

The Coast Ranges of the United States and Canada are somewhat different. In the United States the Coast Range of Oregon and Washington is highest in the north, where it is known as the Olympic Range. This range continues

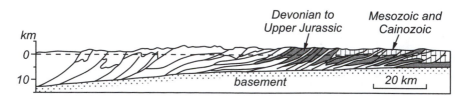

Figure 5.8 Cross-section of the Canadian Rockies showing thrusting to the east over a decollement (simplified after Price and Mountjoy, 1970).

as the Insular Mountain Province in Canada. The Coastal Ranges of Canada are actually a continuation of the Cascades of the United States (McKee, 1972).

The Coast Ranges of Canada consist of a simple arch, breached by a number of rivers that originate along the western edge of the Interior Plateau Province. The Klinaklini, Bella Coola, Dean, Skeene, and Nass are all examples. It seems the rivers were there before the mountains formed: that is they are antecedent rivers.

Bohannon and Geist (1998) describe the Neogene tectonic development of the California continental borderlands, and date the crustal extension and uplift to late Oligocene to Present, giving an earlier start than other workers.

The Californian Coast Ranges consist of many parallel, linear ridges extending from the Klamath Mountains in the north to the Transverse Ranges in the south. Most drainage parallels the structure. Thornbury (1965) suggests that it was during Late Pliocene uplift of the California Coast Ranges that the Valley of California was practically cut off from the sea and came to have essentially its present outline. 'The Coast Range seems to have formed as a large but rather simple arch.'

Western British Columbia is a collage of exotic terranes, much utilised in plate tectonic explanations of the evolution of the area, but the present high relief of the Coast Mountains is a product of rapid late Cenozoic uplift (Muhs *et al.*, 1987). 'A mature erosion surface of moderate relief existed over most of southern British Columbia, including the southern Coast Mountains, in early to middle Miocene time' (Clague and Mathews, in Muhs *et al.*, 1987). The Miocene landscape was characterised by broad low valleys separated by gentle slopes rising to narrow divides and summits 300 to 500 m above the low ground. Remnants of lavas erupted onto this surface when it was near sea level are found today on a few summits up to 2.5 km elevation. Much of the elevation is therefore post-Miocene. In contrast, north of 52°N Miocene lavas or the erosion surface on which these lavas were deposited extend into valleys 1 to 1.5 km below mountain summits, suggesting that some of the present height of the central and northern Coast Mountains is pre-Miocene in age. Fission-track dates also suggest that the Coast Mountains were raised 2 to 3 km in the last 10 million years (Parrish, 1983). Nevertheless for most of the Canadian Cordillera most tectonic uplift occurred in the Pliocene, establishing the main mountain belts and causing streams to incise hundreds or even thousands of metres below the former low-relief, low-elevation, Late Miocene surface (Mathews, 1991).

Additional support for rapid recent uplift comes from an analysis of the present physiography of the Coast Mountains. Over a large area in the south, summits form a broad plateau-like envelope at 2 to 3 km elevation that may represent the Late Miocene erosion surface (Parrish, 1983). To the north the summit surface is lower and more irregular, possibly suggesting greater antiquity of relief.

Many antecedent rivers cross the Coast Ranges south of San Francisco, including the lower Sacramento, the lower Russian, Los Angeles and Santa Ana (Oakeshott, 1971). The Golden Gate outlet at San Francisco is at the mouth of an antecedent river. It was established as a valley before the last uplift of the Coast Ranges, and its present form and that of San Francisco Bay are due to a very recent sinking along the Pacific coast (Atwood, 1940).

The Cascade Range

The Cascade Range is geographically a continuation of the line of the Sierra Nevada (discussed in Chapter 3), but geologically is very different. Volcanic rocks dominate between the Sierra Nevada and the latitude of Seattle, where granites again appear.

But although the Cascade Range is largely a huge pile of volcanic rocks, and includes many famous volcanoes such as Mt Shasta, Mt St Helens, Crater Lake and Lassen Peak there has also been large regional uplift. This is most evident in the north, between Seattle and the Canadian border where the bedrock is mainly non-volcanic, and a plateau-like upland surface indicates a planation surface.

The Cascade Range is one more example of a dissected plateau. Fenneman (1931) wrote, 'the most striking feature of these mountains is the approximately uniform altitude of its peaks and ridges'. According to McKee (1972), 'The modern Cascade Range has been built by Late Cenozoic uplift along a north–south axis. Whatever the reason, this uplift is only part of a truly grand orogeny (the Cascade Orogeny) that has produced much of the present physiography of the North American Cordillera.'

Lowry and Baldwin (1952) thought uplift could be as much as 900 m, greatest near the axis and less on the eastern and western flanks. Evidence of the uplift comes from warped lava flows along the Columbia River gorge which Lowry and Baldwin consider to be an antecedent gorge. 'The Cascade Arch grew across west flowing rivers. The Columbia River is therefore an antecedent gorge.' 'The Columbia River Basalt is exposed along both walls of the Columbia River Gorge all the way through the Cascade Range. It is a huge arch with uplift here of about 900m – but this is a minimum figure since this is the lowest part of the range.'

'The course of the Columbia is plainly antecedent to the last uplift. The mountains rose across its course in a complex fold or series of folds; the channel was lifted and rapids resulted but the stream held its course. Even now the channel is far from graded. At the Cascades (from which the range takes its name) 40 miles [64 km] east of Portland, the Columbia descends to sea-level by a series of rapids 50 ft [17 m] high.' (Fenneman, 1931).

Fenneman (1931) believed the Cascade Peneplain was formed in the Pliocene. Priest *et al.* (1983), reported that uplift of the western Cascade Range occurred about 4–5 Ma. The situation is identical to the rise of the Coast Mountains in British Columbia.

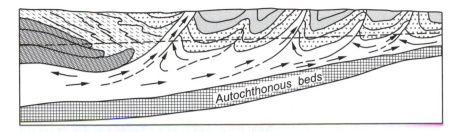

Figure 5.9 A cross-section of the Ouachita Mountains, Arkansas. Movement along a decollement zone accompanied by loading and spreading has caused variations in the thickness of strata and imbricate thrusting.

The Ouachita Mountains

The Ouachita Mountains and the nearby Arbuckle Mountains of Arkansas and Oklahoma are parts of an east–west Palaeozoic fold belt. The structures may be related to those of the Appalachians far to the east, but any connection is hidden under the Mississippi embayment. The Ouachita Mountains are of interest because they are in the middle of the continent, and cannot be related to either subduction (commonly invoked in the west) or passive margins as in the east. The structures are classic gravity slide features (Figure 5.9), but we stress again that these Palaeozoic structures are much older than the uplift of the present mountains.

The Ouachita Mountains reach only about 800 m on the crests. There is a summit level which apparently decreases in elevation to the south and passes beneath the sediments of the coastal plain. If this is so the summit surface may be dated as an Early Cretaceous peneplain (Bridges, 1990). This seems reasonable, as a second, lower erosion surface has Eocene beds resting on it.

Central America

The basic units of central America are shown in Figure 5.10. In Mexico there is a plateau with the Sierra Madre ranges on either side. The Sierra Madre Oriental is essentially a dissected escarpment between the plateau and the Gulf of Mexico lowlands, consisting mainly of folded Triassic and Cretaceous and sedimentary rocks. The highest peaks reach about 2700 m.

The Sierra Madre Occidental reach about 3500 m and include a rhyolite plateau at about 2000 m with horizontal flows which are deeply dissected.

The northern plateau of Mexico is lower than the bordering ranges at an elevation of about 1300 m. It is a continuation of the Basin and Range Province of the United States. The Mesa Central is a continuation of the northern plateau at an elevation of about 2000 m.

The neovolcanic plateau at about 2800 m consists of Quaternary volcanic products. Some large extinct strato-volcanoes including Popacatapetl

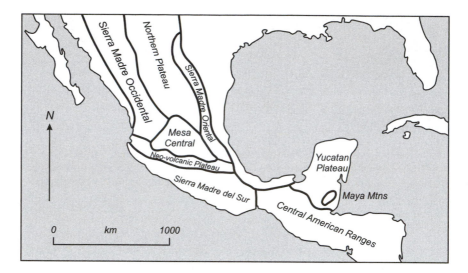

Figure 5.10 Mountains and plateaus of central America and Mexico.

(5440 m) and Orizaba (5747 m) rise above the plateau. The younger volcanic products consist mainly of olivine basalt.

The Sierra Madre del Sur is a complex region geologically, with a rugged landscape. Palaeozoic metamorphic rocks crop out along the coast, and the northern part has Cretaceous limestones.

The mountains of the Central American Ranges result from block faulting and volcanicity. As in the Andes, volcanoes tend to be on top of horsts. Within this group the Sierra de las Cachumatones is a limestone horst with a summit plain at over 3800 m.

The Yucatan Plateau is a low platform of flat lying Miocene limestone.

Plate tectonic explanations

The literature of the last 30 years on the Rocky Mountains has plate tectonics as the dominant paradigm. Plate tectonics has been used in many ways to explain the North American mountains, and much of this work is summarised by Miller and Gans (1997). They look favourably on the effect of magmatic intrusion, and believe that, although long-cited as a classic example of continental growth by lateral accretion of allochthonous terraines, the mechanism is probably not the most fundamental or important process of crustal growth (they equate orogeny with crustal thickening). They write: 'Although "subduction leads to orogeny" has a neat ring to it, true mountain building in the [North American] Cordillera appears to have occurred during finite time intervals of rapid convergence, magmatism and absolute westward migration of North America.'

Some of the problems associated with plate tectonic explanations of the Rocky Mountains are:

1   They are rather far inland to be explained by plate tectonic subduction, as are all the mountains of western North America except the Coast Ranges.
2   Plate tectonic explanations of the Rockies invoke subduction at the edges, but because of the symmetry around individual blocks it is necessary to have subduction in opposite directions.
3   The Rockies are separated from the ocean and any ocean–continent subduction site by the extensional Basin and Range Province and the Coastal Ranges, which are also explained by some as due to subduction.
4   The most general problem is that although the plates have allegedly been colliding for over 100 million years the uplift is confined to the last few million years.
5   With subduction of the Pacific Plate, one might expect tectonic features to be arranged north–south. This is fine for the Front Range and a few others, but the thrusts (or gravity slides) occur in all directions, and the problem is especially acute where ranges run east–west, as the Uinta Mountains.

Gabrielse (1975) summarised the plate tectonic theory for the Canadian cordillera as having a Proterozoic and early Palaeozoic trailing edge, changing to a collision phase. 'The compressional phase was accompanied by subduction zones, volcanic arcs, and plutonism during the Palaeozoic, Mesozoic, and Cenozoic.' None of this seems to explain the actual mountains.

In summary, the main problem with plate tectonics in North America is that so many variations on the theme have been used, including subduction, accretion of plates, strike–slip faulting along north–south lines associated with east–west compression, and various aspects of magmatism.

6 The Andes

The Andes run along the entire western side of South America (Figure 6.1). For most of their length they are divided into the Eastern and Western Andes (Cordillera Oriental and Cordillera Real, or local variations), with an Inter-Andean Depression between. In Colombia they divide into three ranges. They consist of folded and faulted Palaeozoic and Mesozoic rocks, with Cretaceous granite intrusions and volcanics of Mesozoic and Cenozoic age. Tectonic basins such as the Lake Titicaca Basin lie between the East and West Cordillera They are grabens formed by normal faulting, and are filled with thick sediment. This depression is flanked on both sides by volcanoes – the 'Avenue of Volcanoes' of von Humboldt. The eastern slope of the northern Andes drops abruptly to the Amazon lowlands, and several great tributaries of the Amazon have cut through the Andean front.

The Andes provide the prime example of mountains supposedly formed by subduction and compression. However, the geomorphology suggests other explanations.

Planation surfaces and uplift

Planation surfaces provide a major method of working out the amount and timing of uplift of the Andes. If planation surfaces are formed graded to base level (usually sea level), then their present height shows the amount of uplift since they were formed. For this reason, much effort goes into dating the surfaces. Radiometric ages of volcanic rocks, or fossil ages of sediments on the plateau surface are major methods, and fission tracks provide a physical method.

Planation surfaces eroded across folded bedrock dominate much of the Andean landscape (Figures 6.2, 6.3). The planation surfaces are preserved as slightly dissected remnants extending for tens of kilometres. Most authors agree that they represent erosion surfaces created close to sea level and later uplifted. The rocks affected by planation were previously folded, faulted and sometimes overthrust for tens of kilometres. The planation surfaces, preserved on top of the mountain chain, clearly indicate that mountain building processes – that is the uplift – occurred after the creation of these older structures.

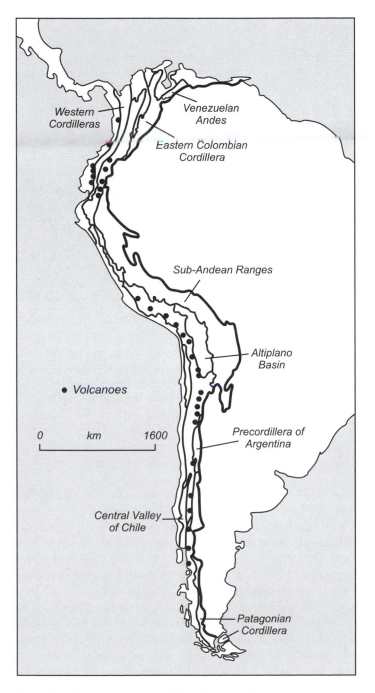

Figure 6.1 The main geographical features of the Andes.

Figure 6.2 The Andean planation surface near Cuenca, Ecuador (photo C.D. Ollier).

Large flattened or gently rolling surfaces have been recognised in the Andes since the beginning of the century, and were often described as the puna surface (e.g. Bowman, 1916) because they are frequently preserved at the elevation of the 'puna', the upper vegetational belt of the Cordillera. The high plains are sometimes known as the Altiplano, but this term is often confused with the upper surface of the depositional material in the Inter-Andean Depression. Many authors recognise multiple planation surfaces.

In the north-east Andes there were at least six phases of uplift and tectonic quiescence between the Late Cretaceous and the Pleistocene (Hoorn, 1995). The main uplift was in the Plio-Pleistocene.

In Colombia planation surfaces have been recognised by many workers (Khobzi and Usselman, 1973; Ruiz, 1981; Soeters, 1981; Page and James, 1981; Kroonenberg et al., 1990). Most agree that their genesis pre-dates the creation of the Inter-Andean Depression. Planation surfaces of Plio-Pleistocene age are reported from Colombia (Kroonenberg, 1990). The Cordillera rose during the Plio-Pleistocene, with more than 3000 metres of uplift since the formation of the puna surface. The progressive transition from tropical lowland to mountain vegetation is revealed in Pliocene deposits whose base is about 2.7 Ma (Radelli, 1967, Van der Hammen et al. (1973), Hooghiemstra (1989), Andriessen et al., 1994). Pliocene deposits are located in the Inter-Andean Depression of Colombia indicating uplift during the Upper Pliocene and the Quaternary.

Figure 6.3 The Andean planation surface north of Quito, Ecuador (photo C.D. Ollier).

In Ecuador planation surfaces cut across both the Eastern and Western Cordillera which are both underlain by folded bedrock. The planation surface is dated by ignimbrite, which covers parts of the planation surface but does not fill the depression, at about 6 million years. West of the Ecuadorian Andes is another dissected plateau, about 500 m high, called the Costa (not a coastal plain). It cuts folded Lower Pliocene sediments. The planation process therefore extends to the Earliest Pliocene.

Coltorti and Ollier (1999) recognise only one major planation surface in the Ecuadorian Cordillera. Near the Peruvian border it is located around 3000–3500 metres. East of Quito it is preserved at 3500–4000 metres and similar elevations are recorded in the direction of the Colombian border. The Western Cordillera is similarly flattened at 3000–3500 metres in the southern part of the country. West of Rio Bamba it is close to 3900 metres. The mean elevation decreases towards the Costa and the Oriente, and large flattened parts are preserved at about 2000 metres. Sometimes the planation surface is tilted or displaced by younger faults. In places it is also dissected by glacial erosion and may be reduced to 'accordant summits'. However, all these modifications, which occurred later in the history of the Cordillera, fail to mask the evidence of planation.

The succession of erosion surfaces preserved on the western flank of the Andes in Peru indicate sporadic uplift of a gently undulating landscape formed by erosion of the Coastal Batholith and its volcanic envelope (Myers, 1975). The planation surfaces increase in altitude with increasing age and distance from the western edge of the Paramonga Block. There is also an increase in the slope of the erosion surfaces with age, indicating uplift by arching of the Chavin Block and westward tilting of the Paramonga Block. The uplift and erosion took place between the emplacement of the last plutons of the Coastal Batholith, about 30 Ma ago, and the eruption of an ignimbrite about 6 Ma ago. Cobbing *et al.* (1981) also described planation surfaces in Peru. Since Middle Miocene time the Andes of central and northern Peru have undergone several distinct stages of uplift. A major erosion surface, the puna surface, closely postdated Middle Miocene deformation. In some places the puna was deeply dissected before deposition of volcanics 10 Ma ago. Renewed uplift led to the Canyon erosion stages, dated in northern Peru at 5–6 Ma ago.

In Bolivia, Walker (1949) recognised a puna surface at about 4200 m which he thought was Pliocene, and a Valle surface at about 4000 m with ignimbrite. He regarded the ignimbrite and subsequent uplift as Pleistocene. The planation surfaces were later dated by ignimbrites at about 7 Ma (Kennan *et al.*, 1997; Lamb *et al.*, 1997). Dresch (1958) regarded the puna surface as Miocene, with uplift in the Plio–Pleistocene. Sonneberg (1963) wrote that 'The present Bolivian Andes are the product of a late Tertiary orogeny' and 'some geologists now strongly postulate that mainly vertical uplift created the present picture of this impressive system in Bolivia.'

Kennan *et al.* (1997) describe low relief palaeosurfaces preserved between 2000 and 4000 m. They were formed between 12 and 3 million years ago. The plains have been uplifted about 2 kilometres and deep dissection occurred since only 3 million years ago. These authors believe that the palaeosurfaces were probably uplifted to within a few hundred metres of their present altitude by 3 million years ago. They note that the complete lack of regional palaeosurface tilt is inconsistent with late, ramp-related folding or subduction – in other words the uplift was simple and vertical. Much earlier Walker (1949), also working in Bolivia, although finding four erosion surfaces showing intermittent uplift, concluded that 'Physiographic and paleontologic evidence lead to the conclusion that the area was relatively low-lying toward the end of the Pliocene and that most of the uplift to the present elevations occurred during the Pleistocene.'

Almendinger *et al.* (1997), writing of the central Andes, distinguish the puna from the Altiplano, the former being 1 km higher, and believe the difference is the result of different crustal thickness. They maintain that the 'timing of deformation, . . . and age distribution of patterns of Cenozoic magmatism suggest that the Central Altiplano region began its principal phase of uplift about 25 Ma, although some uplift could have begun as early as the Eocene (53–34 Ma).' This dating places these events much earlier than most workers now believe to be the case.

Hollingworth and Rutland (1968) describe the evolution of part of northern Chile. They report ignimbrite on plateaus and planation surfaces, and uplift by block faulting. Uplift possibly started in the Late Miocene, but most uplift occurred in the Pliocene and Pleistocene. In the southern

Atacama Desert, Chile, Mortimer (1973) recognises four erosion surfaces. The first began with the elevation of northern Chile above sea level in the Late Mesozoic. Dated volcanic rocks show that the younger surfaces range from Lower Eocene to Upper Miocene, since when the present phase of canyon cutting commenced.

At the southern end of the Andes in Tierra del Fuego the ranges bend to the east as an orocline, around the plains of Patagonia. The structure is shown on Figure 6.4. The mountains consist mainly of old granites, but there is a major thrust of Cretaceous sediments across a basement, apparently with pyroclastics as the lubricant on the decollement. This looks very like a typical plate tectonic diagram of subduction, but if this interpretation were applied here it would be the Atlantic subducted under South America, not the Pacific. As usual, the thrusted structure did not make the mountains of today.

In the Argentina–Chile Cordillera several surfaces have been dated as Miocene. (Segerstrom, 1963; Viers, 1967; Clark *et al.*, 1967). The Sierras

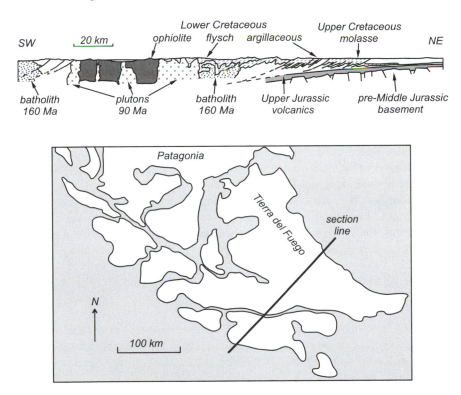

Figure 6.4 Map and cross-section of Tierra del Fuego. The steeply dipping thrust, which would conventionally be interpreted as subduction, would imply subduction of the Atlantic under the continent, whereas it is generally assumed that the Pacific is underthrust under South America to create the Andes. Simplified after Milnes, 1987.

Pampeanas of north-western Argentina are large fault block mountains of Precambrian and Palaeozoic basement, bounded by north–south faults, and rather isolated from the main Andean Range. A pre-uplift regional erosion surface is widely preserved in the eastern slope, tilted slightly to the west. Costa *et al.* (1999) mapped remnants of this surface and analysed the results to reconstruct pre-uplift topography and identify minor blocks and neotectonic faults. The ranges were uplifted in the Neogene, with up to 1300 m in the Plio-Pleistocene.

Drainage patterns

Potter (1998) reviewed the Mesozoic and Cenozoic palaeodrainage of South America, describing the effects of what he regards as the mid-Miocene uplift of the Andes. This work relates to the changes from an old depositional area or geosyncline to a mountain region. Hoorn (1995) studied the change in drainage pattern in the Cordillera and the Amazon Basin. In the Early Miocene the drainage of western Amazonia was directed along a palaeo-Orinoco to the Caribbean. Substantial Andean uplift in the Late Miocene caused major changes in geography, with the Orinoco changing course and the Amazon reaching the Atlantic, drowning older carbonate platforms.

Numerous rivers cut through the eastern and western Cordillera. It is not always clear whether this is by river capture, or if the rivers are antecedent to uplift. In Ecuador the Santiago River crosses the eastern Cordillera to flow to the Amazon Basin. It is very deeply entrenched, and could even have isostatic effects on uplift of the Cordillera like the rivers crossing the Himalayas described in Chapter 7. Potter (1998) describes many short Chilean rivers that start in the Andes and flow across the Great Valley of Chile, and then cross the lower range of the Cordillera de la Costa. From the description the rivers appear to be antecedent to uplift of the Cordillera de la Costa. Indeed, antecedence seems more probable than river capture throughout the Andes. There are no rivers that cross the entire Andes in antecedent gorges, like the antecedent rivers that cross the Coast Ranges of North America.

Volcanoes

There are two main volcanic products in the Andes, stratovolcanoes and ignimbrites. The large stratovolcanoes such as Chimborazo, Cotopaxi, and Aconcagua are mainly Quaternary and modern. The volcanic plateaus with ignimbrites are generally Pliocene.

The volcanic plateau

In Ecuador a series of volcanic deposits lies on top of the planation surface creating a volcanic plateau (Litherland *et al.*, 1993). The Cuenca area was the centre for the volcanic spreads in south Ecuador. The horizontally layered materials cover an area over 80 km across and include ignimbrites, agglomerates, tuffs and lavas of rhyolitic to andesitic composition. Volcanic plateaus did not cover the entire planation surface along the whole Cordillera, and many non-volcanic flattened parts remained exposed. The thin volcanic sheet has been likened to the icing on a cake.

These volcanics must be younger than 6.0 Ma because all along the Cordillera they cover folded rocks of this age. It is possible that they are of Lower Pliocene age, thus giving the age of the planation surface they cover. This, then, is the span of time available for the uplift of the Cordillera and its deep dissection.

The Pleistocene stratovolcanoes

A series of stratovolcanoes erupted during the Quaternary and many of them are still active. Most of the active volcanoes are along the edge of the Inter-Andean Depression and only a few are located in the middle. In Ecuador Pleistocene and Recent volcanoes cover parts of the planation surface. The volcanoes rise to 6000 m, well above the mean elevation of the planation surface which is about 3500 m.

D'Angelo and Le Bert (1968) relate Quaternary volcanicity in Chile to the presence of the Central Valley (longitudinal depression), and show that where the depression is absent there is also no volcanicity (Figure 6.5). They suggest that the depression is caused by normal faults that reach down to magma chambers, while the non-volcanic Norte Chico area between 27° and 32° has thrust faults which they suppose prevent the rise of magma.

Quaternary events in the Andes

The Quaternary of South America is well described by Clapperton (1993). In the Andes glaciation was important in re-shaping the planation surface, the volcanic plateau and the stratovolcanoes. Only two glacial events are well recorded in some parts of the Cordillera. It is possible that in older times the higher parts of the mountain were still too low to be able to generate glacial morphologies although climatic fluctuations are well recorded in the Inter-Andean depression of Colombia (Hooghiemstra, 1989; Andriessen *et al.*, 1994).

Along the coast a series of marine terraces lying at progressive elevation above sea level (up to 225 m in Ecuador) reveal continuing uplift of the area.

Faults

Faults in the Andes includes normal, thrust and strike–slip faults. Many are pre-Andean, and our interest is in the younger faults and their role in the origin of the mountains.

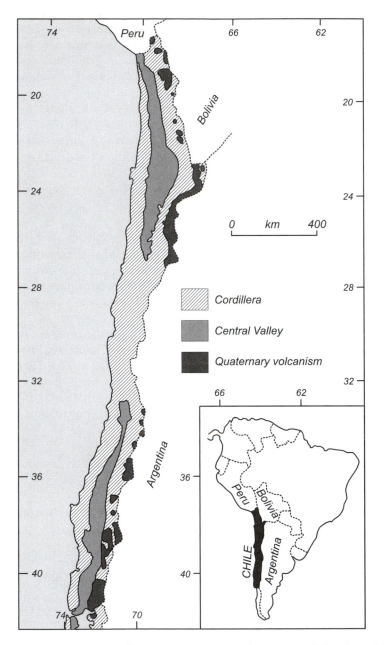

Figure 6.5 Relation between Quaternary volcanicity and the Central Valley in Chile. Where the depression is absent there is also no volcanicity (after D'Angelo and Le Bert, 1968).

The throw of some faults can be calculated, like the vertical displacement of the faults bounding the Cordillera in Ecuador of 1500 to 2000 m. Other major faults might extend right through the lithosphere, an example being those separating the fault blocks of Peru. An east–west section of the Peruvian Andes reveals a series of blocks (Myers, 1975). From the coast these are the Paracas Block, the Paramonga Block, the Chavin Block, the Maranon Block, the East Peruvian Trough Block and the Brazilian Shield (Figure 6.6). The Maranon Block is a geanticline or horst separating two belts of subsidence that were the West and East Peruvian troughs. The Paramonga Block is the site of the Coastal Batholith. Folding of relatively surficial sediments was related to essentially vertical movement of the major blocks, which were long, ribbon-like belts parallel to the coast. The folding appears to be of decollement type, with open folds and no large recumbent folds or nappes, and no ophiolites or melanges.

Block faulting, as postulated by Myers for Peru, suggests normal faults. But many studies reveal low angle thrust faults. The question is, are these faults due to some form of subduction or underthrusting, or are they formed by spreading after uplift of large blocks? Katz (1971) believes the field evidence in Chile shows that the Andes, far from being buckled by subduction and compression, have undergone extension since at least the Miocene, in a belt 300–400 km wide.

It is important to note that the thrusts diverge from the centre of the Andes (Figure 6.7). Both sides of the Cordillera are therefore bounded by thrusts. The thrust faults on the Pacific side are easily explained in plate tectonic terms by subduction. The thrusts on the eastern side of the Andes are explained by subduction of Brazil under the Andes, or by less specific 'underthrusting' from the east, or by thrusting that is in some way connected with the Pacific subduction as in Figure 6.7b. An alternative is that the Andes are spreading like the uplifts in North America described earlier, with a sort of 'mushroom tectonics'. On the other hand, the middle part of the Cordillera is characterised by normal faults with huge displacements, forming typical simple fault escarpments and characteristic of extension. The Inter-Andean Depression is more like a rift valley than a simple graben.

We suggest that the divergent thrust faults are the result of gravitational spreading as the uplifted block relaxed laterally, a situation seen in many other mountain regions. Many authors prefer the hypothesis of underthrusting. On the coastal side subduction of the Pacific Plate is an obvious 'explanation', but the symmetrical underthrust of the Eastern Cordillera presents problems.

Plate tectonic explanations

To express the matter simply, plate tectonics explains mountains on active margins as the result of subduction of one plate under another, as described in Chapter 1. The Andes have become the classical example.

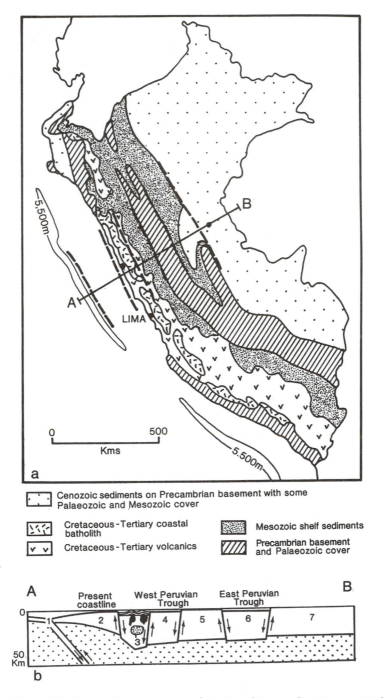

Cenozoic sediments on Precambrian basement with some
Palaeozoic and Mesozoic cover

Cretaceous-Tertiary coastal
batholith

Cretaceous-Tertiary volcanics

Mesozoic shelf sediments

Precambrian basement
and Palaeozoic cover

Figure 6.6 Map and cross-section of Andes of Peru (after Myers, 1976).

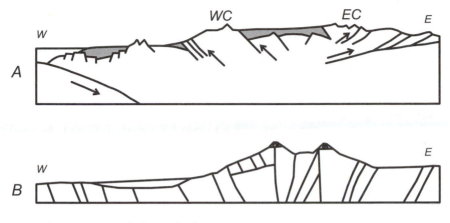

Figure 6.7 Divergent faults in Andes.
    a.   Sketch cross-section of the fan-like arrangement of faults in the Chilean Andes, as shown by Muñoz and Charrier (1996). Subduction is thought to cause the faults in the west, but most think that overthrusting causes faults in the east, though a minority of authors have subduction there also.
    b.   Sketch cross-section of the fan-like arrangement of faults in the Ecuadorian Andes, simplified from Litherland *et al.* (1993).

Zeil (1979) was an early critic, and explained the situation very well as follows:

> James (1971) [and others] drafted a model for the central section of the Andes which was extrapolated by other researchers to apply to other parts or to the whole of the range. In the process a number of geophysical and geological data were inadequately considered or were suppressed in favour of the grandiose tectonic picture.

To this we could add that most geomorphology is omitted or suppressed. Some objections listed by Zeil are:

1   The Benioff zone is not uniform along the Andes. Where it exists it varies in depth. There is no Benioff zone between 250 and 500 km. A clear Benioff zone exists only in the central section of the Andes.
2   The crust in the range varies in thickness. In Colombia it is 45 km; in the Central Andes 70 km; in southern Chile 40 km.
3   The time factor in the genesis of the range is completely ignored.
4   Individual parts of the range exhibit either enormous amounts of uplift or subsidence. Their position in the range and their structural shape cannot be reasonably explained as the results of a process of uniform subduction.

As Gansser has pointed out (1973), plate tectonic theories that use the Andes as a model adopt simplified assumptions that neglect the fact that only the recent morphogenic uplift made the apparently uniform Andes, masking a very complicated geological history. The Andes as a marginal chain are influenced by the shields to the east, which display resurgent tectonics by remobilisation along old fracture zones, and by the Pacific plates in the west, which Gansser thinks more complicated than generally supposed.

It must be stressed that although the area is located on one of the most active margins of the world, where subduction is supposed to be continuous since Cretaceous times, there was a tectonic still-stand period permitting great planation, and the real mountain chain was created only during the Plio-Pleistocene.

The Sub-Recent and Recent volcanic belts in the Andes are approximately 250 km inland from the marginal oceanic trench, a figure that is remarkably constant except in middle Colombia. The rhyolitic–dacitic volcanism can have nothing to do with the melting of a downgoing ocean slab of basalt, but originates from some other source of different composition.

Jordan *et al.* (1997) present a very straightforward account of the plate tectonic approach to determining the uplift of the Andes. They start with two stated premises, summarised here as follows:

1   Surficial uplift correlates in time with thickening crust or thinning mantle.
2   The main mode of crustal thickening in the Andes is shortening.

In view of these premises their approach to determining the topographic uplift of a region is to examine constraints on ages of crustal shortening. From this they conclude that parts of the Central Andean Plateau were near sea level in the Cretaceous, Eocene deformation accounted for a quarter to one half of the uplift of some areas, especially the Western Cordillera, and late Oligocene and younger thickening accounted for most of the elevation. They give dates of uplift that vary from place to place, but this is completely surrogate information: what they really determine is crustal thickening.

Lamb and Hoke (1997), wanting to increase crustal thickness to cause uplift but noting the perfection of planation surface in Bolivia, have an ingenious solution: '14 km of lower crustal thickening beneath an essentially rigid "lid", can be explained by about 100–150 km of underthrusting of the Brazilian shield and adjacent regions beneath the eastern margin of the Central Andes.' Isacks (1988) also noted the planation surface and realised that the volcanoes were merely 'icing on the cake', yet still wanted underthrusting of the foreland on the eastern side to produce plateau uplift.

Allmendinger and Jordan (1997) describe the plate tectonics mainly in terms of high angle or low angle subduction with all the usual plate tectonic attributes including arbitrary thermal effect. For the Andes between 15°

and 24° S they write, 'During the Miocene the crust beneath the current plateau was thermally weakened, resulting in horizontal shortening of the entire crust. At about 10 Ma, shortening mostly ceased in the plateau and in the eastern Cordillera and migrated eastward into the Subandean Belt, a thin-skinned foreland fold-thrust belt. At that point, the cold lithosphere of the Brazilian Shield began to be thrust beneath the mountain belt.' This seems strong on assertion, but short on geomorphic detail.

One surprising example of comparison with another region is that of Isacks (1988) who recognised 'a widespread Basin and Range, Laramide-like shortening', whereas most workers recognise the Basin and Range Province as a classical extensional area, as explained in Chapter 5.

Okaya *et al.* (1997) repeat the usual plate tectonic claim: 'The central Andes are a prime example of compressional deformation and mountain building.' 'For northern Bolivia, fission track data . . . document an increase in uplift of the Eastern Cordillera since ca. 3 Ma'. so far, so good, but they then add, 'i.e. when the Subandean crust was thrust to the west beneath the Eastern Cordillera'. How do they know it was thrust under? They also realise that the Altiplano stays flat, but write, 'The Altiplano rose uniformly by lower crustal thickening.' How can they know the lower crust was thickened? And presumably it had to be thickened uniformly, or have a rigid lid as described by Lamb and Hoke.

Some supporters of plate tectonics relate the uplift of the Andes to subduction of sediments which melts to form granite, which rises, pushing up the Andes. The mostly Mesozoic plutons constitute the largest mass of plutons on Earth, covering about 465 000 km^2, about 15 per cent of the Andes surface. Their alignment parallel to the coast (Figure 6.8) certainly suggests some control on the location and possibly the origin of the Andes, but the plutons made their way upwards over 70 million years and ceased about 30 million years ago – long before the uplift of the Andes which is mainly in the last 6 million years.

In several ways the Andes do not conform to the plate tectonics models that they allegedly illustrate. In particular, there is no evidence of compression along the ocean–continent interface. In the coastal belt, block faulting is the most important tectonic process. Sediments in the Peru–Chile trenches are not compressed and show horizontal bedding. Fractures and graben in the Cocos plate suggest extension, and there is no indication that oceanic plates are being pushed from the widening oceanic ridges towards a subduction zone.

It is interesting to contrast the young ages of uplift of the Andes with the 200 million year span for which alleged subduction has been occurring in South America (Burg and Ford, 1977, p.13). The conventional wisdom is that the Andes were made by subduction of the Pacific, but why did uplift wait for millions of years? How did a pause in tectonics occur to allow planation? And what really caused the young, symmetrical vertical uplift?

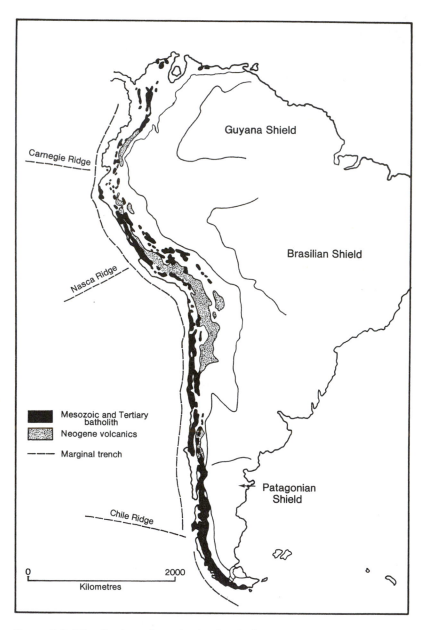

Figure 6.8 Distribution of granite in the Andes.

Summary

A planation surface was cut across the western side of South America during the Early Pliocene, and widespread remnants are preserved on top of the Cordillera. It cuts the older formations including Palaeozoic and Mesozoic rocks, and in places also truncates Late Miocene–Earliest Pliocene rocks. Much of the folding of strata is earlier than and quite independent of mountain formation, which is a result of vertical uplift after the formation of the planation surface. The uplift is mainly Plio-Pleistocene.

Volcanic plateaus formed at the beginning of the uplift processes in the Early Pliocene, covering large parts of the planated Cordillera. Volcanic activity halted for a while, and was reactivated some million years later in the early Pleistocene, possibly connected with faults bounding the Inter-Andean Depression which was relatively lowered, while the surrounding areas were rising. The depression was filled with hundreds of metres of conglomerates interlayered with volcanic products.

The Andes massif was essentially lifted as a great horst, but when lifted beyond the height that the rock strength could bear the horst started to spread laterally, forming divergent faults and an extensional graben (the Inter-Andean Depression) in the middle. The extension is also responsible for the activity of the volcanoes.

7   Asian Mountains

The Tibetan Plateau (Qinghai–Xizang Plateau)

Lying between the Himalayas and the Kunlun Mountains, the Tibet Plateau (Qinghai–Xizang Plateau) is the largest plateau on Earth, with an area of about 2 million square kilometres and an average elevation of about 5 km. Something quite exceptional must be at work in the Earth to create such an extraordinary landform. Here we shall concentrate on the landforms of Tibet, and their evolution. Later we shall discuss the related mountain rims, and only then speculate on the significance of this vast, high plateau (Figure 7.1).

Planation surfaces

Tibet has two major planation surfaces. The higher is called the Peak Surface, and the lower the Main Surface (Li, 1995).

The Peak Surface was complete in the late Eocene. The distribution of many plant and animal fossils shows there were no mountain barriers at that time. Altitude was low and the climate warm. Of the fossils, particular interest lies in the Giant Rhinoceros, which was possibly the largest terrestrial animal of the Cenozoic era. As early as 1927 Grabau pointed out that 'this wide distribution in Asia of an animal of this size and bulk, clearly indicates that during its existence barriers between these different regions were wanting'. Remnants of the Peak Surface are now preserved as plateaus, some of which have ice caps.

Uplift occurred in the Middle Miocene and a new planation surface was formed, the Main Surface, graded in part to Pliocene intermontane basins. Again warm and humid conditions prevailed, associated with tropical to subtropical forest and grassland, and there were no significant barriers (such as mountains) to migration of hipparion, rhino and giraffe – not noted alpinists. Burke and Lucas (1989) described sediments in the Lunpola Basin on the Tibetan Plateau, and wrote that the Lunpolan floras suggest a much lower elevation for Tibet until only a few million years ago. Throughout most of the Pliocene the Tibet region was not a highland but a peneplain with elevation not more than 1000 m. Today the Main surface is

Figure 7.1 The Tibet Plateau, a vast, uplifted pediplain (photo C.D. Ollier).

well-preserved but dissected into plateaus with areas of tens to hundreds of square kilometres.

Shackleton and Chang (1988) note than 'An erosion surface . . . across the Tibetan Plateau was cut across mid-Miocene granite . . . its age is thought to be late Miocene.'

Uplift of the plateau

A period of intense uplift of the Tibet Plateau, known as the Quinzang movement, began about 3.4 Ma. The movement is divided into three phases:

A   beginning 3.4 Ma
B   beginning 2.5 Ma
C   beginning 1.8 Ma

Phase B is perhaps the most significant, and corresponds to the initiation of a monsoon climate and the start of loess deposition in China. On the plateau itself the humid, hot climate was replaced by a dry and cool climate. In a later publication (Li, 1995) the 1.8 million years of stage C is refined to 1.7–1.66, but this may be pushing accuracy too far.

After these major uplifts there were minor uplifts. Regional uplift is postulated to make the seven river terraces of the Yellow River (Huang

Ho). Detailed work on sediments in the Zoige Basin indicates three accelerated uplifts at 800, 360 and 160 Ka (thousands of years ago).

These dates can be contrasted with the date of the initial collision of India with Asia, about 50 Ma, which is far too early to be the cause of the uplift.

A significant feature of the two planation surfaces is that they have least difference in elevation in the middle of the plateau and greatest (up to 1000 m) at the borders. This shows that the uplift of the edges is not the same as that of the interior of the plateau and may have a different cause. The uplift of the Tibet Plateau is a joint operation of the overall uprising of the plateau and local rapid rising of its periphery.

The Himalayas

A further complication in mountain building is isostatic response to erosion. This is best illustrated by the Himalayas, but only when these mountains are considered together with the Tibet Plateau.

The southern edge of the Tibet plateau is marked by the great arc of the Himalayas and several major boundaries (Figure 7. 2). The Indus–Tsangpo suture zone in southern Tibet is believed by many to mark the site where the Indian continent has been subducted under Asia. Two significant lines further south are the Main Central Thrust and the Main Boundary Fault. The High Himalayas lie north of the Main Central Thrust, and the Lower Himalayas between it and the Main Boundary Fault.

The Himalaya Range, from Nanga Parbat to Namcha Barwa, is 2500 km long and between 250 and 400 km wide. The average height of the main Himalayan range is over 6000 m, and it rises above the Indo-Gangetic plain by up to 8500 m.

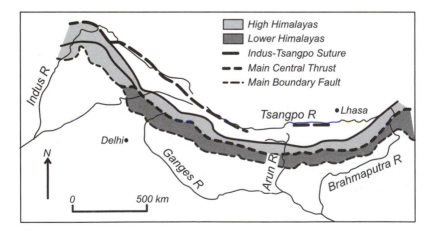

Figure 7.2 Map of mountains and major faults in the Himalayas.

Geology of the Himalayan region

The Himalayas consist of nappes which moved from a northerly direction. The rocks are highly metamorphosed, but some workers have detected tectonic dolomite between the major nappes, such as the Kathmandu Nappe and the Nawakot Nappe. If the nappes are gravity structures, they imply uplift of the Tibet plateau or other high ground as a geotumour from which nappes spread to a lowland further south. It is important to realise that the nappes are older than the uplift of the present Himalayas. There has, however, been renewed movement on some nappes in more recent times, and in places bedrock is thrust over unlithified sediments. Maltman (1993) illustrates (his Figure 7) high-grade Precambrian Nanga Parbat gneiss thrust over unlithified glaciofluvial sediments.

Between the crystalline Great Himalayas and the Siwalik sediments lie the Lesser Himalayas. Nappes of the Lesser Himalayas are thrust for distances of 20 to 25 km over the Siwalik Molasse along the Great Boundary Fault.

A very striking feature of the Himalayas in plan is the perfection of the frontal arc. It is quite comparable with an island arc, yet is firmly located on land, and furthermore the arc can be linked with the festoon of arcs that run from Indonesia, through the Himalayas, and on to the Mediterranean. It is not understood why simple underthrusting should create an arc in plan, let alone why it should approximate to the size and radius of an island arc and be continuous with the island arc chain.

A thick sequence of sediments north of India was deformed 35–45 million years ago to form the first Himalayan mountains, the Tibetan Himalayas. Later, 25 to 15 million years ago, a further rock mass was deformed to make the Great and Lesser Himalayas. Marine sedimentation finished in the Upper Palaeogene. The primary subaerial relief was inclined to the rapidly narrowing shelf sea of the southern foredeep (Kalvoda, 1992).

Observations in southern Tibet agree fairly well with data from the Himalayas. The early stage of tectonic development ended at the close of the Eocene. The greater part of the Tibetan highland emerged from the sea in the Cretaceous.

During the Miocene the Siwalik Depression developed to the south of the present day Himalayas. The pre-Miocene sediments of the southern Himalayan foredeep are derived from the Indian Shield and lie on a crystalline basement (Kalvoda, 1992).

Geological data show that India was once united with the southern continents of Gondwanaland, and until the Mesozoic shared their fauna, their ice-ages and other features. Since then it has separated and drifted north at a rather high speed (16 cm per year) to collide with the Asian mainland. This geological information is confirmed by the evidence of seafloor spreading in the Indian Ocean. For a while part of a great sea, the Tethys Sea of Mesozoic and early Tertiary times, lay between the two continental slabs. But despite the data from seafloor spreading and palaeomagnetism, palaeontological evidence shows that India was never very far from Asia, and Gondwana fossils have been found as far as the Tien Shan.

Rivers and drainage patterns

Major rivers from Tibet cross the Himalayas, so at an earlier stage the Tibet Plateau must have been higher than the Himalayas. Later uplift along the Himalayas created higher ground, but the rivers kept pace with the uplift. This means the rivers are antecedent.

There is now no trace of a planation surface in the Himalayan region, but the landscape evolution of the region can be worked out from the study of the rivers that cross it. Rivers such as the Indus, Brahmaputra and Arun rise on the Tibet Plateau and cross the much higher Himalayas in deep gorges. The Himalayas are not just the world's highest mountains, but also have the deepest valleys – deep enough to cause isostatic compensation. This suggests that as the rivers cut down, the Himalayan Range will rise. Since erosion is greatest in the valleys, uplift is also concentrated in the valleys, and anticlines grow in response to erosion rather than to tectonic thrusts.

The hypothesis of isostatic uplift of the Himalayas (Wager, 1937) starts with a simple uplifted plateau, an extension of the Tibet Plateau (Figure 7.3a). Erosion was greatest on the plateau edge and cut huge valleys, deep enough to cause isostatic compensation. Considered as a whole, the belt of valleys will be uplifted (Figure 7.3b). Since the uplift is general, the inter-fluves, and individual peaks, will rise above the height of the original plateau (Figure 7.3c). In this way the Himalayas rise to be higher than the Tibet Plateau.

Linking this story of isostatic uplift of the Himalayas to the Tibet Plateau, it is interesting to recall that the erosion surfaces on the plateau diverge in elevation towards the Himalayas, which fits exactly the concept of erosion accompanied by isostatic uplift.

This model can be tested by topographic analysis. If the uplift of the Himalayas results from isostatic uplift in response to erosion, the average height of the Himalayas should be the same as that of the Tibet Plateau: if there is an external force causing uplift (such as granite intrusion or pressure from a northward-migrating India) there would be no such accord. Remarkable confirmation comes from the morphometric work of Bird (1978). He carefully derived a composite mean elevation profile across the Himalayas and Tibet (Figure 7.4). The result was a remarkably uniform mean elevation of 5 km throughout. As he expressed it (p. 4979), 'This uniformity suggests that some deeper layer of the crust is weak . . . It can be seen that the high peaks of the Himalayas result from adjacent deep erosion of its valleys [Wager, 1937] rather than from any maximum in the average elevation.' A plan version of the Himalayan region in the form of a Summit Level Map of Nepal is provided in Plate 1 in Ohta and Akiba, (1973).

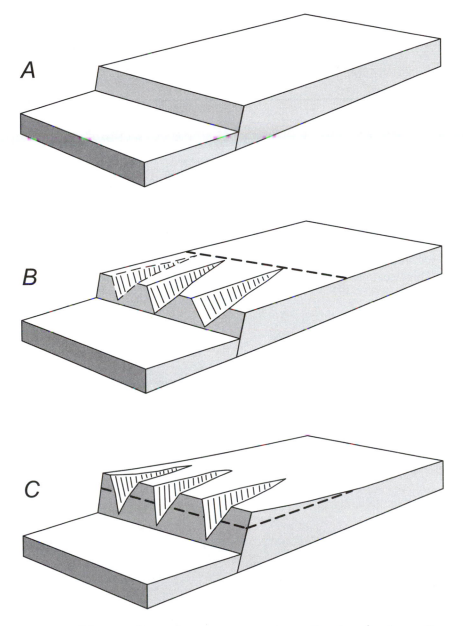

Figure 7.3 Diagram of isostatic response to erosion at the edge of a plateau. For explanation see text.

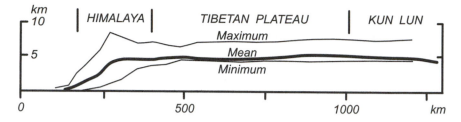

Figure 7.4 Composite mean elevation profiles of the Himalayas and Tibet (after Bird, 1978). The continuation of the same mean level from Tibet to the Himalayas suggests simple isostatic uplift.

An individual large valley might have its own isostatic compensation, as shown in Figure 7.5. Isostatic uplift will cause more uplift along the valley than on the plateau as a whole. If the present topography is extrapolated across the valley, it appears that the valley has cut through a ridge.

This situation was noted by the observant Scottish novelist John Buchan (1924) who wrote:

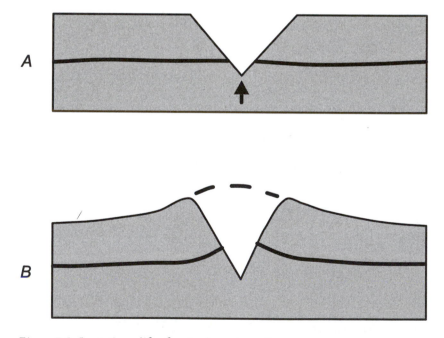

Figure 7.5 Isostatic uplift of a single major valley.
    a.   Before isostatic response.
    b.   After isostatic response. Because of uplift of the edges, the valley appears to flow along an anticline and across the highest part of the topography.

Curiously enough, the rivers which break through the Himalaya chose the highest parts of the range through which to cut. South of Pemakochung is the great peak of Namcha Barwa, 25, 445 feet high; north of it is the peak of Gyala Peri, 23, 400 feet. The distance between these mountains is only some fourteen miles, and through this gap, at an altitude of just under 8,000 feet, flows the great river.

(The gorge is about 5 km deep and 23 km wide, which can be compared with the Indus Gorge which is 6 km deep and 21 km wide at Nanga Parbat).

Isostatic uplift along a valley not only affects topography, but affects the geological structure, creating an anticline along the valley as shown in Figure 7.6. Ohta and Akiba (1973, p. 5) describe the Arun valley as 'a deep valley, cutting along the axial zone of the Arun anticline'. Ishida and Ohta (1973, p. 60) described the structure of part of Nepal where the gneiss strikes NW–SE and dips N at 30°. On this broad structure another fold trend of NE–SW is superposed. 'The antiforms are located along the large river, such as the Tamba, Kosi, the Khinti Khola, the Likhu Khola and the Dudh Kosi.' Furthermore the anticlines are narrow and sharp, while the synforms between the major valleys are all open folds (Figure 7.6). This is precisely what would happen with isostatic uplift along the valleys. Tectonically it is very difficult to make folds that are *parallel* to the main direction of tectonic movement. Folds are normally perpendicular to the direction of thrust. Furthermore it is unusual to have anticlines with a different style from the intervening synclines. Geomorphically it is extremely difficult to explain why all the large rivers flow along tectonically formed anticlines, unless the valleys were there first.

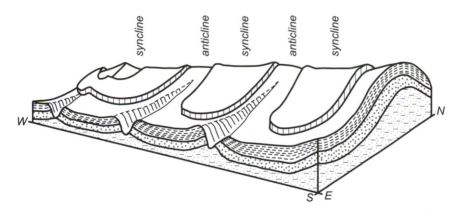

Figure 7.6 Anticlines along valleys (after Ohta and Akibra, 1973). Wherever the rivers cut deeply, a sharp anticline is found. Between the anticlines are very broad synclines. This type of folding is very difficult to produce by tectonic pressure, apart from the difficulties of rivers being located along anticlines.

Buchan also noted that the Brahmaputra 'breaks through the highest range on the globe . . . by means of a hundred miles of marvellous gorges, where the stream foams in rapids, but there is no fall more considerable than can be found in many a Scottish salmon river'. In geomorphic terms, there are no major nick points, or in scenic terms no great waterfalls. With major scarp retreat, or river capture, these are to be expected, but with an antecedent stream they are unlikely to form. This is why the world's greatest waterfalls occur on rivers that fall over Great Escarpments (described later) and not on antecedent rivers.

Recent work on the Tsangpo river has added a lot of detail, but the essential story remains the same. The middle reaches of the Tsangpo have a course with alternate straight stretches through gorges, and meandering stretches across sediments. This is caused by numerous faults across the course, with aggradation on backtilted blocks and gorge incision across the uplifted block edges. The faulting is Pliocene to present, fitting in with the idea of general Pliocene and younger uplift (Zhang, 1998).

Vertical uplift of the Himalayas

Many scientists have concluded that the Himalayas are formed by vertical uplift rather than compression. Crawford (1974, p. 373) wrote, 'The great height of the present-day Himalaya is quite unrelated to their original formation as a geological structure. Least of all is it directly related to Tertiary collision.' Ohta and Akiba (1973) writing of Nepal noted: 'The present height of the Great Himalayas has resulted from block upheavals of this range in the younger geologic periods since Cretaceous' (p. 255). 'The Great Himalayas...have resulted from...block movement with vertical displacement and not from the Alpine folding. No nappe structures can be found in Central and Western Nepal.' (p. 258). 'Even the Main Central Thrust and the Main Boundary Thrust were generated in relation to this vertical block movement.' (p. 257).

Further detail of the vertical movement is provided by the long profiles of 16 trans-Himalayan rivers (Seeber and Gornitz, 1983). North of the Main Central Thrust, in the High Himalayas, the rivers have steep gradients. Downstream, they have gentle gradients, and are not affected by the Main Boundary Fault. This shows that the increased gradients result from differential uplift of the High Himalayas, as we might suspect from the mountains themselves. We thus have a picture of the long and narrow arc of the High Himalayas being uplifted vertically during the Quaternary, and the Main Boundary Fault has little topographic importance.

Ohta and Akiba (1973, p. 6) wrote, 'All these topographical characteristics illustrated in the summit level map are reflecting well the geologic structures of this large area. This simply means that the development of the Himalayan range is a very young event in geologic time.'

Age of uplift of the Himalayas

The model suggested here is of isostatic uplift of the edge of the Tibet Plateau to form the High Himalayas. If this is correct, a further conclusion can be drawn. Since the Himalayan uplift relates to the uplift of Tibet, it must be the same age or a little younger. This means the uplift started at 3.4 Ma, and is largely Quaternary. This has enormous impact on climatic change, both locally and globally, and may be the trigger to the Ice Age.

An outside limit is provided by the Upper Cretaceous marine limestones that cover most of the Tibet plateau, which must therefore have been below sea level in the Upper Cretaceous. Another limit is placed by the molasse deposits in the Siwaliks (south of the Himalayas) which are at least 18 Ma old, so some high ground must have existed at that time.

Estimates of the uplift have been derived from comparison of the altitudes at which plant fossils of the Pliocene and of the interglacials are found now and at which the kind of vegetation they represent is now found on the southern flank. Thus *Cedrum deodar* and *Quercus semicarpifolia*, found in Pliocene gravels at 5900 m on the north side where the mean annual temperature is $-9°C$, grow today in evergreen forest at 2500 m on the south side with a temperature of $10°C$. Most of this difference is attributed to an uplift of about 3000 m. An overall conclusion is that most of the substantial uplift since the Pliocene took place in the Late Pleistocene.

The relatively small volume of sediment of Quaternary age derived from the Himalayas (and Karakoram) in comparison to the huge uplift supports the view that the uplifts are very young (Kalvoda, 1992). The destruction of the mountains by geomorphic processes has not been active for a sufficiently long time to modify significantly the vertical dissection of relief. It is possible that at present the rate of uplift is greater than the rate of denudation. Evidence includes correlated terraces, gradient changes in watercourses and repeated geodetic measurements. According to some investigators only the latest glaciation is found in the Nan Shan and Tsaidam Basin, suggesting ground was lower during earlier glacial periods.

The main period of orogenic movement in the Himalayan zone began during the Middle Miocene, but the Pliocene–Pleistocene period of folding controlled the main types of mountain relief in the foreland. North of the Himalayas sediments were deposited in the Zangpo Basin, deformed and eroded, and by the Pliocene vast tropical lakes covered the area (Kalvoda, 1992). 'Only as late as the onset of the Quaternary does evidence of significant tectonic uplift in the Himalayan region appear.'

The enormous Bengal Fan has collected sediment derived from the Indian craton, the Himalayas and the Indo-Burman ranges. Eocene and Oligocene sediment is dominated by quartz, with very little feldspar or rock fragments, and appears to be derived from a deeply weathered craton. The Miocene is very different, being rich in feldspar and rock fragments. The younger sediments are rich in potassium feldspar relative to plagioclase, suggesting a granitic source (Uddin and Lundberg, 1998). The variation indicates uplift of mountains and unroofing of granites starting in the Miocene, rather earlier than the geomorphic evidence suggests, though the Miocene sediment may have been derived from the Indo-Burman area.

Copeland and Hanson (1990) also used evidence from the Bengal Fan, but came to different conclusions. Some sediment deposited in Early to Middle Mesozoic times they attributed to a pulse of uplift in the Himalayas. They claim at least part of the Himalayas was suffering rapid erosion throughout the Neogene.

Cervenny *et al.* (1988) used fission track ages in the Siwalik sedimentary pile derived from the Himalayas and concluded that for at least the past 18 million years there have been areas in the Indus River catchment with uplift rates and erosion rates comparable to those observed in the Himalayas today.

Derivation of sediment does not, of course, mean that mountains were present. The lowering of the huge area of Tibet from one planation surface to another would itself provide vast amounts of sediments. The common description of the Siwalik sediments as 'molasse' rather begs the question, for the term is generally taken to mean 'sediments associated with mountain building'. If the Siwalik lithofacies are examined in detail, 'it can be found that the piedmont breccia and fanglomerate representing the intense uplift of the mountains, the so-called Boulder stage, did not begin to appear until the Quaternary period' (Li, 1995). Most of the Neogene sediments are fine grained.

Li (1995), well aware that most Western researchers favour a Miocene age for the uplift of the Himalayas and the Tibet Plateau, insists that the Tibet Plateau, the Himalayas and the Kunlun Mountains were all uplifted during the Quaternary. 'The view that the plateau has attained its present height by 14 Ma or 8 Ma can not be backed up by the evidence.'

Plate tectonic explanations of the Tibet Plateau and the Himalayas

At its simplest, the tectonic story is of the collision of India into Tibet, with the Himalayas between (Figure 7.7). But the continent–continent collision is supposed to do three things: first, the collision was by under-riding, to produce the double thickness of crust and the uplift of Tibet; secondly, the collision is supposed to create great nappes at the collision site; and thirdly, collision is supposed to drive the uplift of the Himalayas to become the highest mountains in the world.

The simplest explanation of the Tibet Plateau is that the Indian plate has been subducted under the Asian plate, giving a double thickness of crust. This very thick and light crust would rise, giving the plateau its extraordinary height. Holmes (1965) described it as follows:

> That an area so vast as Tibet should stand at so extraordinary a height remained a geological enigma until the glacial evidence for continental

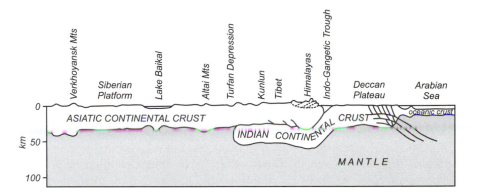

Figure 7.7 Section across Asia from the Verkhoyansk Mountains to the Arabian Sea, after Holmes, 1965. This is meant to illustrate the plate tectonic view of the collision of India with Asia to create the Tibetan Plateau by making a double thickness of crust, the Himalayan nappes by compression, and the uplift of the Himalayas. It is not clear what Holmes thought was going on between the oceanic crust and the Deccan Plateau.

drift suggested that the crust of the ancestral Indian shield had somehow migrated from Africa to Asia. Instead of staging a head-on collision, the shield was drawn downwards to continue its unfinished journey beneath the area that has become Tibet, leaving the distorted sides – Afghanistan and Baluchistan on the west, and Szechwan, northern Burma and Yunnan on the east – to accommodate themselves around the invader as best they could. The isostatic uplift of Tibet was the simplest of the many tectonic consequences of this stupendous encounter.

Johnson *et al.* (1976) go so far as to estimate the areal extent of the Indian plate that is subducted under Tibet as 2.4 million km². Alas, the story did not remain so simple, though it still seems to be the mental model of many geophysicists. If a subduction model is used to make a double thickness of crust, the subduction must have been remarkably flat. The location of the Indus suture ophiolites, the generally agreed site of continent collision, rather suggests that thickening and thrusting were more confined.

Because the underthrusting hypothesis is so entrenched, a few non-geomorphic extra details are elaborated here. The conventional wisdom is that the uplift of Tibet and the Himalayas results from the underthrusting of India under Asia, a thrust of over 2000 km (Burchfiel, 1983) in 50 million years. Why did uplift wait 47 million years to happen, and why were there two still-stands to allow formation of planation surfaces in Tibet? The underthrusting of India under Asia would surely require a low angle contact, but Hirn *et al.* (1984) found that the fault limiting the Lhasa block [the Indus–Tsangpo suture] is a vertical contact. They wrote, 'the exact coincidence of the Moho step with the edge of the Gandise granitic belt ... points to a vertical contact of crustal segments'. Even the India–Asia collision is in some doubt, because Gondwana species are found in Asia, suggesting that Asia and India were never too far apart. In Xinjiang and the Turfan Basin Permo-Triassic vertebrate fossils of Gondwana type have been found. The Late Permian deposits have a *Dicynodon* fauna which is widespread in strata of the same age in India, South Africa and Australia. Indeed the faunal mixture in Xinjiang 'reflects the large scale convergence of every continent in the late stage of Early Permian' (Ministry of Geology and Mineral Resources, 1982). Abundant large dinosaurs in India in Early Jurassic and Late Cretaceous times 'are positive proof that there were over-land communications between peninsular India and other continental regions at times during the Jurassic and Cretaceous periods and perhaps throughout the extent of these periods' (Colbert, 1973). Based on palaeontological rela-tionships, the Tethys Ocean between Eurasia and Gondwana might have been a narrow epicontinental sea, and the vast ocean usually postulated may be an artefact (Crawford, 1979)

Royden and Burchfiel (1997) also present a classic plate tectonic explanation of the Tibet Plateau, starting their essay with the sentence, 'The collision of India with Asia about 50 Ma. ago and their subsequent convergence has produced a spectacular example of active continent–continent collision.' But active convergence of India and Eurasia does not manifest itself in Tibet by crustal shortening, and the dominant tectonic feature at present is east–west extension, with the formation of normal faults in a roughly north–south direction. They point out that in the Tibetan Plateau a series of north–south striking grabens in southern Tibet indicates extension of the plateau at about 10 mm per year. The grabens started about 2 Ma ago. There is also evidence for large scale north–south extension.

According to Molnar and Tapponnier (1978) Tibet is underlain by weak material (in contrast to the double crust idea mentioned earlier) and the hydrostatic head that causes the high altitude of the plateau appears to be maintained by pressure applied by India to the rest of Eurasia. Tibet trans-mits this pressure to the regions to the north and east, and acts as 'the pressure gauge of Asia'. Active convergence of India and Eurasia does not manifest itself in Tibet by crustal shortening, and the dominant tectonic feature at present is east–west extension, with the formation of normal faults in a roughly north–south direction (Molnar and Tapponnier, 1978).

Similarly, Ni and York (1978) explain the extensional tectonics as resulting from the relative eastward motion of the Tibet Plateau as it is wedged from between the converging Indian plate and the stable Tarim Basin (Figure 7.8). Beyond the Himalayas there is little sign of the compres-sion that might be associated with the India–Asia collision.

Chang (1996) presents a modern account of the Tibet Plateau which is almost entirely orthodox plate tectonics. He claims, 'The surface uplift of

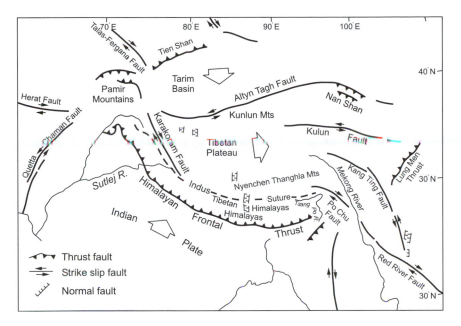

Figure 7.8 Simplified geography and tectonics of the Himalayan–Tibetan Plateau region. Convergence of the Indian Plate and Asia north of the Altyn Tagh Fault leads to a resultant eastward movement of the Tibetan Plateau Plate (after Ni and York, 1978).

the Xizang Plateau *due to isostasy* began in early Miocene' (our italics). How he knows the uplift was isostatic is not clear, and his dating is not in agreement with other workers. He admits that the planation surface cuts across mid-Miocene granite, and that the Pliocene and Quaternary was a period of east–west extension with north–south graben, with about 2 km of plateau uplift 'explained by the catastrophic delamination of the thickened lithospheric root beneath the plateau'. This seems to be a classic case of plate tectonic rationalisation.

Coleman and Hodges (1995) determined an earlier date of uplift of the Tibet Plateau at over 14 million years. 'One way to determine age of uplift is to establish the initiation of NS normal faults – these are due to gravitational collapse of plateau.' They determined a $^{40}Ar/^{39}Ar$ age of 14 Myr for hydrothermal mica from an extensional fracture belonging to such a fault system. So, they concluded, the Tibet Plateau attained its high mean elevation well before Late Miocene times.

This tortuous reasoning assumes that:

1   early uplift was the cause of gravitational collapse of the plateau (which may be spreading for other reasons, as are other mountains all over the world);

2   hydrothermal activity should accompany gravity collapse (hydrothermal activity is normally associated with igneous activity, not mechanical collapse);
3   the specimen of mica dated by argon isotopes represents the age of uplift of the whole plateau

As for timing, Burg and Chen (1984) note that 'significant crustal short-ening was completed by the Palaeocene'. In other words crustal shortening had finished long before vertical uplift started. With even more significance to ideas of India being thrust under Asia they write that 'the crust of the Lhasa Block might have been thickened *before* the India–Asia collision' (our italics). The same idea is expressed by Chang (1996) who wrote 'the age of the thickening of the Qinghai–Xizang Plateau crust is end-Palaeogene.'

A plate tectonic explanation of drainage is given by Brookfield (1998), too complex to summarise here. He believes that the drainage history of southern Asia can be reconstructed by restoring the gross movements of the plates and the tectonic displacement, uplift and erosion of individual tectonic units. Nevertheless he finds the most important changes in drainage took place in Pliocene to Quaternary times.

Other Asian mountains and tectonic features

The Kunlun Mountains

On the northern side of the Tibet Plateau are the Kunlun Shan or Kunlun Mountains. This is an uplifted plateau rim like the Himalayas, but smaller. Marine Late Oligocene sediments are found at 1.5 to 2 km in western Kunlun. It appears that most of the uplift has taken place since the Pliocene. The sides of the uplifted block have spread out laterally, giving the appear-ance of being thrust outwards, particularly along the northern front.

Further north there is a belt of stony desert between the Kunlun Mountains and the Tarim Basin. Here, near the oasis of Pulu, there is a broad plain cut by a deep gorge, exposing the geological sequence. The bedrock was planated, and covered by thick pebbly alluvium. Then there were two exten-sive horizontal lava flows, about 20 m thick and with crude columnar jointing, with gravels between the flows, and covered by yet more gravel (Figure 7.9). The basalt has yielded a K/Ar date of 1.1 million years. The significance of this is that the area had been planated more than a million years ago; that highlands already existed to provide the vast amount of coarse but well-sorted debris; that the area was one of tension, permitting the ascent of volcanic rock; and that there has been no significant folding or other disturbance in the past million years.

Chang (1996) notes that: 'Coarse Pliocene to recent clastics in basins around the plateau suggest fast late Tertiary–Quaternary uplift.' Rapid uplift is still occurring. Repeated levelling over a 20-year period indicates uplift

Figure 7.9 One-million-year-old basalt in horizontal gravels at the base of the
Kunlun Mountains (photo C.D. Ollier).

by as much as 10 mm per year. Terraces on the Kunlun Mountains and knick points in rivers at the edge of the plateau indicate fast recent uplift. Li (1995, p. 190) quotes Cui Zhiju who found that part of the Kunlun was lifted from 1500 m at 2.5 Ma to 4700 m at present.

The Tarim Basin and the Tien Shan

The Tarim Basin is a huge sunkland, now mainly covered with sands of the Taklimakan Desert, and north of this are the ranges of the Tien Shan, almost 2000 km long.

The conventional view of the Tien Shan is that it is essentially a horst, with spreading. Holmes (1965) wrote

> Late Jurassic folding was followed by further erosion and a cover of late Cretaceous and Tertiary deposits. Then came the Pliocene and later block faulting and tilting, with flanking subsidences, giving structures not unlike those of the Rockies in North America. Pleistocene erosion surfaces have been unwarped by rising blocks and downwarped by sinking intermontane basins.

As early as 1904, Davis wrote:

> The best example of an ancient, greatly uplifted former peneplain in a mountain region is the Bural-Bas-Tau, in the Tien-Shan Mountains of central Asia, where massive granite has an evenly planed surface, standing at altitudes of 10,000 or 12,000 feet [about 3,500 to 4,000 m]. This surface is obviously a peneplain, and at the ends of the range it slopes down off the sides, thus indicating the configuration of the uplift which produced the modern range.

Davis' sketch is reproduced as Figure 7.10. On the south side thrusts are to the south, and on the north side to the north. Yin *et al.* (1998) believe the range was formed by compression and crustal shortening around 21 to 24 million years ago. They wrote, 'As no Cenozoic normal faulting has been identified in the Tien Shan region, its denudation may be entirely attributed to contractile uplift and erosion.' It seems quite possible that this is a situation with a horst bounded by normal faults which have spread, like the Rocky Mountains and other uplifts of the Laramide orogeny in North America.

Late Oligocene marine sediments occur at 1.5 to 2 km in southern Tien Shan.

The Turfan Depression

The most spectacular of the downfaulted blocks is the Turfan Depression, which lies between two branches of the Tien Shan and has a lowest point 168 m below sea level, not too far from a peak at over 6000 m. Lake Aiding

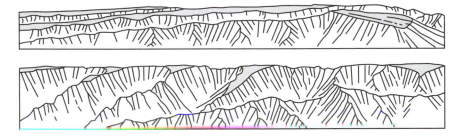

Figure 7.10 The uplifted peneplain of the Bural-Bas-Tau in the Tien Shan, sketched by W.M. Davis, 1904.

lies on the floor of the depression (Figure 7.11). The area has been subsiding throughout the Quaternary, and Quaternary sediments are over 1000 m thick in the northern Turfan Basin and over 700 m in the southern basin. Subsidence has outpaced deposition, causing part of the basin to be below sea level. The bounding ranges appear to be thrust towards the Turfan Depression (Figure 7.12).

The Altai Mountains

North of the Tien Shan lie the Altai Mountains. This area was peneplained in the Lower Tertiary and then rose above the level of a Siberian sea. By

Figure 7.11 Lake Aiding on the floor of the Turfan Depression. The elevation is 168 m below sea level, and the location is a sunken block within the Tien Shan mountains (photo C.D. Ollier).

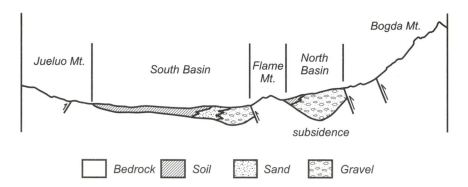

Figure 7.12 Cross-section of the Turfan Basin (from Prof. Zhong Junping, University of Xinjiang).

the second half of the Tertiary epeirogenic movements were intensified, and in the Late Tertiary and the Quaternary there was much block faulting. As a result of the displacement of large blocks the surface of the Tertiary peneplain (and the Tertiary deposits lying on it) was raised to different levels ranging from 500 m to 3600 m (Suslov, 1961).

The Transbaikal Mountains

'East of Lake Baikal the remnants of a summit plateau were disrupted by mid- and late-Tertiary block-faulting into a series of north–east south–west trending horst and graben features' (Bridges, 1990). Lake Baikal itself is a half-graben, with a huge fault on the western side and a down-tilted land surface on the east. The westerly fault scarp remains fresh-looking, but is dissected into triangular facets by many valleys draining into the lake. Ufimtsev (1990) described the geomorphology and tectonics, and produced many diagrams showing the faults and erosion surfaces of the region, such as that shown in Figure 7.13.

East of the Transbaikal Mountains are the Sayan Mountains, which also were eroded to a peneplain during the Tertiary and uplifted in the Late Tertiary. West of the Transbaikal Mountains are the Stanovoy Ranges which make the great divide between drainage to the Arctic and drainage to the

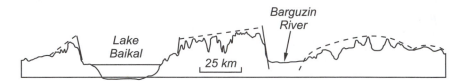

Figure 7.13 Diagrammatic cross-section across Lake Baikal, showing the warped and faulted planation surface (after Ufimtsev, 1990).

Pacific, and north of this is the Aldan Plateau, about 1000 m above sea level, and described as a warped erosional plain (Bridges, 1990). Ufimtsev (1994) describes most of the Mongolian–Siberian region in terms of a warped and faulted planation surface.

The Karakoram

The Karakoram is a mountain range at the north-west end of the Himalayas, about 600 km long and 150 km wide, and the highest peak is K2 at 8760 m.

The region started to be uplifted in the Palaeogene with successive stages which were followed by denudation. The youngest granite in the Great Karakoram is Late Miocene, and was uncovered before the Pliocene. The uplift of the present-day Karakoram began in the Late Neogene and continued through the Quaternary and into the present. The extreme uplift of the Karakoram ranges during the Quaternary was indicated by the appearance of sediment deposits such as the Mirpur gravel and the Gongha conglomerate by 500 000 years ago (Kalvoda, 1992).

The Indus–Tsangpo ophiolite zone, regarded as a suture between the Indian and Tibetan continental masses, continues between the Himalayas and the Karakoram. Two massifs are of major significance in the Karakoram – the Nanga Parbat and the Haramosh Massifs. They consist of metamorphic rocks derived from the Archaean rocks of the Indian Shield, and they appear to be wedged into Miocene to Eocene sediments on the south of the Karakoram.

The landforms of Pakistan are varied and complex, but the essence has been expressed by Schroder (1993) who wrote, 'Most of the landforms of the mountains and plains of Pakistan have been produced within the last 1–2 Ma.'

The mountains of Japan

The major relief and structural framework of the arc–trench system of Japan have been formed during the latest Cenozoic, particularly in the Quaternary. This was proposed by Otuka in 1933, and supported from a plate tectonics view by Sugimura and Uyeda (1973).

The Kitakami Mountains of north-east Honshu are said to be typical of the Japanese islands (Hoshino, 1998). The summit is marked by a raised peneplain which was formed in the Late Miocene and the gently inclined erosion surface and gravel beds were developed along the flanks of the range in the Late Pliocene. Uplift of the Kitakami Mountains took place during the Pliocene to Early Pleistocene (Chinzei, 1966). The raised peneplain, which makes the skyline of the mountains, is 1000 m in the north and falls gradually to 500 m to the south. The peneplain is also very evident in southern Japan (Figure 7.14). The conventional plate tectonic explanation of Japan as an island arc created by subduction completely ignores the very obvious planation and vertical uplift.

Figure 7.14 The peneplained surface of southern Japan (photo Takao Yano).

Hoshino believes the gravels that cap the southern part of the peneplain are of Late Pliocene (Villafranchian) age, and are terrestrial but were deposited very close to sea level. The description sounds very like that of the Apennines as described by Coltorti and Pieruccini (2000) (see Chapter 4).

Overview of ranges of continental East Asia

Gao (1998) rejects the common plate-tectonic model of compression and subduction, and suggests active tension associated with mantle diapirism. He claims that an important Neogene tectonic gap, marked by a widespread planation surface, is widely developed in east Asia from Tibet to the Japanese islands. It almost levelled off all topography on Middle Miocene rocks. This planation is supported by other workers. For example even in Yunnan Province, near Burma and Vietnam, late Neogene planation was well developed and the area was near sea level (Yano and Wu, 1999). The planation surface separates two generations of deformation with quite different mechanisms. Compression dominates before planation, and extension afterwards. The extension is marked by block faulting, associated with occasional volcanism. Gao sees four phases in the general model:

1   Block faulting creates major structures (Figure 7.15). Any basins so formed fill with Late Miocene or Plio-Pleistocene deposits.
2   Differential movements of blocks are accelerated in the Early Pleistocene, and there is accelerated deposition in basins.

Figure 7.15 Morphotectonic profile from the Tibet Plateau to the Yunnan Plateau. A once-continuous plateau, correlated by fossil fauna and flora, was broken up by steep normal faults to form multiple plateaus with total displacement of over 3000 m (after Gao, 1998).

3 The general pattern is dominated by a continental divide that separates drainage to the Arctic from that to the Pacific and Indian Oceans.
4 There is then further normal faulting and drainage modification.

The plate tectonic hypothesis of the India–Asia collision being responsible for so many features becomes more and more improbable with distance from India. It may conceivably have something to do with the Himalayas and the Tibet Plateau. Some workers extend its effects to the Kunlun Range, rather as the Pacific subduction under South America is held responsible by some for the thrusting of the eastern Andes. But then comes the Tien Shan, and some extend the collision story that far. But the story of range and basin, horst and graben, continues to the Lake Baikal depression and beyond (Figure 7.13). An investigator starting at Lake Baikal would find the story of vertical tectonics continues at least as far as the Tien Shan without any need to invoke continental collision.

Besides normal faulting, the tectonics of north-east China is characterised by strike–slip faulting, with right-lateral motion on north–north east tending planes, left lateral motion on west–north west-tending planes, and extension in approximately a north-west direction. South-east China, in contrast, is relatively stable. Molnar and Tapponnier (1978) interpret these as a result of the India–Eurasia collision. No simple plate boundaries can be recognised in north-east China. The strike–slip motion is generally older than the modern topography, which is dominated by vertical tectonics.

The major problem with plate tectonic explanations is that while plate movements have been going on for at least the past 50 million years, much of the vertical uplift is in the past 3 million years. The biggest problem with the story of Plio-Pleistocene uplift of the Tibet Plateau and associated mountains is that there is evidence of derivation of sediments from land masses at earlier times.

8 Mountains with gravity structures

Introduction

The role of gravity in folding rocks and creating landforms is much disputed. Whole books have been written on gravity tectonics (e.g. de Jong and Scholten, 1973), but the role of gravity has often been denied, and with the advent of plate tectonics largely ignored. Here we bring together some examples of mountains where we think the evidence for gravity tectonics is particularly strong. We also touch on some general principles of gravity tectonics, and there is further discussion in Chapter 12.

Underwater folding and faulting

The concept of folding and mountain building being synchronous is so engrained that it is easy to think that folding takes place on land, where the mountains are. In reality considerable folding takes place under the sea, as indicated by the following examples.

The Agulhas Slump

The Agulhas Slump (Figure 8.1) is a large submarine slump on the continental margin of south-east Africa, 750 km long, 106 km wide, with a volume of over 20,000 km^3. It is post-Pliocene in age (Dingle, 1977). It has many of the tensional and compressional structures found in mountains, but has never been above sea level.

The Bengal Fan

Another huge delta is that where the Ganges and Brahmaputra Rivers have combined to build a submarine fan, the Bengal Fan, over 2,000 km long. The oldest known sediments are thought to be of Upper Cretaceous to Paleocene age. Over most of its enormous area the sediments appear to be unfolded, but the distal parts are folded (Figure 8.2) and thrust to make a fold belt. Wezel (1988) regards this folding as possibly the most distant

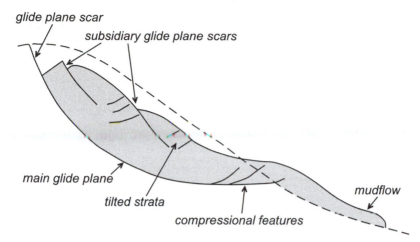

Figure 8.1 The Agulhas submarine landslide, offshore southeast Africa (after Dingle, 1977).

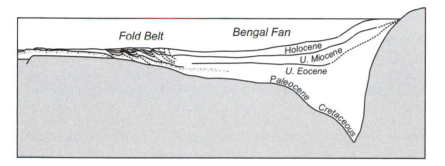

Figure 8.2 Cross-section of the Bengal Fan, showing folding at the distal end. Could this be by gravity spreading? Compare with the Niger Delta (Figure 8.3).

effects of the Himalayan orogeny, but it is more plausibly caused by gravity sliding and thrusting.

The Niger Delta

The Niger Delta is much faulted (Evamy *et al.* 1979), with some major faults that define distinct provinces within the delta region and control sedimentary deposition to a large extent, and many smaller faults (Figure 8.3). These are significant because they demonstrate that not all faults are associated with topographic features on the land – some major faults never did produce fault scarps on the ground surface, and by analogy wherever we see faults in old delta sediments in older rocks we should not think of

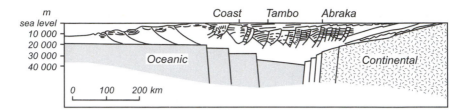

Figure 8.3 The Niger Delta, showing many features of an orogen, with normal
faulting at the back and thrusts at the front, entirely due to gravity
(after Evamy *et al.* 1979).

faulting as being related to younger tectonics and uplift, for they may be
penecontemporaneous with the deposition as in the Niger Delta. Note the
normal faults near to shore, and thrust faults on the distal portion. In many
ancient settings this thrusting at the front would be interpreted in plate
tectonic terms as the result of subduction, but clearly the Niger Delta is
on a passive margin and subduction is impossible.

Other examples

Seismic sections from all over the world reveal similar deformation of marine
sediments. Two further examples are shown in Figure 8.4, from the South
China Sea and from the Baram Delta off north-west Sabah. Both show thrust
structures and decollement, like the distal part of the Niger Delta, and
flow under gravity is the simplest explanation for the structures.

Even in the backarc basin of an island arc, a classic site of plate tectonic
collision, folding of sediments may be by gravity sliding. Yano and Kunisue
(1993) record that in the backarc basin of Japan, Late Cenozoic sediments
accumulated and were folded by gravity sliding after north-west tilting of
the basement blocks.

Folding, nappes and decollement

The structure of many mountain ranges, including the European Alps, is
dominated by great nappes, huge sheets of rock that have clearly moved
over fault planes at low angles, commonly bringing old rocks to lie over
younger rocks. Underlying rocks are not deformed, and the low angle plane
is known as a 'detachment' or 'decollement'. The unfolded unconformity
beneath the folded rocks clearly shows that the mountain mass was not
pushed up from below, but that some sort of lateral force is responsible for
the folding. The nearly horizontal movement may be of about 100 km. In
most instances it is marked by a layer of particularly mobile material that
acts as a lubricant for the slide. Salt (halite) seems to be particularly suitable, as in the Jura Mountains and the Zagros. Dolomite seems to be also

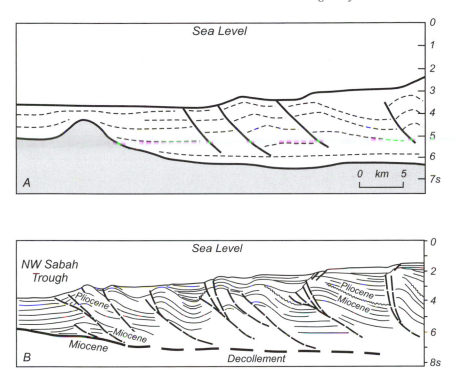

Figure 8.4 Sections derived from seismic survey showing thrusts and decollement, in marine sediments
a. South China Sea
b. Baram Delta, NW Sabah.

very suitable, as in the unconformity dolomite of the Naukluft Mountains.
Gypsum and anhydrite also make good lubricants. Shales may provide
enough lubrication, as in the Oslo Graben.

The nature of the lubricant may affect the behaviour, as illustrated in
the Apennines of Italy, where the base of the Triassic is evaporites in the
north and south and dolomite in the middle, and the resulting tectonics is
different in the three areas.

Some geologists think the nappes result from great squeezing of a sedimentary basin. In plate tectonics they are usually regarded as thrust by
plate collision; in gravity tectonics the nappes are seen as the result of
gravity sliding. The following examples are thought to be especially good
illustrations of gravity tectonics.

The Apennine nappes

The Apennines have been treated in Chapter 4. Here we merely remind readers that the stack of nappes comprising the Apennines has a total thickness

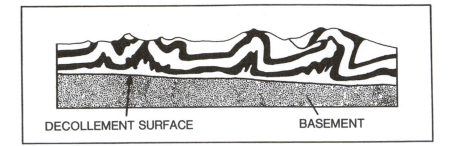

DECOLLEMENT SURFACE   BASEMENT

Figure 8.5 Cross-section of the Jura Mountains, showing rather idealised 'Jura type' folding, and decollement along Triassic evaporites.

of over 10 km, very much greater than the height of the Apennine Mountains, so much of the faulting occurred when the sediments were beneath the sea.

Jura

The Jura Mountains gave their name to Jura-type folding (Figure 8.5), a generic term that refers to folding that does not involve the underlying rock. The folded rocks of the Jura could not be folded by any mechanism involving the whole crust; for the folding is surficial, even though the folded layer is several kilometres thick. The unfolded unconformity beneath the folded rocks clearly shows that the Jura folds were not pushed up from below, but that some sort of lateral force is responsible for the folding. The strata are mainly Jurassic.

The Jura are a crescentic range in plan, separated from the Alps by the Swiss Plain which is underlain by unfolded rocks. At the ends of the crescent there are single anticlines, but the number of folds increases towards the middle. Unfolded tabular areas intervene between some groups of folds, separated by thrusts to the north-west. On the outer side of the crescent are the tabular Jura, with the folded Jura behind, and behind this the almost unfolded sediments of the wide Swiss Plain. The idea that they were pushed by lateral pressure from the Alps is thus untenable. The folding was assisted by layers of Triassic anhydrite and salt along which the overlying beds could glide and fold.

Folding occurred mainly in the Upper Miocene and Lower Pliocene, after which the area was levelled by erosion. Uplift, with some renewed folding, is mainly Pleistocene (Holmes, 1965).

Chuan-Jin Fold Belt

The Chuan-Jin fold belt lies east of the Yangtze, centred about 30°N 107°E, and is about 150 km wide and 400 km long, aligned roughly south west–north east. Anticlines are dominant in the fold system and branch in places. In contrast synclines are flat and broad, almost undeformed. This is interpreted as a 'German type' decollement, with Triassic evaporites as the detachment layer or lubricant. In plan many anticlinal traces of the fold belt gently curve convex to the north-west, indicating the north-westward transportation of the detached strata (Yano and Wu, 1995).

The Pelvoux Massif

The Pelvoux Massif in the south of France is surrounded by nappes. This massif had emerged from the sea by the end of the Oligocene, and so presented an obstacle against which nappes were brought to a standstill when they collided, buckling and imbricating the strata of the toe. Nappes from the main axial region of the French Alps moved westwards, and piled up against the eastern side of the Pelvoux massif: nappes from the north piled up on the northern side; nappes from the south piled up on the southern slopes (Holmes, 1965). Thus there is a centripetal pattern of structures, and a centripetal pattern of forces. No scheme of lateral compression can account for this pattern, but gravity sliding presents an easy explanation. Pelvoux was an island-like obstacle surrounded on three sides by slopes on which gravity sliding took place. The nappes of Pelvoux now reach heights of more than 3000 m, higher than the source area of the nappes.

The Papuan Fold Belt

Papua New Guinea provides a further complication in mountain building and gravity sliding. The highlands of Papua New Guinea are generally envisaged as rugged mountains, but this is not always so. The core of old rock that was uplifted to make the spine of Papua New Guinea is an arched plateau. Part of this is known as the Owen Stanley Range. Relict surfaces are widespread in the Owen Stanley Range, mainly on the principal watershed but also on some offshoot divides. The old erosion surface appears to be Pleistocene or perhaps Pliocene. Thus we have the familiar story of

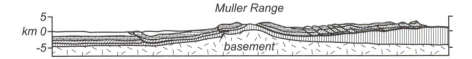

Figure 8.6 Cross section of the Muller Range, part of the Papuan Fold Belt, Papua New Guinea. Note that the basement and the lower part of the Jurassic strata (vertical shading) are not folded, but the upper strata are deformed into complex folds and faults. Strata appear to have slid from uplifted areas, and come to rest against obstacles. (Simplified from Jenkins, 1974).

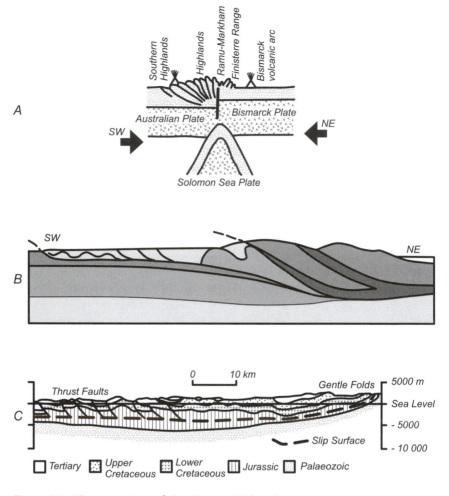

Figure 8.7 Three versions of the Papuan Fold Belt.
   a.   Collision of the Bismarck Plate and the Australian Plate,
        pushing down the Solomon Plate completely and squeezing out
        the Papuan Fold Belt (top left) (after Ripper and McCue, 1983)
   b.   Subduction of the Australian Plate under the Pacific Plate (after
        Burchfiel, 1983).
   c.   A cross-section of the Papuan Fold Belt based on drilling (after
        Findlay, 1974). The folds result from gravity sliding after uplift
        of the highland zone on the right.

planation and young uplift, even in an area which most authors still regard as a classic site for subduction-made mountains.

Deep drilling has shown that the folds are confined to the upper few kilometres (Figure 8.6), the basement is largely unaffected, and the main glide plane is within Jurassic strata. Where the strata could slide into basins they are deformed into complex folds and faults, and folding is greatest where a sliding mass came to rest against a basement obstacle. The sliding was consequent on vertical uplift of the central axis of country. The ranges of the Papuan Fold Belt are hard to explain by any mechanism other than gravity sliding.

Plate tectonic models of the Papuan Fold Belt are shown in Figure 8.7. Ripper and McCue invoke deep seated thrusting driven by subduction on the distant north coast, transmitting the stress across several morphotectonic units. Another plate tectonic model, by Burchfiel (1983), likewise has totally imaginary rock structure in the Papuan Fold Belt (Figure 8.7b). Such 'cartoon' versions can only be accepted by ignoring the results of detailed deep drilling which clearly showed the basement is unfolded as shown in Figure 8.7c (Jenkins, 1974; Findlay, 1974).

Taiwan

The structure of Taiwan, as shown in Figure 8.8, seems to leave no option but gravity tectonics. The main mountain range of the island is due to vertical uplift, and surficial sedimentary layers have slid, by gravity to form the Western Foothills and the East Coastal Range. The basic tectonic force was uplift of a tilt block. The Neogene sediments and thrusts seem to interfinger with Plio-Pleistocene sediments, but the later Pleistocene

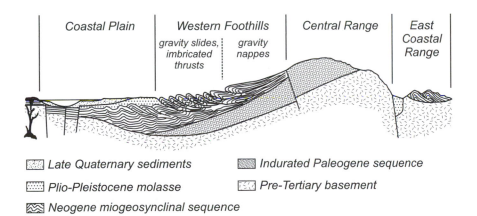

Figure 8.8 A cross section of Taiwan showing a decollement and overturned folds and thrusts in the western foothills of the Central Range (simplified after Chai, 1972). Uplift to the right seems to be the cause of the slide.

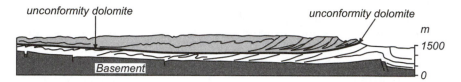

Figure 8.9   Cross-section of the Naukluft Mountains, Namibia, from NW to SE. The unconformity dolomite is undeformed though it lies between two thick sets of intensely folded and faulted rocks. The basement rocks are disturbed only by block faulting (simplified after Korn and Martin, 1959).

sediments appear to be unaffected by gravity tectonics, though affected slightly by extensional tectonics and normal faulting.

Naukluft Mountains

Perhaps the finest example to demonstrate these gravity-slide ideas is provided by the Naukluft Mountains in south-west Africa, described by Korn and Martin (1959). The situation is shown in Figure 8.9. The bedrock of Precambrian rocks is unfolded, but is overlain by a series of intensely folded rocks which evidently moved from north west to south east. The intensity of deformation increases to the south east, where the repeated thrust faults known as imbricate structure are found. This is the effect of breakers-at-the-nappe-front described in Chapter 12. A lateral push in the east would cause greatest deformation where applied, and its effects would die out to the west, just the reverse of what is found. A sliding mass of rock, however, would be most deformed at the front end, where it crashed into the obstacle that brought it to rest. This is what is found.

But there is an even more remarkable feature in the Naukluft. After the faulted and folded rocks were emplaced they were uplifted, eroded to a plain, the area then subsided beneath the sea and a new series of sedimentary rocks were deposited unconformably on the lower, folded and faulted series. The upper series would have been nearly horizontal originally. But this upper series later underwent another phase of folding and thrusting. It slid down the unconformity and was intensely deformed without any effect on the underlying rocks.

The unconformity between the two sets of folded rocks is marked by a distinctive yellow dolomite only 5 to 10 metres thick. This evidently provided the lubricated layer allowing decollement of the overlying mass. Even so the lower part of the dolomite is virtually undisturbed, and the upper part becomes gradually deformed towards the upper adjacent rocks. A slab of rocks several kilometres thick slid on a lubricated plane over a dolomite layer only 10 metres thick, and without disturbing any of the underlying rocks at all. There can be no question of crustal shortening here. Everything points to gravity sliding on a huge scale.

Of course neither of these sets of gravity structures has anything to do with the present Naukluft Mountain topography. The last set of gravity structures was planated, and the erosion surface raised to form a plateau. The present mountains are part of a typical passive margin mountain topography.

Post-uplift gravity structures

Gravity structures that occur after planation and uplift take several forms, which may work in combination:

> gravity collapse structures
> gravity spreading
>> fault spreading
>> mushroom tectonics
> differential loading
>> cambers
>> valley bulges
> lateral tectonic extrusion

Gravity collapse structures

A series of gravity collapse structures in sedimentary rocks reported from Iran were perhaps the first such structures to illustrate the great importance of

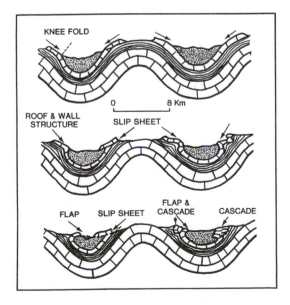

Figure 8.10 Diagram of gravity collapse structures of Iran (after Harrison and Falcon, 1934, 1936).

gravity in folding rocks (Harrison and Falcon, 1934, 1936). The structures are shown in Figure 8.10, and they range in complexity from simple visor and knee folds, to structures as complex as cascades. The present vertical relief may be over 2000 m, but the secondary gravity folds start to form when the relief is no more than 600 m. These structures could easily have been interpreted as 'normal' geological folds caused by 'compressive shortening', but the lack of folding in the underlying rock clearly shows that they are surficial features, not related to compression of the entire Earth's crust. Furthermore the folds and related structures were formed subaerially, after incision of valleys.

Gravity spreading

Fault Spreading

The idea of gravity spreading after vertical uplift has been applied to uplifted blocks in Australia by Ollier and Wyborn (1989), who reasoned as follows.

When a fault scarp runs almost straight for tens or even hundreds of kilometres it has to be a high angle fault, because a low angle thrust fault in an area of high relief would have a sinuous outcrop. Such straight faults are the Tawonga Fault and Long Plain Fault in south-west Australia. Yet detailed examination in tunnels cut through these faults shows them to be low angle thrusts, with granite or Palaeozoic bedrock thrust over Quaternary alluvium in places. The simplest explanation is gravity spreading of an uplifted block over the alluvium.

Similarly in the Andes, the straightness of the Cordillera margins on the large scale indicates dominantly vertical faulting, but detailed studies indicate thrusting, which is a late-stage modification as uplifted blocks spread. In the Rocky Mountains the mountain front is straight on small-scale maps, but detailed work shows thrusting of 100 km (see Chapter 5). Bieber (1983) reported that where thrust the fault may dip at less than 60° but 'elsewhere along the Front Range dips are more nearly vertical'. The situation suggests original vertical faults which spread under gravity and turn into apparent thrust faults.

On an even larger scale the main faults of the Himalayas make great arcs on small scale maps, but detailed mapping shows much thrusting

Mushroom tectonics

If fault spreading is symmetrical on both sides of an uplifted block then the result is 'mushroom tectonics.'

Examples include the southern Rocky Mountains (see p. 101), the Colorado Plateau (p. 99), the Uinta Plateau (p. 100), and the Andes of Ecuador (p. 115).

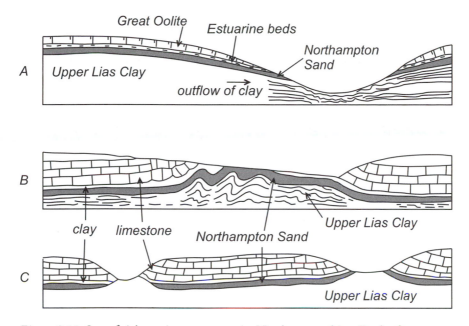

Figure 8.11 Superficial gravity structures in Northamptonshire, England
        a.   Cambering
        b and c.   Valley bulges (after Hollingworth *et al.*, 1944).

Differential load structures

Some classical and informative gravity structures were first reported from the Northamptonshire area of England (Hollinsworth *et al.*, 1944). They are not of mountain proportions, but give clues to the interpretation of larger structures found elsewhere. Essentially the Oolitic limestone cap rock made a series of plateaus separated by valleys cut through to soft clays beneath. However the cap rock did not remain horizontal, but was bent in two different ways (Figure 8.11).

Cambers

In one situation the plateau edges sagged down, extruding the soft clay from beneath the limestone edge and the clay was subsequently washed away. This left the cap rock draped in a gentle arc on the interfluve, a structure known as a camber. The river appears to follow a syncline, but in reality the apparent syncline was formed later than the river valley.

Valley bulges

In other situations the weight of the limestone apparently squeezed up the clay along valleys, where it was eroded, and the edges of the limestone were here bent upwards. The valleys are thus flowing along apparent anticlines – but the rivers were initiated on flat strata and the anticlines developed only after the rivers cut down to the soft underlying clays. These features are known as valley bulges. Rocks which behave in a fairly rigid manner, like the Oolitic limestone in the last example, are known as 'competent beds'; the rocks that flow, like the Lias clay in the last example, are said to be incompetent.

Similar features have been described from many other areas, with different names such as valley anticlines.

Simmons (1966) described stream anticlines in central Kentucky which were only about a metre high (several inches to several feet in his own terms) and confined to limestone and shale. Matheson and Thomson (1973) described very small bulges from Alberta, Canada (but also referring to some in other places including South Dakota), which they attributed to valley rebound. They wrote, 'Valley rebound is a ubiquitous feature in areas where major rivers are incised into flat lying sedimentary rock characterised by a low modulus of elasticity.' In their studies they found rebound up to 10 per cent of the valley depth, but 3 to 5 per cent was more common. The valleys were only 60 m deep, of post-glacial age. The rise is still active at a rate of about 0.01 metre per year. The most significant feature is Matheson and Thomson's identification of rebound as the mechanism, for it is likely to be even more important in the large valleys associated with mountains.

Valley anticlines of the Colorado Plateau

On a larger scale, valley anticlines have been described from Utah that follow the sinuosities of the Colorado River for at least 35 km (Potter and McGill, 1978). The correspondence leaves no doubt that the anticlines formed as a response to valley cutting, but arching is evident as much as 600 m above the valley floor. To some extent the upward flow of evaporites in the underlying rocks is responsible for these valley anticlines.

Valley anticlines of the Himalayas

The anticlines that follow valleys crossing the Himalayas have been described in Chapter 7. A whole suite of anticlines follows the valleys and has a different form and direction from that of other folds in the region.

Valley anticlines of the Zagros Mountains

The Zagros Mountains lie on the south-west side of the Arabian Plateau. They consist of folded rocks, so on plate tectonic explanations are described as the result of subduction. They are also classic plateau edge mountains (see Figure 12.6).

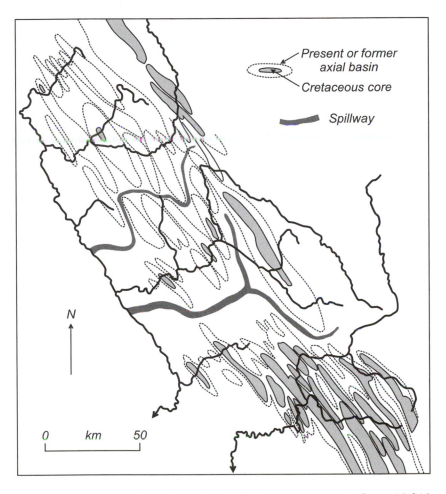

Figure 8.12 Zagros valley bulges. The NW–SE structures result from old folds,
probably gravity slides. The main drainage is approximately perpendicular to the folds. Downcutting by major valleys led to isostatic
compensation and uplift of the valley rims, creating depressions
between major valleys. These depressions look like valleys, but they
never were; they are topographic lows created by rising land on
either side.

Advocates of plate tectonics interpret ophiolites of Iran to represent oceanic
crust, and regard the Zagros Crush Zone as a suture where collision occurred.
Kashfi (1992) has presented much evidence against a plate tectonic interpretation. Basement rocks exposed in deep boreholes indicate that the Zagros
Fold Belt is underlain by continental, not oceanic, crust. He believes the
most compelling evidence against subduction is the continuity of geology,
especially the late Precambrian–Cambrian Salt Range, which extends from

Pakistan, across Iran, and crosses the Persian Gulf into Oman and South Yemen. He believes the Iranian plateau and south-western Iran and Arabia have been a single geologic zone since the beginning of Proterozoic time.

The rocks are tightly folded and thrust. Massive salt layers provide excellent opportunities for gravity sliding, and the structures are complicated further by salt diapirs.

Of special interest here are some complications of the folding, and of the drainage pattern. This has been described in great detail by Oberlander (1965). The significant feature is the dry valleys, which have valley form but no alluvium. Oberlander has a complex explanation, but a very simple explanation is that the through valleys with streams give rise to isostatic compensation and formation of a valley bulge. The valley rims are raised. When two valleys are close together, the uplifted rims give rise to 'apparent valleys' between the active valleys, but these never were valleys originally – they are simply depressions bounded by valley bulges (Figure 8.12).

The North-west Dolomites

The Dolomites, a mountain group in northern Italy, provide a spectacular example of gravity tectonics induced by differential loading, often following erosion.

In general the sedimentary column consists of three parts:

1   an upper, more or less plastic zone;
2   an intermediate zone of reefs, which are heavy and rigid, surrounded by incompetent strata;
3   a lower zone with plastic strata and gypsum.

This arrangement is unstable because the heavy reefs (specific gravity 2.5 to 2.7) overlie gypsum with a specific gravity of 2.3, and because the lower zone may move sideways under the load pressure.

The overburden pressure is at its greatest beneath the dense reefs and causes the lower zone to flow towards the areas with lower load, that is between the reefs. Erosion of the softer rocks accentuates the differences, and causes the reefs to founder. The extruded rock bulges up in domes between the reefs, and may push up the edge of the reef like giant valley bulges. As the reefs are bent into basin shapes, the upper zone may flow towards the depressions, causing summit folding. This is shown diagrammatically in Figure 8.13.

When the reefs are very large, as in the Marmolada complex, the behaviour is different. Here the reef sags at the edges, so the major shape is a dome. It is rather like a camber, but with no long direction. With increasing internal pressure the top of the dome may be pierced, and the plastic rock may rise diapirically through it.

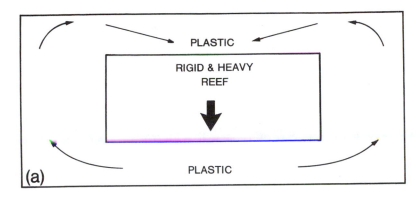

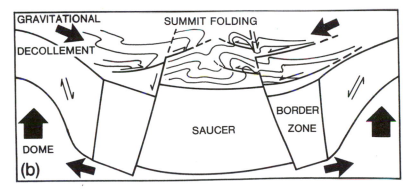

Figure 8.13 Diagrammatic representation of the foundering circuit in the Dolomites.
a. Original relationship of plastic and rigid rock masses.
b. Structural effects of foundering (after Engelen, 1963).

There is a relationship between the size of the rigid body (reef) and the type of deformation it undergoes when foundering. Three types were observed in the Dolomites:

1 small rigid units fall apart into blocks (e.g. Sasso Lungo, 3 × 3 km (Figure 8.14));
2 rigid units of intermediate area assume a saucer shape (e.g. Sella, 7 × 6 km);
3 the largest rigid unit assumes a dome shape and is diapirically pierced later on (e.g. Marmolada complex, 12 × 8 km).

It is also possible for the rigid plate to be breached on a valley side. Once breached, the inner plastic rocks flow through, making a horizontal diapir that looks like a broken-up recumbent fold (Figure 8.15).
All these structures have been described in detail by Engelen (1963).

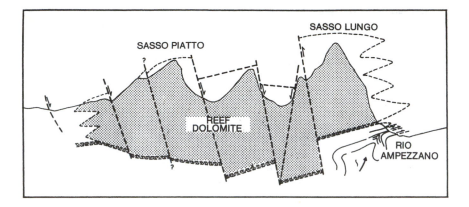

Figure 8.14 Cross-section of the mountains near Sasso Lungo, Dolomites. The break-up of the reef by large faults appears to be entirely due to outflow of plastic rocks from beneath, and not to be deep-seated faulting (after Engelen, 1963).

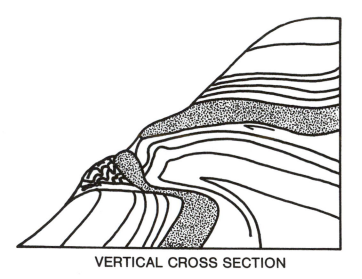

VERTICAL CROSS SECTION

Figure 8.15 Lateral diapir on a valley side. The shaded bed is rigid, the other rocks are plastic (after Engelen, 1963).

Lateral tectonic extrusion

The process of lateral tectonic extrusion appears to be largely a post-uplift type of spreading, but with some plate tectonic associations.

This was proposed by several authors as a mechanism to explain some features of the European Alps (Ratschbacher *et al.*, 1991; Frisch *et.al.*, 1998).

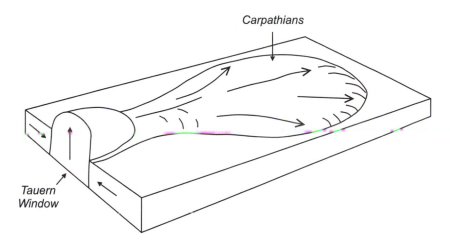

Figure 8.16 Lateral Tectonic Extrusion. Uplift of the Tauern Window (thought
        by the authors to be caused by plate tectonic compression) led to
        gravity sliding of stacks of rocks to the Carpathians (after
        Ratschbacher *et al.* 1991).

An example is shown in Figure 8.16. Uplift of the Tauern window (thought
by these authors to be due to compression in the plate tectonic manner)
led to gravity flow towards the Carpathians. The authors suggest that the
crust was thickened by stacking of wedges as the more rigid Southern
Indenter and the Bohemian Block approached, and the stacked material
flowed by gravity collapse towards the eastern unconstrained margin where
space had been created in the Pannonian Basin.

The timing of the onset of lateral extrusion is unclear, but a late Oligocene
is favoured. Among other reasons is the formation of an erosion surface (the
Augenstein peneplain) in the Oligocene, and its destruction at the Oligocene/
Miocene boundary.

Foothill growth by gravity push

A large mountain exerts a load which may cause upward movements on
lower slopes, in the way that a landslide may cause uplifted blocks at the
toe of the slope. Some of the foothills of the Alps or the Himalayas may
be of such an origin rather than resulting from deep tectonic forces. In the
Sub-Himalayas rivers are deviated by growing anticlines (Gupta, 1997).

Some general aspects of gravity and mountain building are reviewed in
Chapter 12, including criteria for distinguishing gravity structures.

9 Volcanoes and granite mountains

Igneous rocks, magma and lava

Igneous rocks are those thought to have cooled from a hot, molten state, the word being derived from the Latin word for fire. Of the many kinds of igneous rocks distinguished by petrologists, the two broad groups that dominate at the earth's surface are basalt and granite. Basalts are of volcanic origin, but granites originate deep in the Earth's crust.

Granite is produced by massive alteration and melting of rocks of the Earth's crust. Once formed, some granite rises through overlying rock as plutons. The granite may in turn produce dykes of acid rocks, and give rise to acidic volcanic eruptions. Basalt comes from the mantle, from greater depths than the continental rocks.

The igneous rocks produced by volcanoes are often referred to loosely as 'volcanics'. The molten parent material is called 'magma' which becomes lava when it erupts at the Earth's surface. Magma and lava are not synonymous because gases dissolved in the magma are largely lost from the lava.

A simple classification of igneous rocks is shown in Figure 9.1, based on grain size and composition (in this context 'acid' means having more silica). The commonest coarse grained rock is granite. The commonest fine grained rock is basalt. The seafloor consists of basalt; the continents are made largely of granite. Rhyolite is a volcanic rock of granitic composition. It may be erupted as a lava, but most rocks of rhyolitic composition are ignimbrites, described later. Andesite is a volcanic rock of intermediate composition, and is especially common on active continental margins.

Related sets of basalt occur in different environments. Some are:

1   Mid Ocean Ridge Basalts (MORBS). These are basalts erupted from mid-ocean ridges and characterised by very low K and related elements and low rare-earth element (REE) abundance with no enrichment in light REEs relative to heavy REEs.
2   Ocean Island Basalts (OIBS). These include tholeiitic suites as well as alkali basalt suites.
3   Continental Basalts. Basaltic and/or doleritic suites of the continental regions are dominantly tholeiitic rocks although alkali suites may be

		← increasing silica →		
		Acid	Intermediate	Basic
grain size	Coarse-grained	GRANITE	Syenite / Diorite	Gabbro
	Medium-grained	Microgranite	Porphry	Dolerite
	Fine-grained	Rhyolite	Trachyte / ANDESITE	BASALT
	Glass	Obsidian / Pitchstone / Pumice	Pitchstone	Tachylite

Figure 9.1 Simple classification of igneous rocks. Granite is the commonest intrusive rock, often found in the core of fold belts. Basalt is the commonest volcanic rock, especially in oceanic volcanoes. Andesite is found in association with many mountain regions, including the Andes and the mountains of western North America. In plate tectonics subducted material is often alleged to give rise to granites and andesites, but note that these two rocks have different composition.

common. Karroo basalts of southern Africa, Columbia River basalts of the United States, and Parana basalts of Brazil are typical examples. There are isotopic differences between continental tholeiites and MORBS.

When some lavas erupt the rapid expansion of gas breaks them up into fragments called 'pyroclastics'. The finest material is ash; intermediate fragments are known by names including scoria, lapilli, or cinders; the coarsest material is known as blocks or bombs.

4 Island arc basalts. The dominant rocks of the island arcs appear to be basalts with tholeiitic affinities, not andesites as commonly supposed.

Volcanic mountains

Volcanoes do not fit into the main theme of this book, namely that mountains result from dissection of plateaus, but they are mountains in their own right, and have a bearing on many other aspects of mountains that we discuss in the final chapters. The longest mountain range on Earth, and the highest mountain on Earth, are volcanic. Mauna Kea, Hawaii, rises 4205 m above the sea, but it really rises from the seafloor 6000 m deep, so it has a total height of 10 205 m (compare with Mt Everest at 8848 m). The volcanic Mt Elbrus is the highest mountain in Europe at 5633 m. The highest mountain as measured from the centre of the Earth is Chimborazo in Ecuador, at 6272 m (Figure 9.2) Some seamounts entirely below the sea are also high, the biggest being one near the Tonga Trench between Samoa and New Zealand which rises 8700 m from seabed to summit, which is 365 m below sea level.

The seafloor consists essentially of basalt, and there are many oceanic volcanoes of mountain size. Continents consist mainly of granite, but volcanoes occur on continents, especially at favoured sites such as active continental margins, island arcs and associated with rift valleys, but in other situations too.

Types of eruption

Volcanic eruptions may be divided into those that issue from a central pipe or vent, known as 'central eruptions'; those that issue from cracks or fissures, called 'fissure eruptions'; and eruptions scattered over a wide area, known as 'areal eruptions'.

The classification is not always easy to apply. An eruption may start along a fissure but later erupt from a number of separate centres. On the other hand a large central-type volcano may erupt at numerous parasitic centres along a fissure on its flank, as commonly happens at Mt Etna, Sicily.

Some volcanoes demonstrate repeated eruption on the same centre. Valleys eroded on an earlier volcanic cone become filled with younger lava flows, and new cones are built up on the old centre. In recent times a classic

Figure 9.2 Chimborazo, a typical Andean stratovolcano in Ecuador. At 6272 m it has an ice cap on top, despite being located close to the Equator (photo C.D. Ollier).

example is the building of Vesuvius on the site of the old Monte Somma, near Naples, Italy, mainly resulting from the eruption in AD 79.

The gap between eruptions may be millions of years, as on Gough Island in the South Atlantic, which had eruptions about 2.5 million years ago, and about 40 000 years ago (Maund *et al.*, 1988). The island of St Helena is now about 250 km east of the ridge, and probably had an origin like Ascension but it is now extinct and has been carried away from the active site by sea-floor spreading. It was, however, active from 15 million to 8 million years ago, though not continuously.

Major volcanic landforms

The two main products of volcanic eruptions are volcanic cones and volcanic plains. Massive craters (calderas) are another product, sometimes associated with mountains.

Volcanic cones

Volcanoes erupt both on land and under the sea, and this gives a first criterion for classification.

Terrestrial volcanoes

The roughly conical hill or mountain is the most obvious and typical feature of volcanic activity. Cones may be built of lava, pyroclastics, or a mixture of the two.

Shield volcanoes consist mainly of lava and are usually very large, with gentle slopes like the major volcanoes of Hawaii.

Scoria cones consist of generally coarse pyroclastics, are fairly steep sided, and generally small, many less than 300 m in hight. Ash cones are similar but have finer pyroclastics and more gentle slopes.

Stratovolcanoes consist of alternate layers of lavas and pyroclastics. Most of the world's great volcanoes such as Fuji, Vesuvius, Egmont or Rainier are of this type. They are the typical volcanoes of popular imagination and art.

Volcanic cones may also be divided into two types. Monogenetic volcanoes have with a single style of eruption such as scoria cones or lava cones, often resulting from a single series of eruptions from a single centre. Complex volcanoes include both shields and stratovolcanoes, where eruptions of various kinds occur, several eruptive centres may operate and parasitic volcanoes are common, dykes and pipes intrude the volcanic pile, and eruptions may be repeated near the same site for thousands of years.

Oceanic volcanoes

Volcanoes can erupt under water, and, if deep enough, can erupt without explosion. Submarine eruptions often have a characteristic structure called pillow lava, which enables tectonic movement to be recognised when pillow lava is found on land. Many submarine eruptions are huge and prolonged, building the largest volcanoes known, the huge lava shields such as the Hawaiian Islands.

When an erupting volcano reaches sea level the eruption style changes, with many explosions, as was observed on Surtsey. Eventually the volcano can build an edifice clear of sea level, and a 'terrestrial' type volcano is built.

'Seamount' is the name given to volcanoes entirely under the sea. Some may have never reached the surface. Others may have sunk after being subaerial for a while. Large volcanoes contribute, by their own mass, to tectonic adjustment around them. The ocean crust is rather thin, and when a submarine volcano erupts it is presumed that it draws its magma from a considerable area around the vent. When activity ceases a heavy volcano is resting on a fairly limited area of thin crust, and the volcano then starts to sink. That volcanoes sink as they grow older is indicated by coral islands (where the climatic is suitable), which grow up to sea level as the island sinks. Coral therefore attains great thicknesses, and since it cannot grow above sea level or below a limiting depth, the volcanic base must be sinking.

Some volcanic islands are eroded down to sea level, and then sink. This makes flat-topped seamounts known as 'guyots'.

Some Atlantic islands are on or close to the mid-ocean ridges. The active Ascension Island is virtually on the ridge, St Helena 700 km from the ridge and extinct. A very rough correlation of the age of oceanic islands with distance from the ridge led J. Tuzo Wilson (1963) to propose that the Atlantic islands were formed on the rift and then drifted away with the spreading of the ocean. More dates have shown it is not so simple. Some islands remain active for several million years, and for a considerable distance from the rift. Tristan da Cunha (Figure 9.3) is still active (the last eruption was in 1961), although it is 350 km from the mid-Atlantic rift.

In the Pacific there is another pattern of volcanic distribution. Many Pacific volcanoes have a linear arrangement and there is a progressive change in age along the line. The Hawaiian Islands are discussed later in this chapter, but there are many others (Ollier 1988). These observations fully support Darwin's theory of atoll formation by simultaneous coral growth and subsidence – eruption, erosion, coral growth and subsidence.

Volcanic plateaus

Vast outpourings of volcanic products may make plateaus rather than cones. Huge volumes are involved. Two types of very different style occur – basalt plateaus and ignimbrite plateaus.

Basalt plateaus

In the north-western United States the Columbia River and Snake River Provinces are enormous effusions of lava. The two provinces are often treated as

Figure 9.3 Tristan da Cunha, an island volcano in the middle of the Atlantic.
The main cone which erupted about 200,000 years ago has been
cliffed by marine erosion, and the cliffs are about 600 m high.
Younger small volcanoes have erupted later, especially near the coast,
and the last eruption was in 1961. Nearby volcanic islands are a few
million years old (photo C.D. Ollier).

one, but are very different in age and geomorphology. The Snake Province, in
southern Idaho, is essentially Quaternary and covers an area of 50 000 km², and
the Columbia Province is of Miocene age and covers an area of 130 000 km².

On irregular topography lava flows down valleys, sometimes filling them
and spilling over interfluves, displacing river courses, and sometimes
completely altering earlier drainage systems. Sometimes lava production is
on such a vast scale that even the valleys of originally mountainous regions
are completely filled and a lava plateau is produced. The Columbia River
Plateau, for instance, which is of Miocene age, completely buried an original
topography with a relief of over 1500 m, which is deeper than the Grand
Canyon on the Colorado River. New valleys with a similar depth have been
incised into the plateau. The Columbia Plateau basalts average 1000 m
thick. This great thickness of lava was attained by the piling up of many
individual flows averaging 10 m thick, with the largest only 120 m thick.
Some cones are present, but most of the lava was probably erupted from
long fissures. Rather similar, and frequently treated together with the
Columbia River basalts are the Snake River basalts of Quaternary age
covering 50 000 km².

The Deccan traps of India are of Cretaceous to Eocene age with a present
area of 500 000 km² and thickness up to 2000 m. The flows were erupted

over a period of a few million years around the Cretaceous–Tertiary boundary. No original landforms are preserved. Widdowson (1997) has claimed some lateritised surface is the original top of the Deccan flows, but the dendritic plan shows clearly that there has been inversion of relief and the laterite once occupied a valley.

The Cretaceous Parana plateau basalts of Brazil cover an area of over 750 000 km² and the Karroo basalts of South Africa, which range from Late Triassic to Early Cretaceous, cover 140 000 km² but were once probably 10 times as extensive (Cox, 1972).

Ignimbrite plateaus

A mixture of solid particles suspended in a gas can act as a liquid, a principle used in industrial 'fluidisation' for the transport of such materials as cement and coal dust through pipes. The process occurs naturally in some volcanic eruptions, when pyroclastics suspended in volcanic gases flow like extremely mobile liquid. They look like rapidly projected dust clouds, and at night they glow, and they are generally known as *nuées ardentes*, translated as glowing clouds.

An eruption of this type deposits a layer of volcanic ash known as an ignimbrite. Rhyolite and andesite composition is commonest, and many ignimbrites are welded, that is gases are expelled and the hot ash sticks together, to form welded tuff. Welded tuffs often look very like rhyolite, and many rocks that were identified as rhyolites in the past are now known to be ignimbrites.

Pyroclastic flows are very mobile. They flow downhill and fill depressions rapidly, and when they come to rest they have a remarkably flat top. Flows have been traced for 70 km from Mt Mazama, Oregon. The Lake Toba flows of Sumatra cover 25 000 km² and the Taupo–Rotorua ignimbrites have an area of 26 000 km². Calculated volumes of ignimbrites are very great. The New Zealand field has a volume of 8300 km³, the San Juan ignimbrites of California 9500 km³. Individual calderas can produce large amounts such as 90 km³ from Aira caldera and 80 km³ from Aso, both in Japan.

Ignimbrites erupt very quickly. The Katmai eruption in Alaska produced 28 km³ in 60 hours, and it is estimated that the 2800 km² of northern Queensland ignimbrites were erupted in a few days. On the other hand, some ignimbrite areas had many pauses during construction, with sufficient time for forests to be established.

Subsidence and caldera collapse

Calderas are volcanic depressions of large dimensions, much bigger than simple craters and sometimes tens of kilometres across. Once thought to be the result of colossal explosions, they are now generally believed to be created by subsidence. To erupt vast quantities of rock in a very short time, as in ignimbrite eruptions, requires very special tectonic conditions, and the likeliest explanation is that the magma chambers that fed the eruptions were very close to the surface.

Rapid release of magma might empty the magma chamber beneath a volcano and lead to caldera collapse. This seems to be borne out by the very frequent association of calderas with ignimbrites. The so-called Katmai eruption of 1912 led to Mt Katmai losing 240 m from its top and producing a caldera 5 km across, but subsequent investigation showed that an altogether new mountain, Novarupta, was the source of the main eruption. Although the main eruption took place from Novarupta, it was Katmai, 8 km distant and 1500 m higher, that collapsed. Presumably they shared a common magma chamber, and the magma moved 8 km underground from beneath Katmai to be erupted at Novarupta.

Distribution of volcanoes

'Plate tectonics' describes the main features of the Earth in terms of a series of 'plates' bounded by spreading sites (mid-ocean ridges) and collision sites, where plates go down by 'subduction.' These concepts were described in Chapter 1. At the global scale the distribution of volcanoes (Figure 9.4) shows that most of them fit on plate boundaries rather well.

The ocean ridges

The ocean ridges, including the Atlantic Mid-Ocean Ridge, the Pacific Ridge, and all the other oceanic ridges shown in Figure 1.6 constitute the

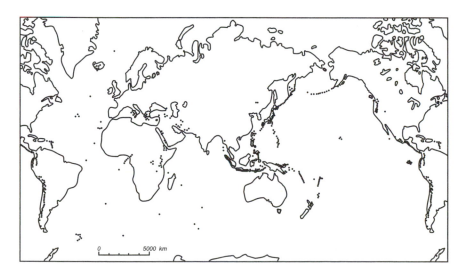

Figure 9.4 Distribution of active volcanoes of the world.

greatest of all mountain ranges. The total length is 65 000 km and the
a greatest height is 4200 m above the base ocean depth.

Iceland is essentially an emerged part of the Mid-Atlantic Ridge. Much
of the volcanic activity takes the form of intrusion of dykes – vertical sheets
of basalt – and Iceland has grown wider by dyke injection. Nevertheless
the island has numerous large terrestrial volcanoes. Presumably the rest of
the mid-ocean ridges have similar dyke injection and occasional central
volcanoes.

Island chains, or hotspot volcanoes

Some volcanoes fall on distinct lines, such as that which starts in the
Hawaiian Islands and can be traced to Midway and then to Meiji Seamount along a line of different direction (Figure 9.5). Volcanoes are still
active in Hawaii, but the other islands are extinct volcanoes. When age
is plotted against distance from the active volcano, the result is almost a
straight line.

This suggests that some migrating point is the source of the volcanoes,
and has been moving at a steady rate. Either the point source is moving
under the crust, or the crust is moving over a stationary or fixed spot. The
line of volcanoes has a kink at Midway. The distances are measured along
the line of islands, not direct from Loihi. The rate does not change along
this line, suggesting that movement was in the direction of the line, and
changed. The hotspot appears to be an intermittent feature, with major
eruptions about every 10 million years, and quieter times in between.

Hot spots can also lead to eruption on continents. The large Central Type
Volcanoes of eastern Australia have a clear relationship of age with latitude

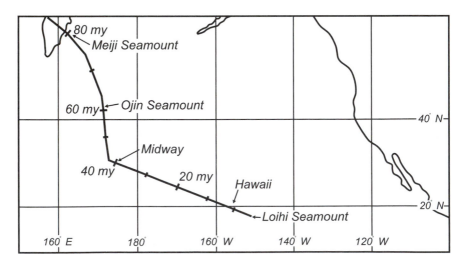

Figure 9.5 Hawaii hotspot trace.

(Figure 9.6). This is explained by Wellman and McDougall (1974) as the result of the passage of Australia over a hotspot. The age ranged from over 20 million years in the north to about 6 million in the south. It should be noted that there are many other volcanic eruptions in eastern Australia that do not fit this picture. Seamounts in the Tasman Sea also appear to be related to hotspots, and show the same rate of plate migration. The rate of northerly movement of Australia over the hotspots agrees with the rate determined independently from seafloor spreading between Australia and Antarctica.

Pacific margin of the Americas

Along the Pacific margin of the Americas andesitic volcanoes are dominant, and they occur along distinct lines or chains. Why the single chain
of southern South America gives way to a double chain at latitude 30°S is
not clear, but there appear to be definite lines at depth governing the location of these huge central volcanoes. It is significant that this area is also

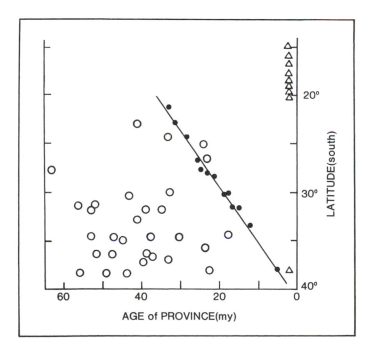

Figure 9.6 A plot of age against latitude of volcanoes of eastern Australia, after
Wellman and McDougall (1974). The open circles are lava field
eruptions; the triangles are eruptions of many small scoria cones and
maars; the black dots are large central volcanoes and show a fine linear
relationship thought to indicate passage of Australia over a hot spot.

intruded by many batholiths of andesitic composition. The line continues into North America, but is less distinct. The Cascade Range of the United States is almost entirely volcanic, and is contemporaneous with the Columbia River basalts to the east. Eruption produced the many large composite cones such as Mt Rainier, Mt Hood, Lassen Peak and Mt Mazama.

Volcanoes of Mexico fall on to two lineaments. One is east–west – Popacatapetl, Colima and Barcena (born 1952) are on this line. The second lineament is north–north west, parallel to the structural grain of the Mexican plateau, and includes Jorullo and Paricutin. In Guatemala there is a line of large composite andesite cones in the west, parallel to the coast. In the east most volcanoes are smaller, basaltic, and more scattered (Williams and McBirney, 1964).

Island arcs

On the western side of the Pacific island arcs predominate, and here again andesitic volcanoes are present, similar in most respects to those of the Americas.

Island arcs are a major topic in tectonic geomorphology, and most arcs are volcanic. There are many theories about their origin and relationships to plate tectonics or other features. There is no doubt that arcs with volcanoes of about the same age are formed today and were formed in the past. Different hypotheses require arcs to migrate in specified ways, but the actual distributions are inconsistent. The distribution of volcanoes of Papua New Guinea (Figure 9.7) shows two main arcs, as well as a scatter of volcanoes with no clear pattern. The northern arc has an inner Palaeogene arc and (on the convex side) an outer Quaternary arc, whereas the New Britain arc has the Quaternary volcanoes on the inside and the Palaeogene arc on the outside.

Island arcs such as Indonesia and Japan have many volcanoes, mainly andesitic, but some volcanoes, including the famous Mt Fuji, are built of basalt.

From a tectonic viewpoint it is interesting that the back-arc basins, lying between the continent and the island arc, are spreading sites like the mid-Ocean ridges and not compressional as might be expected from the presumed subduction under the island arcs.

Andesite line

In the Pacific an area of entirely basaltic volcanoes is separated from those areas around the edge of the Pacific by an 'andesite line', on the shoreward side of which andesitic magmas are commonly erupted, though basalt magmas are also erupted. Andesitic magmas have more silica than basalt magmas and usually erupt with greater violence.

The fact that andesites are absent from the Pacific Ocean within the

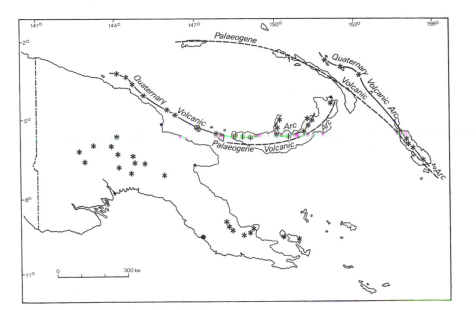

Figure 9.7 Distribution of Quaternary volcanoes in Papua New Guinea (after Ollier and Pain, 1988).

'andesite line' shows that in the Pacific at least andesites are not created by fractionation of basalt magmas, and there seems little reason to suppose they are elsewhere.

Areal volcanism

Areal volcanism (also known as polyorifice volcanism) is characterised by monogenetic volcanoes and the absence of any tendency for eruption centres to be localised at definite points for any length of time. Individual volcanoes are short lived, ranging from a few weeks to perhaps 12 years, and seldom grow bigger than 450 m. Scoria cones, lava cones and maars are the dominant volcanic types and stratovolcanoes are absent or rare. The spatial distribution of the volcanoes is irregular, with some clustering and occasional linear groups. The petrographic composition is basaltic and remains fairly constant within an areal province. Examples of areal volcanic provinces include Victoria (Australia) and the volcanic regions of the Auvergne (France), Armenia and Mexico.

Passive margin volcanoes

Passive margins were described briefly in Chapter 1 and will be covered further in Chapter 10. Of the many such margins, only eastern Australia has significant volcanic activity, with numerous hotspot central volcanoes and hundreds of eruptions that produced small volcanoes and widespread lava flows.

It is not known why volcanic activity did not occur on other passive margins.

Rift valleys

In rift valley areas some volcanoes are basaltic, but there are no andesites or rhyolites. Many lavas are very rich in alkalis, sodium and potassium, and some are rich in carbonate including the remarkable Oldoinyo Lengai in Kenya. Carbonatite is a volcanic rock consisting largely of igneous calcite, and suggests vast accumulations of carbonate at the base of the crust. In the Western Rift Valley of Africa there are many small volcanoes, mostly either in the valley or just east of the fault on the uplifted block, with few on the actual fault. The much larger Bufumbira Volcanoes to the south seem to be slightly off the main rift.

Intraplate volcanoes

Some volcanoes are simply situated in the middle of plates. Mt Elgon and Napak in Uganda are far from any rift valleys or other favoured volcanic sites, but they are huge central volcanoes, 3000 m above their base. Other intracontinental volcanoes include Kilimanjaro in Tanzania, the volcanoes of the Auvergne, and the volcanoes of Hungary.

The Colorado Plateau experienced varied volcanic activity. Extrusive volcanic rocks are scattered around the boundary, but intrusive bodies, such as the laccoliths of the famous Henry Mountains, lie in the central part of the plateau (Figure 9.8).

Erosion of volcanoes

At the end of an eruption many volcanoes have a simple conical shape, and erosion later produces a regular sequence of forms through a planeze stage, a residual stage and finally a skeletal stage. Many variations are possible on this theme, and sometimes planezes are still preserved when the plugs are exposed. Nevertheless the erosion affords a relative dating technique. This was applied by Kear (1957) in New Zealand where it was suggested that stage corresponds to age as follows (Figure 9.9):

Planeze	Mid-Pleistocene to Holocene
Residual	Plio-Pleistocene
Skeletal	Upper Miocene

This mode of erosion can be detected even in some very low angle volcanoes. A 20-million-year-old volcano near Inverell in Australia, Maybole

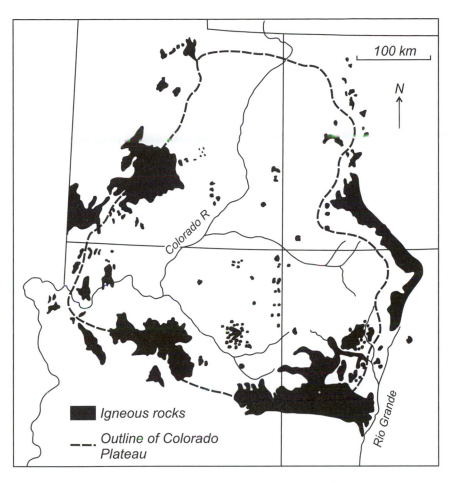

Figure 9.8 Volcanic activity on the Colorado Plateau. Lava flows predominate near the rim and intrusive activity, including laccoliths, is concentrated in the centre.

Mountain – important because of its association with a major sapphire field – exhibits radial drainage and planezes even though the original volcano probably had slopes of only one or two degrees. It fails completely to fit into the New Zealand age–stage sequence. Other older volcanoes do not seem to show the planeze pattern, such as the Miocene volcanoes of Uganda. Mt Elgon is a huge cone about 50 km across, built of a succession of agglomerate deposits separated by ash layers. Radial drainage is present, but instead of planezes the flanks consist essentially of a series of steps or structural terraces. The easily eroded ash is at the base of each cliff, which is in the vertically jointed agglomerate. Further north the volcanic mountains of Torror and Kadam are simply irregular.

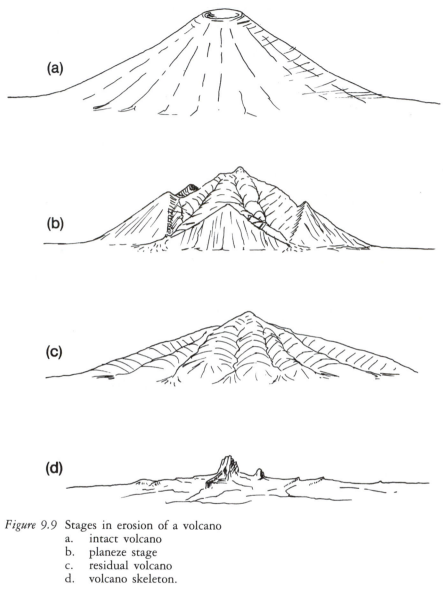

(a)

(b)

(c)

(d)

Figure 9.9 Stages in erosion of a volcano
     a.   intact volcano
     b.   planeze stage
     c.   residual volcano
     d.   volcano skeleton.

Coltorti and Ollier (1999) mapped the volcanoes of Ecuador according to their degree of erosion to see if there was a relationship between age and location, but found no correlation.

Glaciers erode volcanoes too. Mt Rainier, Washington, is a classic example of radial glaciation: the planezes that extend to about 3000 m are known as 'wedges' on lower slopes and 'cleaver' higher up, where they split the descending ice into lobes. Kilimanjaro in Tanzania exhibits a whole range of glacial forms including cirques, U-shaped valleys, striated pavements and *roches moutonnées,* lateral moraines and crag and tail (Downie, 1964)

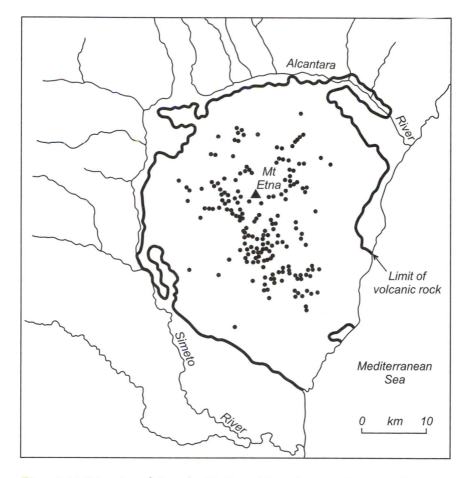

Figure 9.10 Diversion of rivers by Mt Etna. The volcano is about 2 million years old. Pre-volcanic drainage flowed NW–SE. Now the drainage is intercepted and flows around the volcano as the Alcantara and Simeto Rivers. The many black dots are parasitic cones.

Drainage disruption

A volcano may be big enough to disrupt pre-existing drainage completely. Mt Etna is a fine example (Figure 9.10). The pre-volcanic drainage was to the south east, but this was blocked by the volcano and drainage found its way around the north (River Alcantara) and the south (River Simeto). Later lava flows have even diverted these displaced rivers to a lesser extent, as described by Chester and Duncan (1982). Mt Etna was erupted entirely in the Quaternary, and younger eruptions and diversions have occurred until the present time.

A whole belt of volcanoes may disrupt drainage, as happened with the Neovolcanic Plateau in Mexico, which totally blocked the drainage associated with the Sierra Madre and Central Mesa to the north.

In Austria crustal extension in the Neogene created several sedimentary basins such as the Vienna Basin, with associated volcanic activity. Some occurred in the Miocene, but after a period of inactivity a phase of effusive and pyroclastic activity occurred in the Pliocene and into the Pleistocene (ages range from 3.7 to 1.7 million years).

Volcanic intrusions

Volcanic rock may be intruded between strata, forming various kinds of intrusive bodies. The rock type is often dolerite, slightly coarser than basalt because of slower cooling. Volcanic necks are vertical cylinders of rock; steep or vertical sheets of volcanic rock are called dykes; lens-shaped intrusions are called laccoliths, and there are many other varieties. Differential erosion can create mountains out of such rocks and these are not primary volcanic landforms.

The Henry Mountains are laccoliths. In some places the erosion has removed the overlying rocks and exposed the lava core, while in other places the uplifted sedimentary rocks are still intact. In Tasmania many of the mountains are made of dolerite of massive Jurassic sills. The Devil's Tower in Wyoming is a famous example of a volcanic neck.

Granite mountains

Granite is formed and emplaced deep in the Earth, so most mountains on granite are the result of differential erosion. Our concern here is with the possible relationship between granite and mountain building, and with a few special kinds of mountain where the granite intrusion rose beyond the general ground level.

It is quite possible that some granitisation could occur deep in the crust without forming mountains. However if a rigid batholith rises, it seems quite plausible that it will push up overlying strata, and make a topographic high, possibly of 'mountain' dimensions. There appears to be clear evidence that this can happen, and that granites can approach close to the ground surface, and even rise above the regional ground level.

A granite batholith may metamorphose the surrounding rock to make a rock called 'hornfels', which is very hard. Some granites are very prone to deep weathering and erosion, especially those rich in biotite. In this case the granite may give rise to an erosional basin, and the neighbouring hornfels may give rise to hills or even mountains. These are not primary tectonic mountains, but are made by differential erosion.

Granite in fold belts

In the Andes the intrusion of the massive batholiths through the Mesozoic was controlled by major fault lines in the basement (Pitcher and Bussell, 1977). Individual plutons were controlled by transcurrent faults and by smaller scale joint patterns. Intrusion took place over a long time, and the same structures were repeatedly exploited. The Andean granite mass is the largest on Earth, running the length of the Andes (Figure 6.8). As Gansser (1973) wrote, 'It is however clear that any effect the granite has on Andean mountain building is very indirect, for the granite was intruded between 70 and 30 million years ago, and the Andes did not start to rise until perhaps the last ten million years.'

Similarly, a belt of granite runs along the edge of eastern Australia, roughly parallel to the Dividing Range, but the granite is mainly Triassic and the uplift of the mountains Cenozoic.

Examples of granite mountains

Mole Granite, Australia

The Mole Granite in New South Wales was apparently emplaced very close to the surface, with no more than half a kilometre of overlying rocks (Kleeman, 1984). These have been stripped off, leaving the top of the granite as a structural plateau The stripping probably occurred quite early, and the Mole Plateau landform has probably looked very much as it does today since the Triassic. The granite intrusion itself is of interest, being a tabular intrusion (Figure 9.11). Calculations suggest it has a maximum thickness of 4 km, and a more probable thickness of only 1 km.

Mt Duval, Australia

A somewhat similar situation was described from northern New South Wales by Korsch (1982) at Mt Duval. It seems that a large volcano was erupted onto a palaeoplain, over a granitic magma chamber, and then the granite continued to rise until it intruded its own volcano. In doing so the granite reached a height of some 100 metres higher than the palaeoplain. The

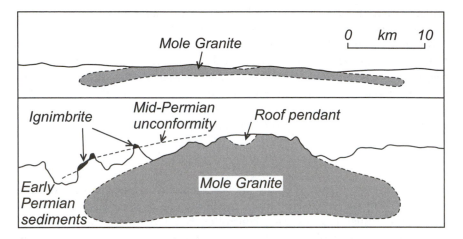

Figure 9.11 Cross-section of the Mole Granite, New South Wales. Top is natural
scale, bottom is 10 × vertical exaggeration. The intrusion is tabular,
and the location of the feeder pipe is not known. This granite was
intruded very close to the ground surface, as indicated by the roof
pendant, the Mid-Permian unconformity and the Ignimbrite. (after
Kleeman, 1984).

ascending granite gave rise to a contemporaneous synclinal rim depression
around the granite in which there accumulated coarse sediment that later
became a conglomerate (Figure 9.12).

Mt Kinabalu, Borneo

This granite mountain rises gently at first to a talus slope, above which are
cliffs up to 1000 m high with occasional gullies, and the top is a bare
plateau about a kilometre across with a few jagged peaks and strange granite
weathering forms rising higher (Figure 9.13). The highest point is 4101 m,
much higher than the sandstone and shale mountains of Borneo which reach
about 2000 m.

Gneiss mantled domes

Some granites have pushed up overlying rocks into domes, and a gneis-
sosity (layering within the gneiss) is found to be concentric with the granite.
Many of these gneiss-mantled domes are Precambrian, but some are of
Cenozoic age and directly involved in mountain building.

Papua New Guinea

Ollier and Pain (1981) described a gneiss dome from Goodenough Island,
Papua New Guinea, which consists of amphibolite facies gneiss, is cored

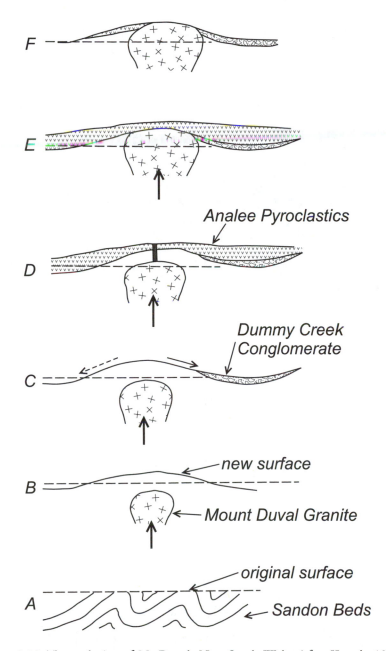

F

E

Analee Pyroclastics

D

Dummy Creek
Conglomerate

C

new surface

B

Mount Duval Granite

original surface

A

Sandon Beds

Figure 9.12 The evolution of Mt Duval, New South Wales (after Korsch, 1982). As the batholith rose a syncline developed around the base. The Analee Pyroclastics were erupted from a volcano, and then the granite rose to intrude its own associated volcano. At this stage the granite was above the general level of the ground surface.

Figure 9.13 Mt Kinabalu, Borneo. The rock is remarkably bare, and odd-shaped granite weathering features prevail (photo J. Wilford).

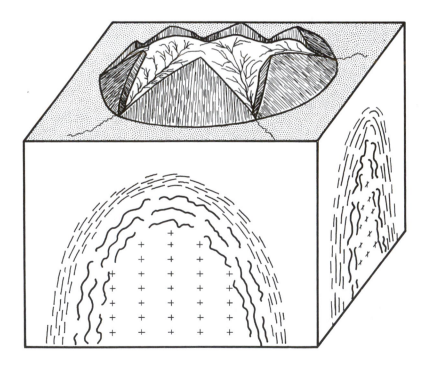

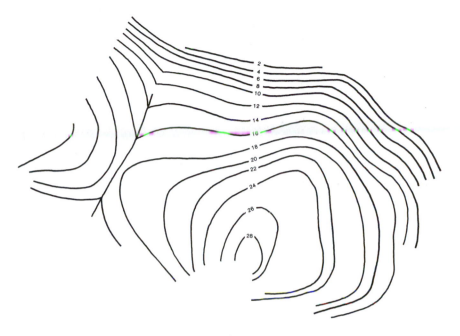

Figure 9.15 Generalised contour map of the Dayman Dome, Papua New Guinea.

by granodiorite only 1.9 Ma old, and which appears to have risen to be a surface landform – a dome 2500 m high – by shouldering aside surrounding rocks (Figure 9.14). The main evidence is that the preservation of facets of the dome is too good to be produced by differential erosion, and if the whole region were uplifted since the granite intrusion it would be impossible to create the dome by selective erosion in the time available.

Another example is the Dayman Dome on mainland Papua New Guinea. This consists mainly of greenschist facies metamorphic rocks, with a foliation everywhere parallel to the surface of a remarkably preserved dome. The shape of the dome is indicated by generalised contours (Figure 9.15) and part of the flank is shown by detailed contours (Figure 9.16) and an air photograph (Figure 9.17). The small section of the north flank looks like a typical but exceptionally well-preserved fault scarp, which it is. Furthermore, the increasing perfection of the slope and the reduction in dissection towards the base suggest that the fault scarp is still emerging.

Figure 9.14 (opposite) Diagram of the gneiss-mantled dome of Goodenough Island, Papua New Guinea (after Ollier and Pain, 1981). The granite, which is only 2.9 million years old, is rising to a level above the general ground surface. The cover of metamorphic rocks has a foliation parallel to the ground surface on the extensive triangular facets that remain after erosion of the dome by radial drainage. The highest point now is at 2500 m, and the base is very close to sea level.

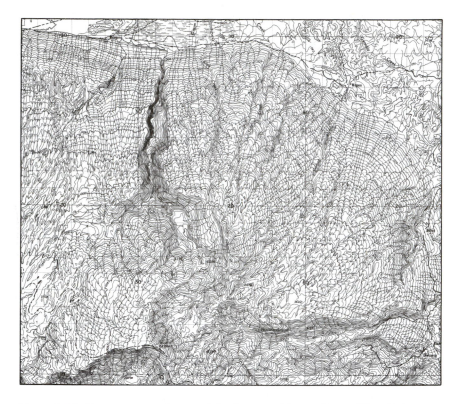

Figure 9.16 Contour map of part of the Dayman Dome, Papua New Guinea.

The air photo even shows bands parallel to the base of the scarp with different drainage density and tone, suggesting bands of different age of emergence. The gneissosity is parallel to the topographic surface or fault plane. The only remarkable thing is that this linear fault scarp can be traced continuously around the dome, and onto the flatter upper surface, so the entire dome surface can be regarded as a fault plane. As with Goodenough Dome, it would be impossible for differential erosion to remove vast quantities of overburden to produce this dome, and it is regarded as a primary tectonic landform. The foliation is produced by shearing and perhaps plastic flow in a zone that defines the shape of the dome.

There are many other young granites in Papua New Guinea. Some are remarkably intact, retaining their mineralised roof or 'gold cap' which give rise to copper and gold mines such as Bougainville and Ok Tedi, which has been dated at only 1.4 million years. The geomorphology of Fergusson Island (2.2 million years old) was described by Ollier and Pain (1981), who note that the rapid rise of the granites is accompanied by very high rates of weathering, and high rates of fluvial erosion, so the mountains are perhaps a genuine example of equilibrium landforms.

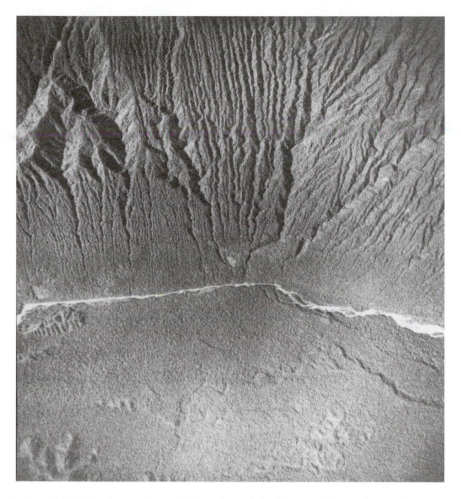

Figure 9.17 Air photo of part of the edge of the Dayman Dome, Papua New
Guinea. The dome is still emerging, like a normal fault, but the
'fault' goes all around the dome. Note how the youngest exposed
fault plane at the bottom of the slope is scarcely eroded, while the
earlier oldest exposed slopes are well dissected. A river flows along
the contact of the steep dome surface and the flat lower plain.

Arizona

The Rincon Mountain dome of Arizona may be similar. It is a granitic
gneiss mass bounded by a dome surface known as the Catalina Fault, which
parallels the attitude of the foliation in the gneiss. Around the base of
the dome are gravity slide structures that have a radial structure about the
dome (Davis, 1975), so presumably the dome was pushed up as a topographic feature beneath the beds that slid from it.

Davis (1975) described gravity folding in another situation, around a gneiss dome, where no other explanation seems feasible. Most of the Rincon Mountains consist of granitic gneiss. It is bounded by a surface known as the Catalina Fault, which parallels the foliation in the gneiss. The gneiss is in the form of a double dome, which Davis interprets as due to folding of the Catalina fault and gneiss. Around the base of the dome, detached isoclinal folds, overturned asymmetric folds, and unbroken cascades of recumbent folds pervade sheets of sedimentary rocks of Palaeozoic and Mesozoic age, resting on the dome surface. The slip line directions define a radial pattern centred on the Rincon Mountains, and Davis concludes that the folding was brought about by local gravitational tectonics. This accompanied the 28–24 Ma uplift.

Pain (1986) describes the geomophology of a number of gneiss mantled domes, or metamorphic core complex mountains, in Arizona, and concludes that they are active tectonic landforms.

10 Mountains on passive margins

We come now to a group of mountains – the mountains on passive continental margins – generally ignored in plate tectonic theory, because subduction is impossible. The Drakensberg, the Western Ghats, the Eastern Highlands of Australia and the Appalachians are just a few examples.

The basic geomorphology of passive margins with mountains is shown in Figure 10.1. There is a broad asymmetrical marginal swell or bulge parallel to the continental edge and the coast. This is separated from the coast by a Great Escarpment, which when seen from the coast looks like a mountain range. These features have been noted in European literature for a long time, but they did not gain much attention because the mountains are not as grand as Alpine mountains, and as generally perceived they did not fit any grand theory.

Marginal swells

A rise or swell along the edges of continents has been noted by many people in many lands, and so has different names in different languages. In English we can call them 'marginal swells' or 'marginal bulges'. Lester King called them 'rim highlands'. In German they are *Randschwellen*, and in French they are *bourrelets marginaux*.

The marginal swell appears to have formed after a period of planation. The whole land surface has been warped into an asymmetrical bulge, with a gentle slope on the inner side, and a steeper slope to the coast, though it must be remembered that the 'steep' slope is still only about 2°. Some have regarded this as a coastal monocline, and in some situations there are suitable rocks and structures to demonstrate that this is so. Others simply note a downwarp to the coast, and as we shall see later some people deny any downwarp on the ocean side of the continental high.

Lester King believed in the coastal monocline, especially in Natal where he lived. He also maintained that where two monoclines meet an area of extra uplift is produced. For example, in Australia the northerly monocline of New South Wales meets the east–west monocline of the Victorian Highlands, with Mt Kosciusko in the corner – the highest mountain in

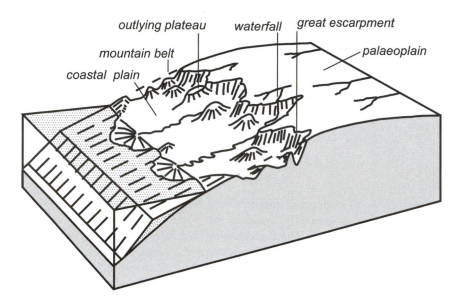

Figure 10.1 The basic geomorphology of passive margins with mountains.

Australia. In South Africa the Natal monocline meets the Cape Province monocline at the Compassberg, the highest point in Cape Province. Likewise the East Brazil monocline intersects the Paraiba monocline at Pico da Bandiera, the highest point in Brazil. As it happens, these three examples have precisely the same orientation, and the monoclines actually meet at about 120°, the angle of the triple junction so significant in plate tectonics.

Perhaps the most significant early account of *Randschwellen* was given by Jessen (1943), but in the middle of the Second World War there was little interest in academic morphotectonics, and few outsiders were reading German literature. His work is made somewhat more accessible by Bremer (1985). Many other authors made contributions, limited to their area of study and a few other areas that they knew. A range of marginal swells have been described by Godard (1982) and Ollier (1985).

In many later studies the emphasis shifted from the broad swell to the Great Escarpment commonly associated with it, and a more spectacular landform.

Great Escarpments

The popular concept of a mountain range is a tent-shaped symmetrical ridge with a sharp, angular top and steep slopes on each side. But the name 'range' has often been given to features which are in reality escarpments – steep slopes on the side of a plateau.

Many continents exhibit spectacular landforms along their passive margins – Great Escarpments. These are landforms on the grand scale, thousands of kilometres long, and often up to 1000 metres high. Some have structural control of horizontal rocks, but many do not. Despite their size, until about the last 20 years the study of Great Escarpments was strangely neglected. In Southern Africa the feature is often called the Great Escarpment in ordinary talk, and has long been a subject of debate. In India the scarp has long been recognised, but the name Western Ghats did not convey the nature of the geomorphic feature to those who have not seen it. Many of the world's great mountain 'ranges' are in fact plateaus bounded on one side by a Great Escarpment: the Drakensberg, the Transantarctic Mountains, the Snowy Mountains and the Serro do Mar are just a few examples. The large number of local names obscured the unity of the Great Escarpment, so that in Australia, for example, the unity was reported only in 1982.

Great Escarpments tend to run parallel to the coast, and they separate a high plateau from a coastal plain. An Australian example is shown in Figure 10.2. In reality they separate two regions of vastly different geomorphology: the tablelands with many palaeoforms, low relief and slow process rates, and the coastal zone with few palaeoforms, moderate relief and rapid process rates. This functional relationship can be discovered even when the escarpment is not very high: the Meckering Line in Western Australia is not a high escarpment but it separates a plateau with ancient drainage lines marked by salt lakes from a 'normal' region of rivers draining to the coast.

The top of the Great Escarpment can be very abrupt (Figures 10.3 and 10.4). The Great Escarpments are undoubtedly erosional, despite the tendency for writers on plate tectonics to draw cross-sections of passive margins showing only normal faults downthrown on the ocean side. The mechanism of parallel slope retreat seems to be responsible for creating the

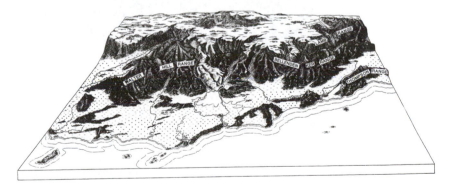

Figure 10.2 The Great Escarpment near Innisfail, south of Cairns, Queensland. In the centre is a lava flow which has poured over the escarpment and now has twin lateral streams. Here the plateau is known as the Atherton Tableland, and it is bounded by various 'ranges' – part of the Great Escarpment and associated ridges.

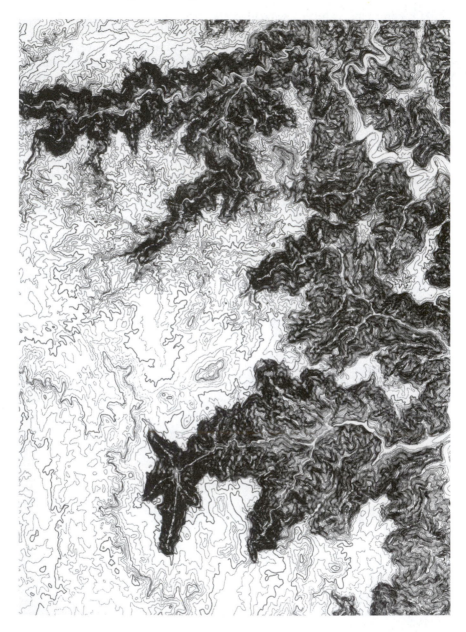

Figure 10.3 The abrupt junction between the New England Tableland and Great
Escarpment, New South Wales. Part of the Armidale 9236
1:100 000 sheet. Contour interval 20 m.

Figure 10.4 The abrupt junction of the plateau and the Great Escarpment, here cut across varied rocks. Bakers Gorge, New South Wales (photo C.D. Ollier).

Figure 10.5 Wollomombi Falls, New South Wales. Here the Great Escarpment is retreating as a wall, and old valleys are perched on the top of the cliffs, similar to hanging valleys at the top of marine cliffs (photo C.D. Ollier).

Figure 10.6 Contour map of the Carrai Tableland, New South Wales, showing an isolated high remnant of an erosion surface, east of the Great Escarpment. Based on 1:100 000 topographic map, reduced. Contour interval 20 m.

escarpments, following drainage lines to the coast which first make great gorges, and then valley widening and coalescence creates a continuous escarpment. Rugged mountainous areas form coastwards of the Great Escarpment, where the old plateau surface has been destroyed. Many of the world's large waterfalls are found where rivers leave the Great Escarpment. In places the escarpment retreats as a wall, leaving old shallow valleys perched at the top of steep cliffs (Figure 10.5). Occasionally a patch of plateau is isolated from the main tableland to form a peninsula or tableland as an outlier from the main plateau (Figure 10.6).

Examples of mountains on passive margins

The Eastern Highlands of Australia

Most maps of Australia show 'The Great Dividing Range' running inland of the eastern coast. There *is* a Great Divide separating drainage going to the Pacific from that going west, but for much of its length the divide crosses a nearly featureless plateau. Much more spectacular is the Great Escarpment. When seen from the coast it looks like a 'range' but beyond it lies a plateau. The Great Escarpment of eastern Australia has been described by Ollier (1982), Pain (1985), and Ollier and Stevens (1989). Its relationship to the Great Divide and to the present coastline is shown in Figure 10.7.

The flat country of central Australia gradually rises to the highlands, arches over and descends towards the eastern seaboard, but erosion from the sea has cut back considerably so scarps or ranges face the sea. Uplift was associated with faulting, minor for the most part, but occasionally defining distinct mountain ranges.

The Eastern Highlands of Australia consist of Palaeozoic rocks that have been planed down by a succession of erosion surfaces, and bowed up in a series of swells. In most places the palaeoplain cuts across folded Palaeozoic rocks, and is referred to by Hills (1975) as the Trias–Jura Surface. The folding of the Palaeozoic rocks did *not* make the Eastern Highlands, but is much older.

To the east the Tasman Sea is an extinct spreading site, and there is no indication of any trench, Benioff zone, or other indication of subduction, or indeed of any pressure from the sea. Further north the Coral Sea is an extinct sphenochasm where spreading has long since ceased, and again there is no evidence of subduction or pressure. Palaeomagnetic evidence and the evidence of hot spots indicate that Australia has moved in a direction about 10° east of north over the past 50 million years or thereabouts, which is roughly parallel to much of the axis of the Eastern Highlands and nowhere perpendicular to it, so the drift of the continental plate is not related to the uplift of the Highlands. Furthermore, the dating of the basalts that are related to the uplift indicates that the sinuous divide was already in existence on the drifting plate and at least some parts were already considerably uplifted, so the northern movement had nothing to do with the creation of the divide or the Eastern Highlands.

Neither the regional tectonic setting nor the details of geology and geomorphology suggest that the Eastern Highlands are caused by compression. The combination of tension, vulcanicity and uplift, with occasional downwarped areas, marks the Eastern Highlands as a region of vertical tectonics (Ollier, 1978). Despite the strong geological and geomorphic evidence for an extensional regime following doming, the situation is not completely understood. Measurements of residual stress in rock of south-

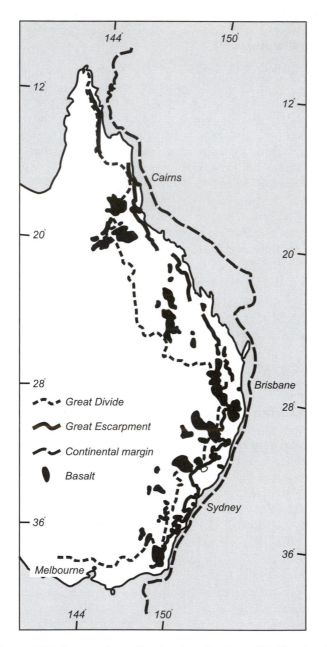

Figure 10.7 Eastern Australia showing the Great Divide, the Great Escarpment, the continental margin, and areas of basalt.

east Australia show a considerable stress in an east–west direction in most places, which would be consistent with compression, though the source and mechanism of such compression remains unknown. Denham and Windsor (1991) report stress oriented north west–south east.

Smith (1982) explained the highlands as the result of passage over a former spreading site. 'Southeast Australia moved onto the former ridge position in the Tasman Sea shortly after it had opened between 76 and 56 m.y. If the Australian lithosphere is similar to Africa I expect strong uplift beginning about 16 m.y. ago.' Nowadays few workers on the Southern Highlands of Australia would believe that uplift was so young (Wellman, 1987, 1988; Bishop, 1986, 1988). Furthermore there is a unity of the Eastern Highlands of Australia, but the entire continental margin does not have the same relationship to spreading sites as south east Australia does.

It is generally thought that the Eastern Highlands uplift occurred around the end of the Cretaceous and start of the Tertiary, as seafloor spreading of the Tasman Sea and the creation of the eastern Australian coast dates back to 80 Ma ago. Erosion grading to the new coast and new continental margin made valleys which coalesced to form the Great Escarpment, which then continued to retreat. In numerous places it intersects volcanic flows around 30 million years old. Some younger volcanoes such as Tweed Volcano were erupted about 25 million years ago on the plains created by retreat of the escarpment, and are now considerably eroded themselves. Only one lava flow goes over the escarpment, and that is about 3 million years old.

The southern part of the Great Escarpment is related to the opening of the Tasman Sea. The northern, Queensland, section is related to an offshore rift which did not reach the stage of seafloor spreading. In places the uplift of the plateau can be dated by sediment still covering the plateau, and some parts result from post-Cretaceous uplift. The escarpment can be dated by volcanoes it intersects (the Great Escarpment cuts across the 19 million-year-old Ebor volcano, so here is post-19 million years); by lava flows that run over the escarpment (a 3 million year old lava flow runs over the escarpment near Innisfail (Figure 10.2) so here the escarpment was in existence at that time, and has not retreated very far since. Other volcanoes erupted after the passage of the escarpment, such as the Tweed Volcano in northern New South Wales which erupted on the coastal plain about 25 million years ago after retreat of the escarpment. Spry *et al.* (1999) report on a basalt that flowed down the Clyde River valley, east of the Great Escarpment in south-eastern Australia, about 28 Ma ago, indicating that the drainage systems in that area were already established at that time. The escarpment is undoubtedly a diachronic feature, and is still eroding. It may have very nearly reached its present position some time ago, and it seems that such escarpments advanced very rapidly across soft rocks and have now reached hard rocks across which they are advancing very slowly.

'What caused the uplift of the Eastern Highlands?' may not be an appropriate question according to Ollier (1995). Ollier and Pain (1994) assembled evidence of widespread reversal of rivers along the east coast of Australia, and concluded that the rivers formerly flowed from Pacifica (land to the east of Australia) to the Great Artesian Basin, and following the Tasman Sea opening the plateau was downwarped to the coast, reversing the rivers and creating a watershed (the Great Divide) on a plateau that was already high.

South-west Australia

In just one part of south-west Australia the typical passive margin prevails. The general situation is shown in Figure 10.8 and details of drainage evolution in Figure 10.9. The broad picture is of a palaeoplain in a semi-arid region, with chains of salt lakes that mark the course of huge ancient rivers. This plain is bounded in part by the Meckering Line, and its continuation marks the edge of the palaeoplain. 'Normal' valleys drain to the coast.

At an early stage of margin evolution (Figure 10.9A) drainage came from the south and flowed to the north. With the separation of Australia and Antarctica by rifting and seafloor spreading the coastal part was downwarped, creating the Jarrahwood Axis, which may also have had some

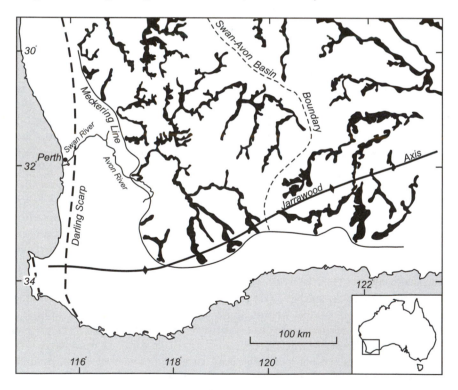

Figure 10.8 Major physiographic features of southwestern Western Australia. The Meckering Line and its continuation mark the approximate edge of the palaeoplain, but there is no Great Escarpment.

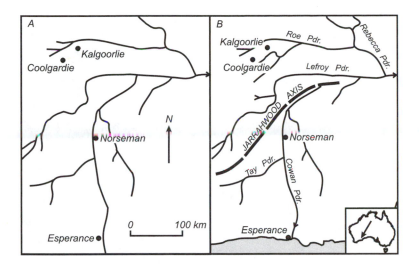

Figure 10.9 Evolution of the Cowan Drainage (after Clarke, 1994).
A. Palaeodrainage. B. Formation of the Jarrahwood Axis and
downwarp to the south, with reversal of drainage. Note the barbed
drainage, explained in Chapter 11.

tectonic uplift. On the seaward side of this axis the rivers were reversed
(Figure 10.9B). Clarke (1994) studied one of these rivers, the Cowan, in
detail. The reversed Cowan river would flow into the Southern Ocean or a
precursor rift and would deposit sediment. These sediments have been found,
and turn out to be Jurassic. This is actually a longer time scale than we
have in eastern Australia, even though less geomorphic evolution has been
revealed. The oldest sediments actually found in the valleys are Eocene.
There is no Great Escarpment, and the Stirling Ranges shown on Figure
10.8 are simply erosional remnants rising above a palaeoplain.

India

The Western Ghats of India comprise a Great Escarpment that runs parallel
to the western coast of India (Ollier and Powar, 1985). In the north it is
cut across the Deccan Basalts and in the south across planated metamorphic and igneous rocks of the Precambrian Shield (Figures 10.10 and 10.11),
with no significant change in form.

   The Great Escarpment (Western Ghats) of peninsular India was described
by Ollier and Powar (1985). The creation of the escarpment was related to
the formation of a new continental margin with the break-up of
Gondwanaland and the opening of the Arabian Sea, which here occurred
at the Cretaceous–Tertiary boundary, the time of extrusion of the Deccan
Basalt. The continent is markedly asymmetrical, and the Eastern Ghats are
younger and lower than the Western Ghats.

Figure 10.10 The Western Ghats between Bombay and Poona. The horizontal
　　　　　　lava flows of the Deccan Traps are clearly evident (photo C.D.
　　　　　　Ollier).

The escarpment has a total length of over 1500 km and is seldom more
than 60 km from the coast. The edge of the escarpment is almost entirely
coincident with the divide between eastern and western flowing rivers, and
there is no evidence of the reversed drainage so prevalent in Australia.

Southern Africa

Much of South Africa is a plateau (the Highveld), bounded by marginal
swells and Great Escarpments (Ollier and Marker, 1985; Partridge and
Maud, 1987; Partridge, 1998) which goes by many local names. The
Drakensberg is perhaps the most spectacular of these 'mountain ranges'.
The Great Escarpment makes a huge arc all around southern Africa and as
far north as Angola. The planation surface may be of Cretaceous age, and
has been related to the sub-Cretaceous unconformity offshore (Partridge and
Maud, 1987).

　　The Great Escarpment of southern Africa was reviewed by Ollier and
Marker (1985). It runs in a large curve from Namibia to the Limpopo River
(and no doubt further if a larger area had been studied), and the Kalahari
Basin is bounded by the marginal swell behind the Escarpment. The high
erosion surface is probably of Mesozoic origin, and the escarpment started
to retreat after continental break-up in the Jurassic. De Wit (1988) suggested
that the Great Escarpment in the west of South Africa is related to a

Figure 10.11 The Western Ghats in southern India, near Kodaikanal. Here the Great Escarpment is cut across Precambrian metamorphic rocks and granites, yet the general form is the same as that where it is cut across the Deccan traps (photo C.D. Ollier).

pre-Karroo period and not to the break-up of Gondwanaland, which would make it a very exceptional escarpment.

The Drakensberg in South Africa is essentially a Great Escarpment facing the Indian Ocean, and above it is a plateau – the Highvelt. The rocks are generally horizontal, and the upper part consists of 1500 m of Triassic basalt. Nowhere are there accurate indications of its age. It is presumably associated with the opening of the surrounding oceans in the Jurassic and Cretaceous.

Partridge (1998) has provided the most recent and thorough review of the morphotectonics of southern Africa. He stresses that:

1   Much of Africa posessed high elevation prior to rifting.
2   The ground surface was lowered by 1 to 3 kilometres during the Cretaceous
3   Neogene uplift re-established high elevations, especially in the eastern half of the subcontinent.

According to Partridge, the African Surface was generated by multi-phase Cretaceous erosion, and is recognised by widespread deep weathering, and massive ferricretes and silcretes. After Cretaceous planation a drainage was established on a well-planed land surface from the Santonian (Mid to Late Cretaceous) on. He wrote: 'The evidence for large-scale Neogene uplift ... is now beyond question ... The largest movements post-date the Miocene and have contributed both to the anomalous elevation of the eastern hinterland and to the strong east–west climatic gradient across southern Africa.'

Both the absence of major incision and terrigenous sedimentation on the continental shelf argue against large-scale uplift in the Middle Cainozoic. Dramatic uplift that elevated southern Africa (and much of interior Africa up to the Red Sea) occurred considerably later. A wide variety of evidence is available. River long profiles are convex up; marine Early Pliocene deposits have been uplifted 400 m; remnants of the African Surface have been warped from 3 m/km to 40 m/km. These and other data show total uplifts along the axis of warping of 700–900 m, within the last 5 million years.

For a mechanism Partridge favours a buoyancy force originating from a massive low density anomaly in the Earth's mantle. Much earlier, Smith (1982) proposed a model for uplift of southern Africa based on migration of the African continent over a source of heat parallel to the east coast: 'those areas affected by plateau uplift now overlie or have passed across the former positions of the oceanic ridges that separated Madagascar and India from Africa.' But the model is based on perceived asymmetry and accounts for uplift only in the east of the region. In reality the marginal swell and the Great Escarpment in southern Africa make a complete horse-shoe around the continental margin. Africa could not be drifting in all directions over conveniently located heat sources.

In the east of southern Africa the Great Escarpment runs through Namibia (where the local name is actually 'Great Escarpment') and Angola. On the western side of southern Africa the escarpment is more complex, and makes detours, perhaps even to the Victoria Falls. But it must be remembered that there is still a lot of the African passive margin that does not have the typical passive margin landforms.

The Appalachian Mountains

Main features

The Appalachian Mountains have been the centre of many controversies. They consist of folded and thrust-faulted Palaeozoic rocks, but the folding of rock generally had nothing to do with mountain building. Hall, the pioneer of Appalachian geology, visualised that folding and faulting associated with mountain belts took place during subsidence of the trough, thus anticipating later ideas of gravity tectonics. As one opponent of this view put it, Hall had 'a theory of mountain-making with the mountains left out' (Spencer, 1965). Later workers have confirmed that sedimentation and deformation were going on at the same time within the Appalachian

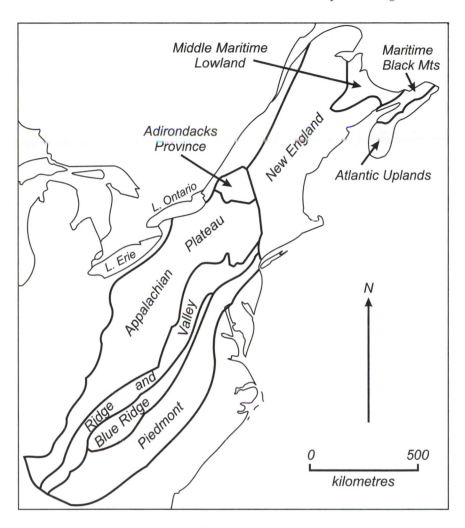

Figure 10.12 Physiographic units of eastern North America.

geosynclines. The structural features were produced while the sediments were being deposited and not late in the geosyncline's history.

The main physiographic units are shown in Figure 10.12. At the simplest approximation to the passive margin model, the Appalachian Plateau (with many subdivisions such as the Allegheny Plateau and the Cumberland Plateau) is the palaeoplain. The Ridge and Valley Province is a dissected plateau or palaeoplain. The Blue Ridge, in part, may be equivalent to the Great Escarpment. The Piedmont and Coastal Plain are typical passive margin features.

All the Palaeozoic structures were planated by the Schooley Peneplain (with bevelled cuestas providing incontrovertible evidence of planation) which possibly dates back to Cretaceous or early Tertiary time. Parts of the Blue Ridge may be equivalent to a Great Escarpment. Davis (1904, p. 214), wrote, 'in southern Virginia and North Carolina [the Blue Ridge] is not a ridge, with a crestline and well-defined slopes on either side; it is an escarpment', and on p. 240, 'The escarpment itself is by no means a straight and simple wall. The ruins of the upland often form a labyrinth of hills and spurs at the back of the piedmont coves.' The present Appalachians have nothing to do with the Palaeozoic folding and thrusting.

The coastal plains of the United States make a physiographic province which is tectonically significant as a wedge of sediment stretching from an inland margin where coast plain sediments lie unconformably on a variety of older rocks to the seaward and Gulf side where the sediments may be traced out under the continental shelves. The oldest rocks of the Coastal Plain sediments are of Cretaceous age. Along most of the margin these rest on Palaeozoic rocks, and the unconformity between them represents anything from 25 million to 500 million years.

The major effects of Palaeozoic folding were over in the Triassic, and no Jurassic sediments are found along the Atlantic seaboard, even at depth, though some shallow sea deposits with salt and gypsum represent the Jurassic in some Gulf Coast rocks. It seems that all previous structures were planed off in the Jurassic. Deposition started again in the Cretaceous when the sea advanced extensively into North America.

The Appalachian region, which might once have had high relief, had been planed down to a subdued surface by Cretaceous times. But since the present mountains are not related to folding in the Appalachian 'orogeny' there is no need to postulate any direct association of folding and mountain building at all. Any uplift that preceded the Cretaceous transgression was not necessarily on a mountain building scale. Detailed studies of river terraces support flexural isostasy as the first-order late Cenozoic deformation mechanism. Total uplift of the central Appalachian piedmont is only about 90 m in the past 15 million years (Pazzaglia and Gardner, 1993).

As sediment accumulated on the plain there was minor movement, warping, and shifting coastlines, but no folding of significance. In general the seaward side sank while the land side rose, culminating in the Appalachian Highlands or Plateau. Accumulations of geosynclinal proportions developed from Newfoundland to Honduras, and the zone of maximum deposition has tended to shift seawards ever since the Cretaceous.

Some students of Appalachian geomorphology have suggested that the Cretaceous sea actually covered a large part of the region which is now the Appalachian Highlands, citing the apparent summit accordance as one line of evidence. Johnson (1931) postulated that the area was covered by a Cretaceous sea, and a series of consequent streams developed when it withdrew. The lack of any remnants of a Cretaceous cover makes this idea questionable. It seems equally probably, if not more so, that the area was a Cretaceous peneplain bounding the Cretaceous transgression to the east.

Meyerhoff and Olmsted (1963) go back in time, and believe that the present drainage lines are direct descendants of Permian streams. They stress the supposed coincidence between present stream courses through ridges and structural sags and fault zones, a situation unlikely if the drainage lines are superimposed from a Cretaceous cover. Thornbury (1969, p. 230) suggests that the drainage may have been along present lines since Triassic times.

With more modern knowledge of the opening of the Atlantic this seems quite reasonable, for the streams would have been draining the Appalachian region to the line of rifts and lakes that were a Triassic precursor to the opening of the Atlantic.

Further south the plain around the Gulf of Mexico shows progressively younger sediments towards the coast. The sea transgressed in Lower Tertiary times, and since then the history has been one of marine regression. The Gulf Coast has an east–west axis, parallel to the edge of the continental shelf along the north side of the Gulf. In general sediments change towards the axis from non-marine to marine. Deep subsidence has taken place between the coast and the continental shelf, and Cenozoic sediments reach 14 000 m, but there is no associated upwarp of the land as in the Appalachians.

Plate tectonic explanations of the Appalachians

The Northern Appalachians are explained by Bird and Dewey (1970) in simple plate tectonic terms. The continental crust is rifted in late Precambrian time, and a proto-Atlantic ocean is formed. Sedimentation occurs on the margins of the opposed continents, and then the continents move together again. At first there was 'Andean type' orogeny on the Appalachian side (though why not on the European side is not explained), and then when the opposed continents collided there was 'Himalayan type' orogeny, with disappearance of the proto-Atlantic Ocean.

Dietz (1972) provides a slightly different plate tectonic model for the Appalachians using lateral compression and plate tectonics (Figure 10.13). Starting with geosynclinal deposits and platform deposits, deformation, vulcanism and intrusion took place about 250 million years ago at a plate convergence, before the opening of the present Atlantic started about 225 million years ago.

In both these models it must be borne in mind that the existence of the present Appalachian Mountains is not the result of plate tectonics. The mechanisms of Bird and Dewey, and of Dietz, produce a fold belt by plate tectonics. This idea is still prevalent, as shown by Pinter and Brandon (1997), who refer to the Appalachians as 'old mountains' in contrast to the 'new mountains' of active continental margins. The fold belt is presumed to have been a mountain range, but the folded rocks have been worn down

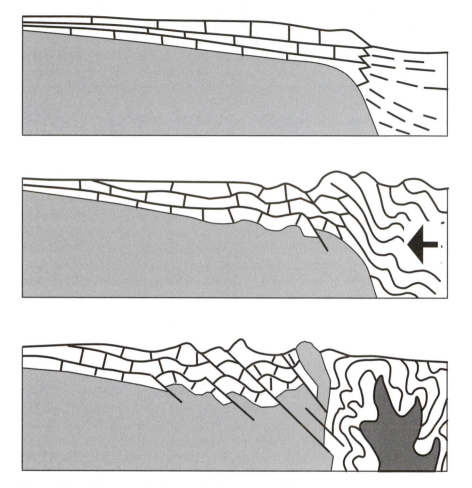

Figure 10.13 Diagram of the Appalachians caused by simple compression due to continental collision (after Deitz, 1973).

to a plain, and uplift of this plain created the present mountains much later in Earth history and at a time when simple plate tectonic compression cannot apply.

Brazil

In South America the only contenders for passive margin mountains appear to be in Brazil.

The Brazilian Plateau is bounded on the Atlantic side by a Great Escarpment, which has been given various local names. The largest section is the Serra do Mar, which extends 800 km with a maximum height of

2245 m (Maack, 1969). It separates a coastal strip from the interior plateau, and all major streams drain westwards on the gentle inland slope to the Parana. In this it is like the Western Ghats of India, and unlike the more complicated drainage situations of Australia, eastern North America and southern Africa. There is little to date the uplift of the plateau, but the escarpment is presumably related to the opening of the South Atlantic, which started in the Jurassic, and is younger than the Parana Basalts which are of Triassic age.

Scandinavia

Scandinavia is commonly described as having Caledonian mountains, but although the rocks and structures may be Palaeozoic (Caledonian orogeny) the present mountains are much younger. It is also generally thought that glaciation was dominant in forming the landscape, but in reality many palaeoforms are preserved.

A broad and generally flat surface (usually called the palaeic surface in Scandinavia) covers much of Sweden and extends over the divide between the Atlantic and the Baltic. It was identified by Reusch (1901), who realised that it had existed before a late upheaval of the land. The surface is described in more detail by Gjessing (1967). Schipull (1974) made a thorough study of a part of the surface near Hardangervidda in south-western Norway. He found remains of old meandering valleys and small intact parts of an exhumed sub-Cambrian surface, and concluded that the Quaternary was a period of conservation of pre-existing relief on the plateaus.

In some places the land suddenly becomes steep at a Great Escarpment (Figure 10.14). The Great Escarpment is much degraded by glaciation in many places, especially northern Scandinavia, but it is still evident in southern Norway. Glaciation deepened the valleys and steepened the slopes, but the Great Escarpment was in existence before ice ever appeared. In some places the rugged mountains of southern Norway are dissected below the level of the palaeic surface (Figure 10.15), but the accordant summit leaves little doubt of the former planation.

The uplift of the palaeoplain may be related to offshore sedimentation. Doré (1992) regarded the envelope surface (roughly speaking the palaeoplain) as equivalent to the base-Tertiary surface offshore, and evolved by late Mesozoic base-levelling. Jensen and Schmidt (1993) paid particular attention to the base-Quaternary surface offshore, which cuts across older strata, and interpreted the uplift of Norway as a mainly Neogene event with a magnitude of 1500–2000 m caused by broad regional warping without faulting. Riis (1996) suggested the palaeoplain was of Jurassic age, but concluded from an overview of the offshore sediments that there were two main uplift phases – Palaeogene and Neogene. The early one had a maximum in northern Scandinavia, the later one in southern Norway. Lidmar-Bergström (1995) follows this opinion, based on geomorphic data.

Figure 10.15 The Palaeic surface lowered by glaciation, southern Norway. This is a classic gipfelflur (photo E. Hagen).

Holtedahl (1953) looked upon Scandinavia as part of a rifted continent – a bold view long before plate tectonics. He also noted what we now see as a passive margin assemblage on the opposed coasts of Norway, Greenland, and Baffin Land in Canada.

Greenland

As they are parts of the same Caledonian orogen broken apart by the seafloor spreading of the Norwegian–Greenland Sea, the Lofoten–Vesteralen (Norway) and the Scoresby Sun area (Greenland) have comparable morpho-tectonics. Both these mountainous areas belong to the steep side of long marginal bulges of Tertiary age (Peulvast, 1988).

Weidick (1976) provides more detail. The present Greenland coastal strip contains alpine landscapes and elevated plateaus, with one high summit level and some minor ones. The highest and oldest planation surface cuts across the Tertiary basalt as well as Precambrian bedrock and so was formed since the deposition of the basalts, that is in Miocene times or later. Several kilometres of the basalt flows were eroded 'and uplift therefore took place in the late Miocene, Pliocene or Pleistocene' (Weidick, 1976). Brooks (1985)

Figure 10.14 (opposite) A typical example of the Great Escarpent in Scandinavia and its contact with the Palaeic surface or paleoplain. The Aurland valley, Norway (photo Fjellanger Wideroe AS).

suggests an age of plateau uplift of about 35 Ma. Since uplift, the margin has been eroded to a considerable depth, to give the appearance from the coast of a steep rugged mountain range. However, inland of the mountains there is a plateau, dissected only by large glacial valleys.

The Miocene climate of Greenland was similar to that of southern Europe today, so at that time glaciation would have been limited to small local glaciers on the highest areas – if any existed. Glaciation appears to have developed quite rapidly about 3 million years ago.

Antarctica

Antarctica consists of two parts: the roughly equidimensional Eastern Antarctica and the tail-like Western Antarctica. Eastern Antarctica is a fragment of Gondwanaland, like Australia or Africa, and has some of the same features. Näsland (1998) briefly reviews earlier work and notes that on the Antarctic continent, exposed palaeosurfaces are uncommon because of the ice sheet cover. However, such surfaces have been observed all around the continent, at the margin of the ice sheet, including Marie Byrd Land, the Prince Charles Mountains, the Dry Valleys, the Shackleton Range, and Dronning Maud Land.

The Transantarctic Mountains run along the edge of eastern Antarctica. The sinking of Victoria Land Basin, and presumed associated uplift of the nearby Transantarctic Mountains, may have occurred in Oligocene and Miocene times, although Behrendt and Cooper (1991) suggest that the rift shoulder along the Transantarctic Mountains has been rising at a rate of ~1 km/Ma since the Early to Middle Pliocene. Glacial sediments off Antarctica also go back to the Oligocene–Early Miocene, and the greatest Cenozoic fall in sea level occurred in the Oligocene, and may be related to growth of the Antarctic Ice Cap (Barrett *et al.*, 1991). Smith and Drewry (1984) make the point that, although the Transantarctic Mountains are one of the great mountain chains of the world, they lack all the thrusting and folding usually associated with subduction and continental collision. On the contrary, Jurassic sediments, now at 3200 m, are still flat lying and undisturbed. These mountains are clearly formed by Late Cenozoic uplift followed by marginal erosion.

In Dronning Maud Land, East Antarctica, a flat, high-elevation plateau was once part of a continuous regional planation surface. Isolated remnants are still remarkable landforms (Figure 10.16). Overlying Late Palaeozoic sediments show that the surface existed in the Early Permian. An 'escarpment that bounds the plateau remnants to the north probably developed as a result of the rifting of East Antarctica and southern Africa' (Näslund, 1998, 1999). However, he later writes that the morphology of the Dronning Maud Land escarpment as a whole 'supports the idea that the escarpment in general was formed in the same way as escarpments along other passive continental margins, that is by scarp retreat'.

Figure 10.16 A palaeoplain remnant in Dronning Maud Land (from Näslund, 1998).

Because of glaciation the geomorphic history is different from many other passive margins, and is summarised by Näslund as follows:

> In the late Mesozoic, the landscape was uplifted in association with the rifting between East Antarctica and southern Africa. After rifting, the palaeosurface and escarpment were subject to intensified weathering and fluvial erosion. In the middle Cenozoic, probably prior to the Oligocene, erosion by wet-based mountain glaciers started to form an alpine landscape out of a presumably mainly fluvial morphology . . . During the late Cenozoic up to present, the ice sheet has been cold-based at high elevations within the landscape, inhibiting glacial erosion and preserving existing landforms.

The Prince Charles Mountains consist of large flat-topped massifs with accordant summits which Wellman and Tingey (1976) postulate are the remnants of a preglacial surface of low relief. An Eocene lava flow on the surface suggests that erosion of the surface, and thus uplift, must have been younger than Eocene (Tingey 1985).

Asymmetry of passive margins

Cross-section diagrams of several continents are shown in Figure 10.17. Southern Africa is the most perfect. Starting in the west there is the coastal

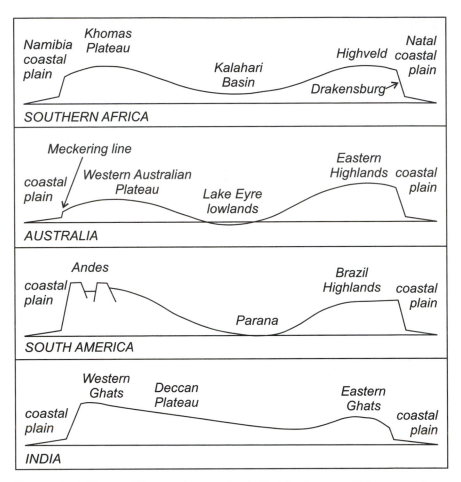

Figure 10.17 The possible morphotectonic similarities between different continents with passive margins.

plain of Namibia which is backed by the Great Escarpment. The Khomas Plateau is the palaeoplain and this gradually sinks to the Kalahari Basin and then rises again to become the Highveld. This ends abruptly at the Great Escarpment of the Drakensberg, and then comes the coastal plain of Natal.

Australia has the high plateaus of Western Australia and Eastern Australia on either side of the central lowlands, which actually go below sea level at Lake Eyre. Despite a common misconception, the western plateau is generally as high as the eastern one except for a triangle of extra high ground in the south east. There is a major Great Escarpment in the east, but in the west only the rather vague break of the Meckering Line.

South America has the swell of the Andes, the central depression of the Parana lowlands and the Brazilian Plateau, with its Great Escarpment, the Serro do Mar. The symmetry is incomplete, of course, because the eastern side is an active margin. The same situation is found in North America, with the Central Plains between high land on west and east, but while the east has plateaus and a partial Great Escarpment, the west is tectonically complex.

India has both Western Ghats and Eastern Ghats, but while the first is a splendid example of a passive margin, the latter is only a gentle warp. Nevertheless there is a lowland between the two.

Antarctica and Greenland both have marginal swells. It is interesting that both have huge ice caps in the centre and mountains around the rim. It is generally thought that the weight of ice depresses the crust, and this is no doubt correct, but since many other continents have central depressions too, this might not be the whole answer. Such basins would provide ideal conditions for the collection of ice if the climate was right.

Absence of typical passive margin morphotectonic features

A general pattern has been identified for passive margins, with a marginal swell and Great Escarpment. These are found in many parts of the world, but not on all passive margins, and the first big problem is – why not?

In Australia there is a superb example along the eastern edge of the continent, an old and in some ways unusual one in the south west, and an argument can be made for a very low passive margin in part of the north. But elsewhere it is missing. This is particularly obvious across most of the southern edge of the continent, especially around the Great Bight, where Tertiary rocks have remained virtually horizontal. In Africa the marginal swell is beautifully simple in the south, but in most of West Africa there appear to be no special geomorphic effects, and the same is true in Tanzania and Kenya. In India the Western Ghats are superb, but the Eastern Ghats are little more than a gentle swell. In South America the passive margin style of geomorphology appears to be restricted to part of Brazil.

Simplified versions of the Great Escarpments of three continents are shown in Figure 10.18, all at the same scale. The Australian Great Escarpment has numerous gaps, sometimes because the rocks are too weak to maintain a steep slope, sometimes because of tectonic complications such as en echelon faults, and sometimes for no obvious reason. The African Great Escarpment is more continuous, but in detail it too has gaps. The Indian situation is of a continuous escarpment from north of Bombay to the Palghat Gap, but south of the gap the scarp retreat from the west is matched by scarp retreat from the east, and a plateau is completely isolated, surrounded on all sides by a Great Escarpment.

Whatever explanation might be proposed to account for the typical features should also be able to account for their absence.

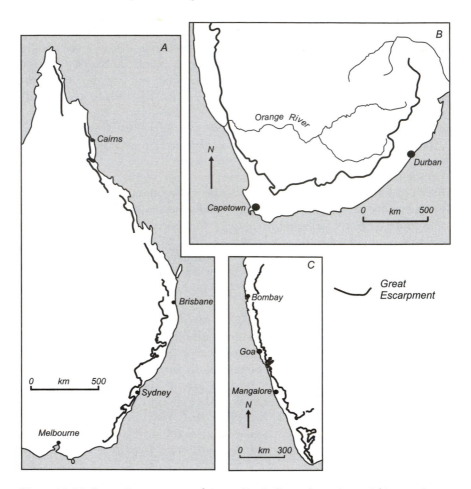

Figure 10.18 Great Escarpments of Australia, India and southern Africa at the
same scale.

Time of uplift

From the brief review given earlier, it is clear that many passive margins
date back long before the Late Miocene to Pleistocene age of many other
mountains. This is consistent with the idea that the uplift related to the
rifting that was a precursor to the break-up of Pangea or Gondwana.

Nevertheless, some uplift may be more recent. Partridge (1998) has
Pliocene uplift of the South African high plains of up to 800 m. In the
Appalachians, although the palaeoplain might date back to the Cretaceous
there is evidence of Cenozoic uplift. In Australia the eastern highlands were
once thought to be formed by the Pleistocene 'Kosciusko Orogeny' but this
idea gave way to a general belief in Early Cenozoic uplift. Nevertheless,

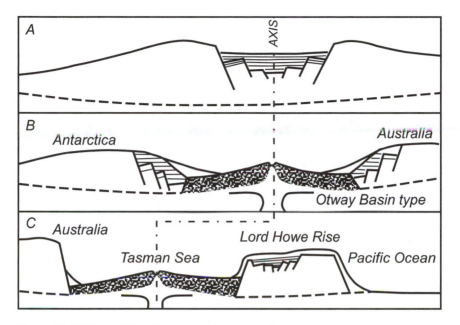

Figure 10.19 From rift valley to sea floor spreading
    a.   The breakup of a continent to form a rift valley, followed by
        extension with normal faulting, and eventually sea floor spreading.
    b.   Symmetrical sea floor spreading, as between Antarctica and
        Australia.
    c.   Asymmetrical sea floor spreading between Australia and the Lord
        Howe Rise.

some movements of up to 1 km may have occurred in the Pleistocene (Ollier and Taylor, 1988). It is quite possible that the age of uplift varied along the eastern highlands, and may range from perhaps 80 million to as few as 2 million years.

Causes of passive margin uplift

New continental margins and coasts came into existence with the break-up of super-continents. The edge of the siallic continent is roughly at the continental shelf, and the coastline is a rather accidental line where the modern sea happens to lap onto the continental mass.

Some think that a rift valley model helps explain the features of passive margins. The first stage is a simple rift valley or graben. This may have a complex of faults in the middle as extension continues. Eventually a basalt floor appears in the rift (the Red Sea stage of evolution) and this is followed by sea floor spreading. These stages are shown in Figure 10.19.

In the simple case spreading is symmetrical. The Atlantic Ocean is roughly of this kind. In the South Atlantic spreading is symmetrical, and passive

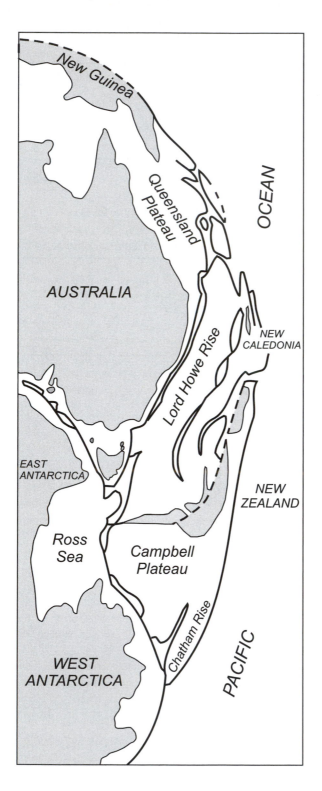

Figure 10.20
The geography of
eastern Gondwana
before the break
up (after Griffiths,
1971).

margins of Brazil and Africa are symmetrically disposed. In the North Atlantic the spreading ridge forks around Greenland, producing passive margins on opposite shores in Baffin Land, Greenland and Scandinavia. On the margins of Gondwana reconstruction is not as easy, but it seems that there was break-up and sea floor spreading, starting perhaps with a situation like that shown in Figure 10.20. South eastern Australia is rather complex, because the final split was very close to Australia, which has a very narrow continental shelf, and the complicated graben with its faults and sediments has drifted away as the Lord Howe Rise (Falvey and Mutter, 1981). The southern coast of Australia has retained its complex offshore faults, but is lacking the usual passive margin features of marginal swell and Great Escarpment.

The details of the continental margin depend on the nature of the original rift. If there is a 'rise to the rift', the area of drainage into the rift is very small. This might be something like the Red Sea Rift of today (Figure 10.21A). The continental margins of Brazil and western India might have started like this. If there is a Lake Albert type rift, like that shown in Figure 3.5, there is a rise to a shoulder which is tens of kilometres inland of the rift faults. Ancient drainage will be reversed in places, as shown in Figure 10.21B. The Scandinavian passive margin and the east Australian margin seem to be of this type.

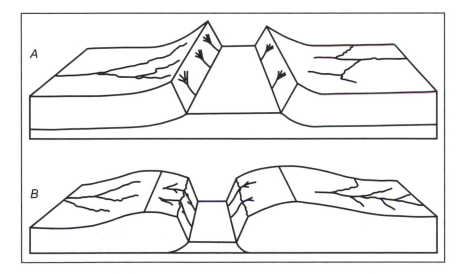

Figure 10.21 Contrasted styles of rifting.
    A. Simple rifting. The edge of the block is uplifted at the fault (the 'rise to the rift') and the watershed is coincident with the top of the fault.
    B. The Lake Albert type. A central graben (or seafloor spreading site) lies between two uplifted blocks, but the axis of greatest uplift lies many kilometres away from the fault scarps. Reversed and disrupted drainage is found on both uplifted blocks.

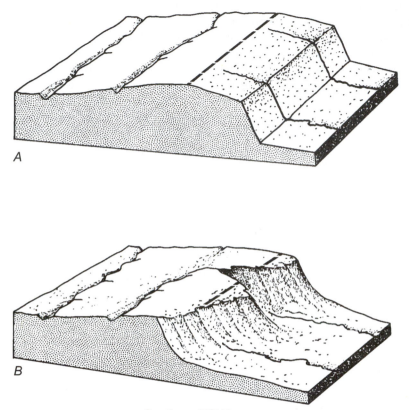

A

B

Southern NSW

Figure 10.22 Retreat of the Great Escarpment where the rivers run roughly
parallel to the escarpment, as in southern New South Wales.
Rivers are captured and diverted to the coast. Retreat of the
escarpment will eventually remove all trace of the original tectonic
axis (dash line) but the drainage pattern may still reveal the
landscape history.

The break-up of Pangea occurred at different times in different places.
The Atlantic opened in the Jurassic, so no coastal features can pre-date
that. Australia separated from Antarctica about 55 million years ago,
although initial rifting may have taken place about 100 million years
ago. The Tasman Sea, between Australia and New Zealand, was created by
sea floor spreading that started about 80 million years ago. The Queensland
Great Escarpment is even younger, related to a rift that failed to open and
is now a submarine channel between Queensland and the offshore and
submerged Queensland Plateau (Ollier and Stevens, 1989).

In simple rifting, the top of the rift fault will be the watershed between
coastal draining rivers and inland draining rivers. The coastal drainage is

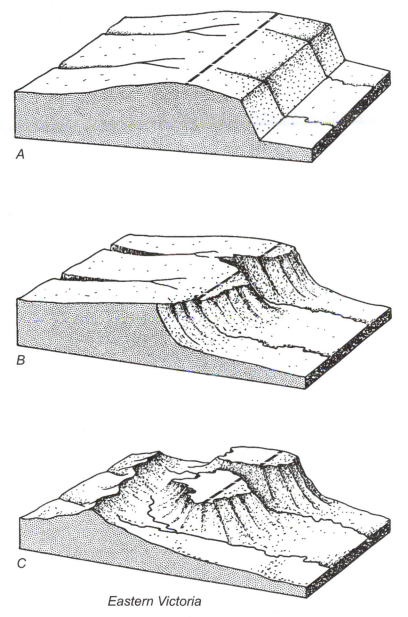

A

B

C

Eastern Victoria

Figure 10.23 Retreat of the Great Escarpment where the divide is roughly
perpendicular to the drainage. Rivers on the steep side cut back
faster than those on the northern side, with headward erosion. By
stage C headward retreat has passed the original tectonic divide
(marked by the dash line), and only isolated high plains remain.
The Great Divide is inland and lower than the high plains.
Western Victoria (Australia) is a good example.

likely to be steeper, and so more erosive, and headward erosion will cause headward erosion or rivers, and when valleys coalesce there will be inland retreat of the Great Escarpment.

In the Lake Albert style of rifting, landscape evolution is more complex. The Great Divide will be well inland, and the situation will be like that in eastern Australia where the divide runs along a plateau. Headward erosion will cause headward erosion of individual valleys, and eventually the creation and inland migration of the Great Escarpment. Two possibilities are shown in Figures 10.22 and 10.23.

If the ancient drainage was roughly parallel to the axis of warping and the coast, headward erosion of coastal rivers will capture plateau rivers, with all the usual features of river capture described in Chapter 11. This will also cause 'leaping of divides', also described in Chapter 11. Eventually the original tectonic divide may be totally destroyed, and the situation might be difficult to distinguish from a Red Sea type rift. However, the complex drainage pattern resulting from repeated capture and changes in catchment size may help to decipher the history.

If the original drainage was roughly perpendicular to the warping and coast, a different set of landforms will result (Figure 10.23b). Headward erosion of coastward flowing rivers will eventually reach the original tectonic divide, and then cut beyond it. Isolated plateaus may be left on the old divide, and headward erosion will reach a new divide, inland of the original tectonic divide. This is the situation reached in eastern Victoria in south east Australia. Numerous isolated high plains are preserved south of the watershed known as the Great Divide, and they are higher than the divide.

These different mechanisms should all produce different effects, probably at different times. If the hypothesis is sufficiently developed to allow prediction, and the age of geomorphic development can be determined, then the hypothesis can be tested. Thus geomorphic features of passive margins should be able to put constraints on tectonic theory, and perhaps even afford crucial tests. This is one reason why we emphasise the importance of drainage patterns in Chapter 11.

Downfaulting or downwarping at the margin

There are two main models for passive margin evolution, both related to rift valleys. With a simple rise to the rift (Figure 10.24A and B) there has to be a massive fault somewhere offshore. Such a fault has not been found on many margins with typical geomorphic features. There would be massive retreat from the fault to the present position of the Great Escarpment, and no old regolith would survive. This model does not take into account, or explain, the offshore break-up unconformity (sometimes called the basal unconformity), which cuts across older rocks and is covered unconformably by younger ones.

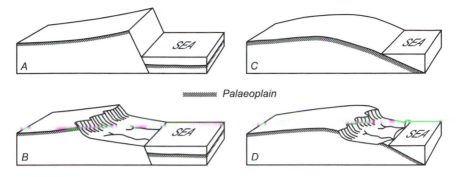

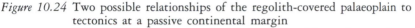

Figure 10.24 Two possible relationships of the regolith-covered palaeoplain to
tectonics at a passive continental margin
Left.   With continuous rise of the palaeoplain to the continental
margin (A), any continuation of the palaeoplain has to be
downfaulted below sea level. When erosion produces a
coastal zone with a Great Escarpment (B), no trace of the
old regolith will be preserved on the coastal lowland.
Right. With downwarp of the palaeoplain from the marginal swell
to below sea level (C), the regolith covered plain outcrops
at the coast, and some palaeo-landforms may be preserved
close to the coast (D) though not nearer to the Great
Escarpment.

The alternative, which we favour, is downwarping of the palaeoplain
below sea level (Figure 10.24 C and D). This simultaneously accounts for
the plateau between the Great Divide and the Great Escarpment, and the
creation of the break-up unconformity, which we regard as simply the downwarped palaeoplain. The amount of erosion required to form the coastal
belt backed by the Great Escarpment is very much less in this model.
Another feature is that the regolith is downwarped to the coast and a belt
of old surficial features can occur near the coast, which is impossible on the
first model. We believe that deep weathering profiles, and some old lavas,
have been warped down to the coast in this way. As a recent example, Spry
et al. (1999) describe the Tertiary evolution of coastal lowlands in southern
New South Wales where basalt flows give numerical dates, and conclude:
'The basalt and sediments in the Clyde River valley indicate that the coastal
lowlands in southeast New South Wales were developed prior to the mid-
Tertiary period.' Furthermore we observe triangular facets of the old surface
near the coast, as shown on Figure 10.1, which invariably retain the old
regolith.

The ultimate cause of marginal swells

There are many possible causes for the formation of marginal swells, as
there are for uplift in general, and a list is provided in Table 10.1. There

Table 10.1 Some possible causes of formation of marginal swells on passive
continental margins

1	Passage of the continental margin over a zone of anomalous high heat flow (e.g. Smith, 1982: Karner and Weissel, 1984).
2	Thermal expansion of a mantle plume beneath the continental margin (e.g. Cox, 1989).
3	Isostatic response to erosion of the continental margin, especially by escarpment retreat (e.g. King, 1955).
4	Uplift to compensate for subsidence offshore, caused in turn by loading of deposited sediments (e.g. Gilchrist and Summerfield, 1990). Mechanisms 3 and 4 usually work together.
5	Underplating of the continental margin by masses of light rock (e.g. Wellman, 1987).
6	Intrusion of large amounts of igneous rock.
7	Delayed response to erosion of earlier orogenic belts (e.g. Lambeck and Stephenson, 1988).
8	Subsidence of basins on each side of an originally high continental margin (e.g. Ollier, 1992).

are theories as different as isostatic recovery after erosion of earlier mountains, underplating of the crust by light material of unknown nature, and
response to thermal phenomena associated with earlier rifting. Lambeck and
Stephenson (1986) proposed a model in which the present elevation of the
eastern highlands of Australia is a result of isostatic rebound due to erosion
since at least 150 Ma ago. This is still supported by modellers who provide
mathematical explanations of the passive margins that leave out such important basic features as Great Escarpments and coastal facets (van der Beek
et al., 1999).

11 Plains and planation surfaces, drainage and climate

Plains are large flat areas that have several types of origin. If they are uplifted they become plateaus. A theme in much of this book is that many mountains are plateaus or dissected plateaus, so we need to elaborate somewhat on plains and planation surfaces. Similarly, most dissection takes place in river valleys, and we have used drainage as evidence supporting some of our assertions, so we also need to cover drainage in some more detail.

Plains

Depositional plains

Alluvial plains may be underlain by planated bedrock at a shallow depth, or they may be underlain by great thicknesses of alluvium (and possible sediments from marine incursions, lacustrine sediments, etc.). The Murray Basin of Australia, for example, has sediment that is up to 600 m thick, and similar great thicknesses underlie the Ganges, Indus and Mississippi. In China the loess plateau has a cover up to 300 m thick of wind blown loess, in the form of individual sheets with palaeosols in between.

Lava plains

These are described in greater detail in Chapter 9. Large flat areas may be covered by basaltic lavas. The lava may be relatively thin, covering an already flat landscape, such as the lava plain province of western Victoria, Australia. It may also be very thick. In the Snake River lava province of the United States lavas were erupted onto a dissected landscape with a relative relief of over a kilometre, rather like the Grand Canyon landscape of today. The lavas not only filled the valleys, but accumulated well above them to form a lava plateau. The lavas of the Deccan Traps in India are also kilometres thick, and the lava plain formed a new basic plain or plateau that was the start of the present landscape. Of course, the lava plateaus have suffered some erosion. Remnants of ancient land surfaces may be found, and Widdowson (1997) believed he had found the top of the Deccan Traps.

However, it is more likely that his ferricretes were formed on old drainage lines, now preserved by inversion of relief (Ollier and Pain, 1996).

Ignimbrite plains and plateaus

The phenomenon of pyroclastic flow deposits was explained in Chapter 9. The resulting plains can completely bury previous landscapes, giving a totally fresh start to regional landscape evolution. The top of the ignimbrite flow is exceedingly flat. The rock itself is highly porous. However, there may be a welded tuff layer at a depth of some metres below the surface which is impermeable (except through joints) and difficult to erode, which helps in the preservation of ignimbrite plains.

Structural plains

Sometimes plains or plateaus are on horizontally bedded rock. It is then hard to prove that the plains are truly erosional plains cut at some particular base level, for they may merely represent the top of an old bedding plain from which softer rocks that once were on top have been removed. In reality sedimentary rocks are rarely horizontal over the areas occupied by broad plains, but small remnants of old plains are very hard to separate from structural surfaces. Regions of horizontal rock are not good for the study of erosion surfaces.

Planation surfaces

Planation surfaces are simply fairly flat topographic surfaces formed by erosion. Erosion by rivers (fluvial erosion) is the commonest type. Such erosion cannot continue below sea level, which is known in this context as the base level of erosion. There is no landscape-forming process that can make a flat surface from rugged country at a high altitude, so plateaus (high level flat surfaces) indicate vertical uplift of a former low-lying plain. Erosion is not instantaneous, so widespread planation indicates a period of tectonic stability when erosion was not offset by uplift.

Evidence for planation surfaces

It is not difficult for most people to tell when they are in an area of relatively flat country: they can recognise a plain when they see one. After this it is relatively easy to determine whether the plain is a depositional plain built up of alluvium, marine deposits, lake sediments or whatever, or whether it is an erosional plain cut across rock. If it is erosional it may be given a technical name such as peneplain, pediplain or palaeoplain, depending on the detailed evidence and the taste of the investigator. In brief it is a plain or planation surface.

An elevated plain can also be recognised without too much sophistication, and 'high plains' is the term given by farmers and foresters to various areas where common sense is enough to show that the country is both high and flat. 'Plateau' is another term used for the same thing, usually when large. 'Tableland' and 'mesa' are other popular terms. High plains may cover areas of thousands of square kilometres or be tiny remnants. They are separated by a zone of steep ground from lower plains, if any exist.

A number of small plain remnants at the same level suggest that they are remnants of a formerly more extensive plain, dissected by subsequent erosion until only fragments remain. This argument is particularly convincing if the surfaces are cut across varied bedrock. In the Apennines for instance, 'The main geomorphic feature . . . is a planation surface which cuts most of the relief of the peninsula resulting in a mountain landscape with large flat parts on top' (Coltorti and Pieruccini, 2000).

In areas of folded or tilted strata the hard rocks form ridges and the soft rocks form valleys. If the ridges tend to reach the same elevation over a wide area it might suggest that the fold had been planated before differential erosion brought out the hills and valleys. On a simply dissected fold a wider range of elevations would be expected. However, there is no agreement on how much variation in elevation might be attained by either process, so simple summit accordance is never more than suggestive.

The best proof of all comes from the presence of bevelled cuestas, as explained in Chapter 2.

Some argue that as rivers cut down the interfluves are also lowered, and numerous 'phantom' planation surfaces may be revealed by accordant summits. This idea is shown in Figure 11.1. But in three dimensions the situation is easier to decipher because simple fluvial interfluves will have a slope downstream, but those approximating to a true planation surface will have level profiles in a direction roughly parallel to the stream courses as well as across the stream profiles. The evidence for former surfaces must be considered in three dimensions. The same is true for reconstruction of old surfaces in folded sedimentary rocks, for if the strike sections of ridges show accordant summits then there is good reason to think there was a planation surface not much higher, even if bevelled cuestas are absent.

Perhaps the weakest evidence for the existence of a former plain is the 'eyeball' method, in which the eye sees a landscape of hills, mountains or ranges, and the mind casts a planation surface over all like a great blanket. However, it might be asked why hills do tend to reach accordant heights, and why sometimes the heights fall into a particular pattern. The Gipfelflur of the European Alps is an imaginary surface that can be seen riding over the summits of the Alps in a huge dome. Cynics may disbelieve, but many experienced geologists feel they are seeing something real.

From afar one always sees how the individual tumble of jagged peaks, so impressively irregular at close sight, approaches a strikingly even

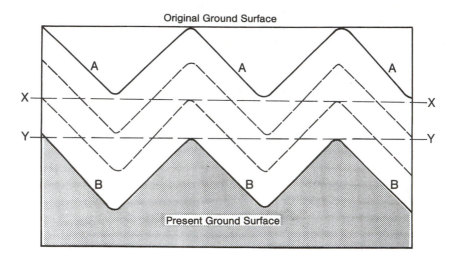

Figure 11.1 A dissected planation surface may be detected at stage A by the
accordant summit levels. Without other evidence this is a risky
method because at later stages of erosion such as B, the present
ground surface, the approximate summit accordance may suggest the
existence of a planation surface, as may stages X and Y, but such
'planation surfaces' never existed in reality. In three dimensions the
problem is not as bad as the simple sections suggest, and with the
assistance of correlative sediments real planation surfaces can often be
detected even after much erosion.

upper limit. This summit level passes over the tops, which are in most
cases residual mountains of some former erosion surface. Many individual peaks are eroded below the surface, and a few stick above it,
and there are wide undulations of the summit plain itself. But as an
overall feature this summit plain is a very real thing indeed.

(Rutten, 1969)

A gipfelflur does not prove a former single peneplain, but it reflects an
integration of former base levels which is today broadly unwarped. In the
case of the Alps the gipfelflur was upwarped during the Late Pliocene and
Pleistocene times.

Varieties of plains and planation surfaces

Accepting the fact that planation surfaces exist, we can move on to examine
briefly the different types. We will have to touch on some of the controversies
that have exercised, confused and wasted the time of geomorphologists for the
past century. There have been long debates about the names for the plains of
different types, arguments about how they were formed, their relevance to
geomorphic history, and their relationship to uplift.

It is certainly very desirable to have a non-committal term for describing planation surfaces whether it be erosion surface, peneplain, pediplain or some other term. The differences between the application of these terms usually depend not on what we can observe of the plain, but on the deduced or observed mode of formation.

Incidentally, some purists of the English language object to the term planation, and especially derived words such as 'planated', but the dictionary definition is:

> **planation** *n.* the erosion of a land surface until it is basically flat
> (*Collins English Dictionary*)

Peneplains

Writers early in the twentieth century used the term 'peneplain' (nearly a plain) to describe planation surfaces that had low relief, were of subaerial origin, and cut across geological structures. The concept was introduced by W.M. Davis (1899), the man commonly regarded as the 'father of geomorphology'. Residuals of higher country standing above the peneplains were called 'monadnocks'. Remnants of more than one peneplain were identified in some areas.

The problems that confuse studies of planation surfaces soon set in. People identified different surfaces, different numbers of surfaces, and correlated surfaces in different ways. Surfaces that some thought to be peneplains were explained by others in different ways. Davis drew diagrams that implied that a peneplain was formed by valley sides becoming ever gentler in slope (slope decline). For some, slope decline is a necessary part of peneplanation. Others use 'peneplain' in the sense in which it was originally used, to mean 'almost a plain'.

Pediplains

A challenge to the peneplain concept came from L.C. King (1953) who propounded his concept of the pediplain. A pediplain is made by the coalescence of pediments, which are footslopes, often backed by hills or mountains. They may be cut across hard rock, or soft rock, and in European literature those on soft rock are called 'glacis'. They are believed to be formed as a surface just steep enough to permit the transport of debris produced by weathering of the steep slope behind, which is actively retreating. A fine example of a pediment is shown in Figure 11.2.

Pediplains have many of the same properties as peneplains – they have low relief, subaerial origin, and cut across geological structures. The big difference between peneplains and pediplains lies in their supposed mode of origin. Peneplains are said to be formed by downcutting of rivers followed by valley widening and slope decline, whereas pediplains are produced by

Figure 11.2 A major pediment at the foot of the Flinder Ranges, South
        Australia. The pediment is itself being eroded to form a lower
        planation surface in the foreground (photo C.D. Ollier).

slope (scarp) retreat after a period of downcutting producing a new plain
near the new base level (Figure 11.3).

An important feature of pediplains is that as the scarp retreats a new bit
of pediplain is created, and so the pediplain is a diachronic erosion surface
getting younger towards the scarp. Landscapes may consist of pediplain
remnants of various ages, giving a stepped landscape.

Etchplains

The etchplain concept, devised in the first place by E.J. Wayland (1934)
in Uganda, adds chemical weathering to the planation process. Basically
the idea is that chemical weathering rots the underlying rock, and in a
succeeding phase the saprolite (rotten rock) is eroded away, leaving a new
plain of fresh rock. One problem with the process as a 'planation' process
is that it starts off with a plain. There is much debate about the depth of
weathering, the degree and frequency of stripping, and the age of etchplains (Thomas, 1989; Ollier, 1984). Büdel (1959) proposed a variation on
the theme, called the double surface of erosion, in which the surface of fresh
rock parallels the ground surface beneath a saprolite many metres thick.
The stripping of the saprolite gives rise to a double surface, one underlain
by saprolite, the other by fresh rock. Other writers (including Ollier, 1959)

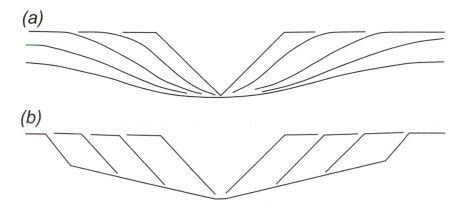

Figure 11.3 Erosion by slope decline leads to the formation of a peneplain, while
slope retreat is said to create a pediment, and eventually a pediplain.
(a) Valley widening by slope decline.
(b) Valley widening by slope retreat.

believe the weathering front (the surface between fresh rock and saprolite)
is very irregular and not parallel to the ground surface. Soil surveys have
shown that some planation surfaces cut indiscriminately across both fresh
and weathered rock (Figure 2.20), as in parts of Uganda (Ollier, 1993).

A variation on the etchplain, tied to climatic change, is described by
Schubert and Huber (1990) who wrote:

> An erosion surface is formed by the alternation of humid and dry
> climates over the surface of the Earth. During the humid periods weath-
> ering (the chemical decomposition of rocks) proceeds up to a certain
> depth. The decomposed rock, or soil, is protected from erosion by the
> vegetation cover. During the dry periods the vegetation cover dimin-
> ishes and, in many instances, disappears (as in the deserts), and its
> protective action for the soil also decreases, the soil is eroded and trans-
> ported by rivers and creeks, whose flow is not normally abundant but
> has a strong erosive effect. In this way each humid–dry cycle reduces
> the Earth's surface and gives way to ever lower erosion surfaces.

Altiplanation surfaces

While most flattening is attributed to erosion graded to a base level, it is
possible that at high altitudes frost weathering and other periglacial processes
might reduce a more rugged area to a flatter one, down to the level at
which cold climate processes work. Such surfaces are known as altiplanation surfaces. Few geomorphologists have accepted this idea, but it was

most forcibly put forward by Mackin (1947) in the Rocky Mountain region, described in Thornbury (1965).

Exhumed plains

A variant on a structural plain is the exhumed plain. This is a plain which has been formed, buried under younger sediment, and then re-exposed by removal of the younger sediment. The sub-Cambrian surface is very important in the landscape evolution of southern Sweden (Lidmar-Bergström, 1995). Sub-Cretaceous surfaces are extensive in Australia (Twidale, 1994). They are also recorded in the southern Urals.

The exhumed plain is an unconformity, but the plain was originally a planation surface. The term 'exhumed plain' should not be applied to just any geological surface, but only to demonstrable instances of exhumation. Nor should 'peneplain' be used for just any structural surface. Sweeting (1955) described a plain separating Carboniferous limestone and shale as a peneplain, but it is actually a surface within conformable marine strata, and was not formed by terrestrial erosion.

Palaeoplains

Hills (1961; 1975) described the oldest surface of which evidence remains in south-east Australia as a palaeoplain. Because we do not know the details of formation of old plains Hills suggests the non-committal term 'palaeoplain' may be preferable to peneplain or pediplain for the final product of planation. 'Palaeoplain is the term for oldlands that have been affected by later events of major kind and have been dissected so that only relics now remain.'

Plains of marine erosion

When the term 'peneplain' was first proposed, debate centred not on the details of whether plains were eroded down or back, but on whether they were really subaerial erosion surfaces or plains of marine erosion. The sea is very effective at carving plains at the coast, and with sufficient time could it not carve plains of continental dimensions? Stratigraphers were aware that most unconformities have marine sediments on top, and the plains of unconformity could most probably have been carved by the same advancing sea that deposited the marine sediments. Were the plains on land carved in the same way?

Marine erosion can only carve a plain of very limited width, for the very existence of a plain of marine erosion causes waves to lose energy and erosive power in crossing it. The only way to get a really broad plain of marine erosion is to have a slow rise of sea level accompanying coastal erosion. This is indeed what often happens in a marine transgression, and many unconformities are created in this way.

Some surfaces that are exposed plains of marine unconformity have been called peneplains. This is incorrect for a peneplain is, by definition, a plain formed by subaerial erosion. The so-called Schooley peneplain of the Appalachians appears to be a stripped sub-Cretaceous unconformity. If this is so it should be called the 'Schooley surface', or 'Schooley unconformity', but not 'Schooley peneplain'.

Argument between proponents of marine erosion and those of terrestrial planation were great in the nineteenth century, but are scarcely heard today, although the question has not always been resolved. A good account of the debate was provided by Davis as far back as 1896.

Multiple planation surfaces

In many parts of the Earth the landscape is made of plains at several levels. Figure 11.4 from central Australia clearly shows two plains: the lower one is an erosion surface of the present day; the upper plain is presumed to be an older one. High and low surfaces can be found on rocks of complex structure where erosional origin is quite definite, and sometimes a succession of several surfaces can be found.

From India, in the region behind Madras, Büdel (1959; 1982) described multiple erosion surfaces in an area entirely underlain by granite. The lower plain, the Tamilnad Plain, broadly follows the coastline and reaches inland an average of 100 km and in places 200 km. It rises to an altitude of 200 m

Figure 11.4 Double planation surfaces at Coober Pedy, central Australia (photo C.D. Ollier).

to 500 m (with a slope of no more than 2 per cent or 0.6°) and is succeeded by a steep step some 100 m in height which leads to the higher (Miocene) Bangalore Plain (750–900 m), surmounted by inselbergs up to 1500m.

The Tibet Plateau is described in Chapter 7. To summarise, this is a composite plateau, with remnants of three surfaces of different age as follows:

A 3.4 Ma
B 2.5 Ma
C 1.8 Ma

Although the overview of Tibet is a huge plateau, in detail there are many smaller, younger surfaces, and repeated etchplanation with a vertical difference of a few metres seems to be the dominant process. Individual straths seem to cut across bedrock while a free face retreats (Figure 11.5).

In the southern Atacama Desert, Chile, Mortimer (1973) recognises four episodes of landscape evolution, each of which led to the development of a regional erosion surface. The first began with the elevation of northern Chile above sea level in the Late Mesozoic. Each subsequent phase started with incision of drainage into the preceding landscape. Radiometric dating of volcanic rocks shows that the earliest elements of the topography existed in the Lower Eocene, and the landscape had formed by Upper Miocene, since when the present phase of canyon cutting commenced.

Figure 11.5 Photo of minor pediment surface in Tibet. The skyline planation surface is cut across steeply dipping strata, and a younger pediment is cutting into the old surface. Note the abrupt angle between the free face (the steep slope) and the footslope on which the man is standing (photo C.D. Ollier).

The morphological development of the north-west Dolomites shows the close interaction between tectonic processes and erosion, and according to Engelen (1963) shows the following stages:

1   After the first phase of Alpine uplift in the Lower Tertiary the area was planated by an extensive erosion level of Oligo-Miocene age. The rather widespread preservation of this oldest erosion surfaces shows the slow rate of erosion of the surface, explained by the fact that the truncated rocks are mainly limestones and dolomites with no surface drainage.
2   A second phase of uplift during the Miocene and Lower Pliocene led to downcutting of valleys by about 1000 m.
3   A new planation surface formed in the Upper Pliocene, now ranging in altitude from 1900 to 2300 m, especially well developed on the less resistant Middle Triassic strata around the reef masses.
4   A third phase of uplift in the Upper Pliocene and Quaternary was accompanied by renewed vertical river erosion of 1000–1500 m.

Multiple plains of the same age

The high plains of today are being eroded actively, but are not related to present sea level. Remnants of an early origin will be steadily removed if the process goes too far, but we do not imagine that high plains with ancient regolith are immune to further erosion and lowering.

This idea has been expressed forcibly by Partridge (1998) with reference to the high plain and coastal plain on either side of the Great Escarpment in South Africa. He wrote:

> throughout this major interval of Cretaceous erosion, land surfaces on either side of the receding escarpment were formed under the influence of two separate base levels of erosion. In the coastal hinterland, sea-level constituted the ultimate control, but inland of the Great Escarpment, the base level for erosion was provided by major river systems such as the Orange and Limpopo at their point of egress from the interior plateau through the Great Escarpment. This created the unusual situation that land surfaces of essentially the same age were cut at different levels above and below the escarpment line . . . The net result of this situation was the creation of two vast erosional bevels, above and below the Great Escarpment.

The dating of planation surfaces

Relative height

If there is a series of planation surfaces the normal age sequence is from oldest at the top to youngest at the bottom.

Dating by overlying rock

If fossiliferous sediments lie on an erosion surface then the erosion surface must be contemporaneous with, or older than, the sediment. The Pliocene Lenham Beds, for example, provide a date for a bevel on the Chalk of England.

Basalt, erupted from volcanoes, is a suitable rock for potassium argon dating and where it lies on a planation surface the age of the basalt gives a minimum age for the erosion surface. If the basalt has been eroded in the creation of a younger planation surface it provides a maximum age for the younger surface.

Some of the East African volcanoes can be related to local erosion surfaces, and it is possible to work out the relationship between volcanicity and landscape development, and also to use the radioactive age of the volcanoes to put limits on the age of the erosion surfaces. An example from the edge of the rift valley in East Africa is shown in Figure 3.7, and an example remote from broad plains between rift valleys in Figure 11.6.

Ignimbrites in the Andes are often around 6 Ma old, so the associated underlying surface is about the same age, or older.

East of Port Elizabeth in South Africa downwarped remnants of the African Surface are overlain by marine deposits of late Cretaceous and Eocene age (Partridge, 1998)

Dating by soils and weathering profiles

Field evidence sometimes shows that particular surfaces have distinctive weathering profiles, soils, ferricretes, silcretes or similar features. These may be used to correlate surfaces and may also give a relative chronology, as in northern Australia where the oldest surface has a deep weathering profile, the next planation surface is produced by partial stripping of the weathering profile down to a silicified horizon, and the youngest surface is created by stripping the weathered profile down to bedrock. Palaeomagnetic studies of ferricrete horizons may, in favourable circumstances, provide a numerical age for the time of laterite formation and this gives a minimum date for the planation itself.

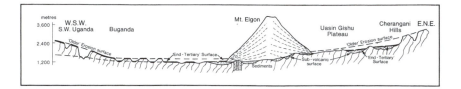

Figure 11.6 Erosion surfaces in relation to the Miocene volcano Mt Elgon, on the Uganda–Kenya border (after King *et al.*, 1972).

Dating relative to drainage patterns

The ages of erosion surfaces and drainage patterns are closely related, and can sometimes be related to stratified deposits.

For example in the Western Rift Valley of Africa, the rifting that dismembered the drainage pattern created a trough which accumulated sediments. Since the oldest sediments are Miocene, the rifting must be at least Miocene in age, and the drainage pattern older still. The plain on which the drainage pattern was formed must therefore be at least Miocene in age too (Gautier, 1965).

The ancient salt lake drainage pattern of south-west Australia dates from before the severance of Australia and Antarctica, which is Eocene (Ollier, 1977). The river system cannot be older than the planation surface it flows across, so this planation surface must be Eocene or older. This date can then be related to the Stirling Ranges which rise above the planation surface.

Correlation with offshore or basin sediments

As the land is eroded to form a plain, the eroded material is deposited elsewhere. This may be in lake sediments, and some of the planation surfaces of Tibet have been traced to Pliocene lake deposits. Attempts have been made in many places including South Africa and eastern Australia to relate planation surfaces to offshore marine deposits. Sometimes even deep ocean deposits have been used, as in attempts to date the uplift of the Himalayas, and hence their precursor planation surface, by sediments in the deep Bengal Fan.

Fission track ages

Fission tracks have been used in many parts of the world in an attempt to date planation surfaces. For instance, Okaya *et al.* (1997) wrote, 'For northern Bolivia, fission track data . . . document an increase in uplift of the Eastern Cordillera since ca. 3 Ma i.e. when the Subandean crust was thrust to the west beneath the Eastern Cordillera.' (The date might be right, but the thrust under the cordillera seems to be pure conjecture and dated by the same measurement.)

Unfortunately fission track data, and other physical methods such as vitrinite reflectance, often give results in conflict with geomorphic evidence. This is especially so in studies of passive margin uplifts. No doubt in time the problems will be sorted out.

Tectonic deformation of planation surfaces

Planation surfaces are related to tectonics in two ways. First, the initiation of new erosion surfaces requires uplift and rejuvenation of valleys before the new surface can be graded to the lower level. A sequence of planation surfaces, carefully interpreted, may provide a history of uplift movements.

Second, planation surfaces may show directly what tectonic movements have occurred since the planation surface was formed. Planation surfaces are approximately flat when first formed and if they are warped or arched then tectonic movement post-dates the planation. In eastern Australia, for example, there is a gentle rise from central Australia, an arching over the Highlands, and a steeper slope towards the east coast. In southern Africa the planation surfaces rise from the centre of the continent towards the rim, and arch over and descend more steeply to the coast. In East Africa the planation surfaces show a rise towards the rift valleys, with the highest part of the arch about 50 km from the rift valley faults. Examination of such planation surfaces enables the geomorphologist to locate axes of uplift and work out the amount of uplift and perhaps even the date. The movements that tilt and warp planation surfaces will also affect the direction of drainage, and the study of planation surfaces is usually supplemented by drainage pattern analysis.

The uplift of a plain to form a plateau can produce several forms, depending on what happens on the edge of the uplifted area.

1   Broad warp, with gentle slopes. This is essentially the *cymatogen* of King (1962). On a large scale it is epeirogeny.
2   Horst, or fault-bounded plateau. The plain is uplifted while remaining largely horizontal, and is bounded by steep faults.
3   Plateaus bounded by monoclines. Between a fault and a low-angle warp there is a steeper slope or monocline. The Natal Monocline, bounding the Highveld (high plateau) in South Africa is a good example.
4   Tilt blocks. A surface may be uplifted asymmetrically, with no uplift on one side and a fault on the other, uplifted side. The Sierra Nevada of California is a good example. Ruwenzori (in part) is another example. Many faults can give a composite or mosaic of tilt blocks, as in the Basin and Range Province of the United States.

Significance of planation surfaces

Planation surfaces mark a relatively confined point in time, which in itself is a useful adjunct to stratigraphy and other ways of approaching geological history. They are particularly useful in providing a starting point for later events. As Coltorti and Pieruccini (2000) wrote:

> The Planation Surface levelled all the topographic contrasts and therefore constitutes a starting point for the following evolution of the landscape of the Apennines. It is a key tool to recognise the neotectonic structures and to differentiate between pre-planation and post-planation activity.

In Chapter 12 we stress the importance of distinguishing between pre-planation and post-planation gravity tectonics.

Planation surfaces become particularly significant if they have very wide-spread, even global, extent, suggesting major events in Earth history. The 'global' planation surfaces of King would be one example. Coltorti and Pieruccini (2000) suggest that the Planation surface of the Apennines is related to a single global event, the late Lower Pliocene transgression, which relates to a general marine transgression that operated very quickly.

General surface lowering

Can a planation surface be lowered by weathering and erosion after uplift? If so it might mean that we are not seeing the original planation surface, and depending on the degree of lowering we might have wrong estimates of the amount of uplift.

In considering the evolution of landscapes it is often loosely said that weathering will cause general surface lowering. In most landscapes this is unlikely. Where weathering profiles are deep, it is found that constant volume alteration features are present to within a metre or so of the top of the weathered rock (saprolite). If the volume remains the same there can be no general surface lowering due to weathering. Lowering can only be by erosion. Only in places where the rock is soluble, as in limestone or dolomite regions, is weathering likely to cause general surface lowering, and proven examples are rare.

General landsurface lowering is by erosion, not weathering. Most fluvial erosion is concentrated in river valleys, which should destroy flat surfaces, and not maintain flat surfaces as they are lowered. Only by slope retreat and pedimentation can a surface be extended in a mainly horizontal way, at a high altitude, for the retreating slope is graded only to its pediment, and not to some distant base level.

Schubert and Huber (1990) propose a complicated scenario, with repeated planation by general surface lowering, and also scarp retreat through Phanerozoic time. Their scheme is shown in Figure 2.5, and the caption explains their ideas.

Drainage patterns and their interpretation

In various places in this book we have described drainage and used such terms as 'antecedent' and 'superimposed drainage'. Here we give a fuller and more systematic account of drainage patterns and their interpretation.

An understanding of drainage patterns is essential to decipher the relationship between tectonic movement and landscape evolution. Since the study of river patterns, once a main theme of geomorphology, is now unfashionable, even most geomorphologists do not understand drainage patterns very well. Most geologists and geophysicists have only a rudimentary knowledge. Even the very idea that drainage patterns might be older than erosion surfaces and mountain ranges is news to many. It is absolutely essential in the study of mountains to understand superimposed and antecedent drainage, and to comprehend not only how river capture really works but what ancillary features should be sought to validate it.

Simple patterns

There are simpler aspects of drainage pattern interpretation to be learned before the more spectacular drainage modifications can be understood.

Simple river systems have a dendritic pattern with tributaries joining a main stream at acute angles that point downstream (Figure 11.7). Tributaries have steeper gradients than the main stream, but their junctions are at precisely the same elevation (Figure 11.8). This relationship is known as 'Playfair's Law'. Simple valleys increase in width and become flatter in the downstream direction. If these conditions are not met we may assume that a valley has been affected by some complicating factor of structure, tectonics or geomorphic history.

Structural control of drainage patterns

Drainage patterns on horizontal sedimentary rock

Approximately horizontal sedimentary rocks tend to form plateaus, plains and stepped topography, with harder rocks making the flats and softer rock being stripped away. The simple dendritic pattern can be well-maintained

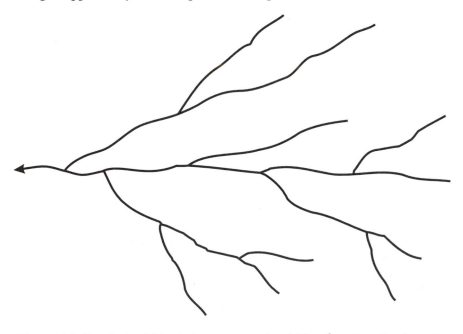

Figure 11.7 Simple dendritic drainage pattern in which tributaries join the main stream at acute angles.

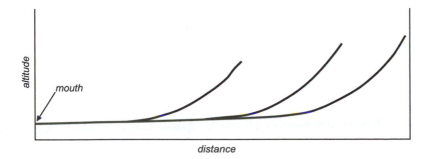

altitude

mouth

distance

Figure 11.8 Long profile of main stream and tributaries. Tributaries are steeper than the main stream, and normally join at just the right level (Playfair's Law), without waterfalls or exceptionally deep pools.

in most instances, but is likely to be modified by joints in the harder rocks. These may form an angular pattern, commonly with two sets of dominant joint direction almost at right angles.

Drainage patterns on dipping sedimentary rocks

In a series of dipping sedimentary rocks, the soft rocks such as clays and shales are eroded faster than the harder rocks, so a pattern with rivers running along the strike is very common. Strike valleys run in the strike direction, and are separated from each other by strike ridges on harder rock.

Tributaries to strike valleys commonly enter almost at right angles. Streams that flow down the dip slopes of the hard beds are called 'dip streams'; those that flow in the opposite direction are called 'anti-dip streams' (Figure 11.9). When the dip is not very high the dip streams will be longer than the anti-dip streams, as they flow on longer slopes. The anti-dip streams also have a greater drainage density. This results in a drainage pattern consisting of parallel strike streams with relatively long and widely spaced dip streams coming in on one side, and relatively short but closely spaced streams on the other. Such a pattern not only clearly indicates the effect of tilted sedimentary rocks, but shows the direction of dip, gives the location of more and less permeable strata, and even provides a rough idea of the amount of dip. Such a pattern (Figure 11.10) is known as a trellis pattern.

Even if the dip of the rock is such that the ridges are roughly symmetrical, commonly known as hogback ridges, the direction of dip can generally be deduced from the drainage density being greater on the anti-dip slopes. Contour maps also show that the anti-dip slope is usually more scalloped and embayed.

An older nomenclature for the streams of a trellis pattern is shown in Figure 11.11. This is established in many textbooks but should be discontinued because it builds into the nomenclature a presumed (and erroneous)

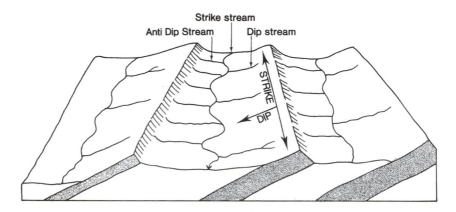

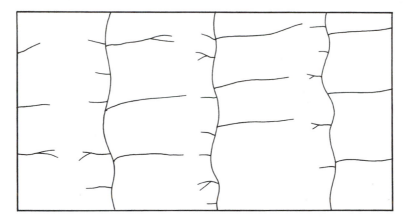

Figure 11.9 Block diagram and map showing the structural relationships of
            streams on dipping strata, and the preferred nomenclature of streams.

time-sequence for the origin of the various streams. It was thought that a
main stream flowed down an initial surface on uplifted folded strata, and
this was called a 'consequent stream', being a consequence of the initial
dip. The strike streams were thought to be developed subsequently to the
consequent streams, so were called 'subsequent streams'. Once these had
carved valleys, further streams would form on the new valley sides; those
parallel to the consequent stream were called 'secondary consequents'; those
in the opposite direction were 'obsequents' (anti-dip streams). In reality the
whole of any land area must be drained, and runoff and initiation of drainage
affect all parts of the landsurface right from the beginning. No part of the
land area awaits the establishment of main drainage before becoming drained
itself, and 'subsequent' is misleading, as well as being less descriptive than
'strike' valley. The early history of a drainage pattern may be difficult to

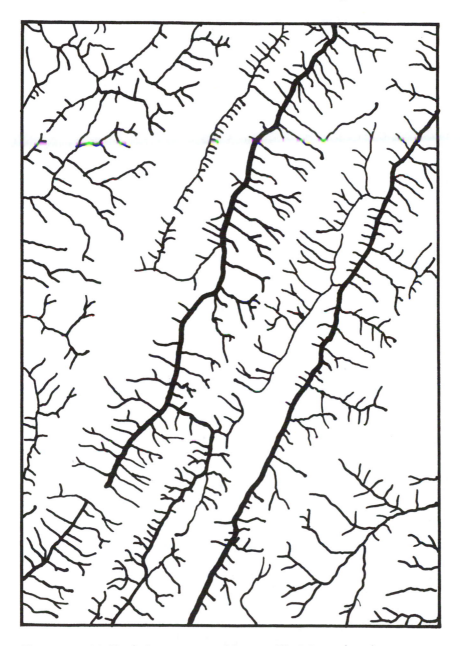

Figure 11.10 Trellis drainage pattern. Montery, Virginia quadrangle.

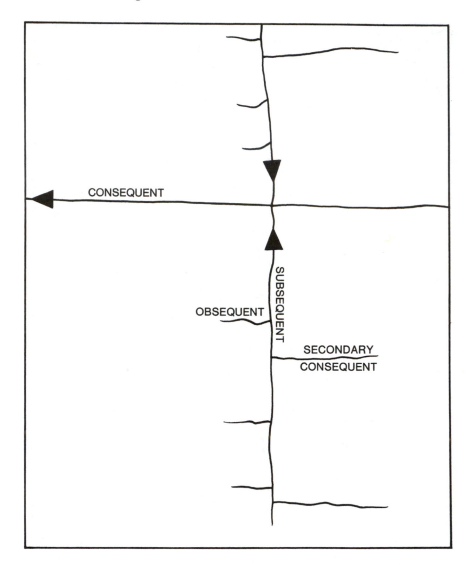

Figure 11.11 Outdated nomenclature for structurally controlled streams in regions of dipping strata.

work out, but the structural relationships are clear, so structural terms for drainage pattern description are to be preferred.

Drainage patterns on folded sedimentary rocks

A fold consists of two limbs, each of which has a simple tilt. On each limb there will be a trellis drainage pattern, but the directions of the dip and anti-dip streams will be different on opposite limbs. This results in a trellis drainage pattern that is symmetrical about the fold axis. An anticline has in-facing escarpments, so the dip streams on opposite limbs flow away from the fold axis, and the anti-dip streams flow towards it. Thus the trellis pattern not only indicates details of hard and soft beds in each limb, as outlined earlier, but clearly indicates the position of the fold axis.

A succession of folds simply provides a succession of tilted beds alternately dipping in one direction and then in the opposite direction.

If fold axes were horizontal the strike streams on opposite limbs of a fold would be parallel, but this is not common. Folds plunge, and the strata swing round the nose of the plunging fold to produce curves in the strike ridges and valleys. Each limb of a plunging fold has a simple trellis drainage pattern, but this trellis pattern is bent around the nose of the plunging fold. This distorted trellis pattern therefore not only indicates dips, strike and fold axis position, but also shows the direction of plunge. The folded rocks of the Appalachian Mountains provide excellent examples of this drainage pattern, which is sometimes known as Appalachian-type drainage.

In areas of folded rocks many valleys are formed along softer sediments and the country has a distinct 'grain' parallel to the strike. On such a grainy basis we can expect overall drainage patterns to have some major stream directions resulting from the original slope of the general ground surface, but the strike of the rock will show through in many tributaries. The drainage patterns of the Upper Lachlan and the Upper Murrumbidgee illustrate the effect of a general slope to the north west in a region of uniform strike to the north (Figure 11.12).

It is important to notice rivers that are *not* controlled by structure. River courses may well be inherited from an old planation surface and have persisted in their course while minor tributaries became increasingly controlled by structure. The Appalachians provide fine examples, such as the Susquehanna River (Figure 11.13).

Modifications of river valley landscapes

Fluvial modifications

From the moment they are initiated river valleys continue to be modified by the action of river erosion. There is downcutting (vertical erosion) and valley widening by slope processes and lateral erosion by the river. Downcutting may lead in turn to extension of the river upstream, that is headward erosion. The river cannot cut below the level of its mouth, which is known as 'base level': the sea is the ultimate base level, but temporary base levels of erosion may be created at rock barriers or lakes. Temporary base levels tend to be short lived, but even the ultimate base level of the sea is varied by eustatic changes in sea level or by tectonic movements.

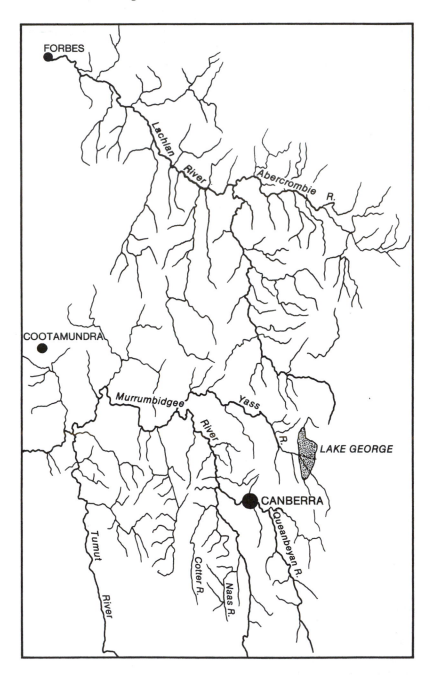

Figure 11.12 Drainage pattern of the Upper Lachlan and Upper Murrumbidgee
Rivers (SE Australia) showing the 'grain' produced by the regional
northerly strike of the bedrock in an area sloping to the north
west.

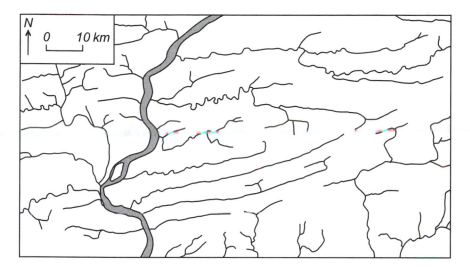

Figure 11.13 Drainage in the Harrisburg region of the Appalachians. The Susquehana River (shaded) has a course that is unrelated to structure, probably superimposed. The tributary streams are strongly controlled by structure, and it is possible to see the location of fold limbs and axes from the drainage pattern alone.

When vertical erosion is reduced (by nearness to base level) valley widening occurs, that is lateral erosion takes over. If vertical erosion increases again (because of uplift of the land or for any other reason) the river is said to be rejuvenated: rejuvenation simply means increase in vertical erosion. Tectonic uplift is likely to cause rejuvenation, but caution is needed in interpretation because such things as climatic change, chance exposure of softer sediments, marine erosion at a river mouth, or eustatic changes in sea level can all have the same effect.

Alternating vertical and lateral erosion may lead to the formation of step-like landforms on valley sides, known as 'river terraces'. Some terraces (known as 'straths') are rock cut, others are formed of thick alluvium laid down when a river is aggrading, that is building up its bed. Terraces are first formed at river level, as flood plains, and only become terraces when vertical erosion cuts through them. They are valuable because they give indications of former river gradients, and successive terraces show successive river positions. In areas of tectonic activity river terraces may be warped in various ways, giving valuable clues for interpretation of tectonic movement.

River capture

River capture is the name given to change in river pattern when a stream which is cutting down rapidly extends headwards into the catchment of a

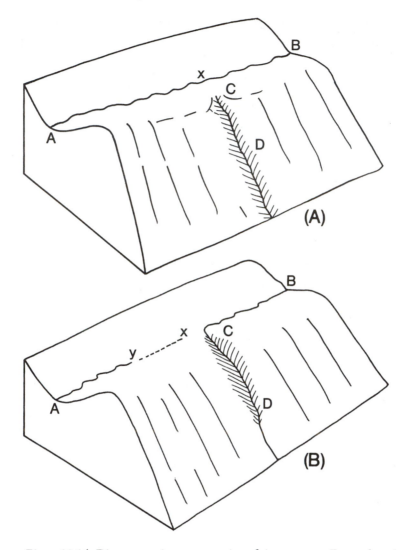

Figure 11.14 Diagrammatic representation of river capture. For explanation see text.

stream which is cutting down only slowly. Thus in Figure 11.14A, stream CD is steep and is eroding downwards and headwards quite quickly. Stream AB has a gentle gradient and is cutting down only slowly. Figure 11.14B shows the result of capture. Stream CD has captured the waters of river AB at x. Stream Ax is reduced in volume, and the waters of Bx are added to CD increasing its vertical erosion even more. The sharp change of direction at x is called the 'elbow of capture', and the sudden increase in gradient at x is called a 'knick point'. The knick point may eventually migrate upstream.

The old section of stream AB at y no longer carries water, and is known as a wind gap. Old river gravels and alluvium mark the former course of the river at y, and if AB was a complex valley before capture, the section y may even have old river terraces or other fluvial features.

Amateur geomorphologists have often heard of 'river capture' even if they do not know any other geomorphic terms, and the term is often bandied about as a loose 'explanation' of drainage anomalies. In reality it requires rather special conditions to provide the juxtaposition of a rapidly eroding and a slowly eroding river. If river capture is thought possible, a search should be made for the following questions:

1   What is the reason for the capturing river to be eroding faster than the captured river?
2   Is there an elbow of capture?
3   Is there a knick point at the elbow of capture, or upstream of the elbow if the capture is old?
4   Can the old course of the captured river be detected by features such as an old channel, a wind gap, old fluvial sediments?

It is quite common for an erodable stratum or structure to give rise to successive river captures An often described example comes from Yorkshire, where the Ouse has successively captured the Nidd, Ure and Swale (Figure 11.15).

Changes in divide location

Divides are not fixed in position for all time. The divide is the intersection of slopes from adjacent valleys, and if one valley erodes faster than the other, the divide will slowly migrate. This slow migration is *creeping*.

Some theoreticians speculating about landscape evolution have suggested that a tectonic axis may slowly migrate, and be reflected in changing positions of divide. Mazzanti and Trevisan (1978) thought such a wave was crossing the Apennines in Italy, as explained in Chapter 4. Such a tectonic wave moving across topography is unlikely to have a direct effect, but by changing the steepness of rivers it may have secondary effect. In reality there seems to be little evidence for such tectonic waves.

When river capture occurs the divide suddenly moves to a new position, as shown in Figure 11.16. The first diagram shows the original drainage, with two north-flowing streams, A and B. The second diagram shows the drainage after capture of the headwaters of stream A by a tributary of stream B. There will be a water gap, elbow of capture, and beheaded stream. There

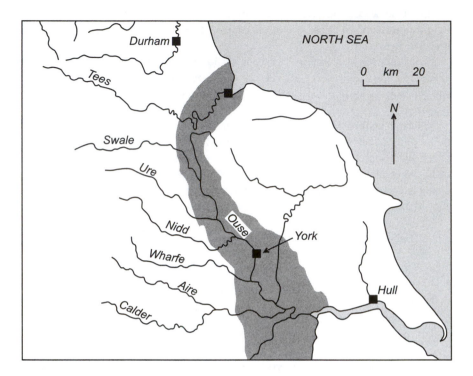

Figure 11.15 Successive river captures. The Ouse, eroding along the outcrop of
soft Triassic sediments (stippled) has captured the Nidd, Ure and
Swale.

will also be a new divide, as shown, between the old headwaters of stream
A and its beheaded continuation. At the moment of capture the old divide
has jumped from position a to position b: it is a *leaping* divide. The capture
adds a new area to the catchment of stream B, so it will probably erode
fast, and compete even more with stream A.

Although this concept is very necessary in the study of the evolution of
divides it is seldom treated in modern geomorphology texts. A good account
is provided by Cotton (1918).

Superimposed drainage

Suppose a river is flowing across a plain that has been cut across varied
rocks, and then starts to cut down vertically. The river will cut down to
the new base level, and with its tributaries will excavate its drainage basin
towards the new base level. Softer rocks such as clays and shales will be
removed readily, but any harder rocks may resist reduction to the new level.
However, the main river may retain its course where this accidentally crosses

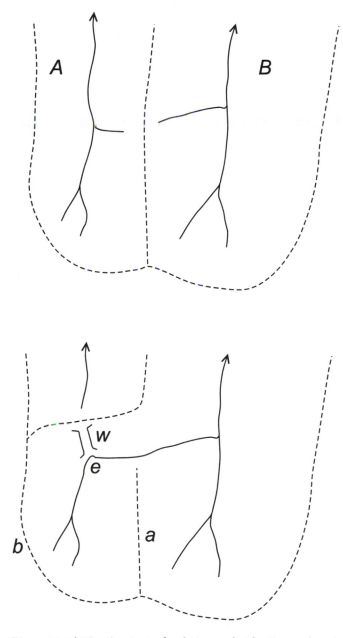

Figure 11.16 The 'leaping' of a drainage divide. For explanation see text.

a band of hard rock. When much of the area has been reduced to the new lower level the river course will have gorges where it goes through hard rock. The river course may seem incongruous, as it leaves a wide open plain to plunge into a gorge.

For example the Bristol Avon flows across a broad plain that continues out to sea, but the river leaves the lowland and takes a remarkable course through a ridge of carboniferous limestone to form the Clifton Gorge. This river course is inherited from a higher planation surface, and is an example of superimposed drainage.

Some superimposed drainage is initiated on a cover of rocks over a more varied assemblage of basement rocks, but this is not essential and some superimposed drainage originated on a plain cut directly across complex basement rocks.

Antecedent drainage

Another way to get remarkable gorges is for tectonic uplift to raise a strip of highland across a river by warping or faulting. If the uplift is sufficiently slow for river erosion to keep pace the river will maintain its course. This is called 'antecedent drainage'.

One of the best indications of antecedent drainage is if the ridge that the river crosses is higher than the source of the river. The Fitzroy River of Queensland rises some considerable distance from the coast and crosses a range of high coastal mountains. How does the Fitzroy cross ranges that are higher than its source? River capture of the inland drainage by a headward-eroding coastal river seems improbable, as there is no suitable soft rock in appropriate places, and none of the features such as wind caps or elbow of capture. More probably the Fitzroy is antecedent where it crosses the ranges. Similarly in the Burdekin system long strike rivers unite and these flow across a gorge in the Leichhardt Range (said to be impassable to four-footed animals) to the coastal plains. Again antecedence seems to be the best way to account for a river that takes a course through an impassable gorge across a mountain range higher than the source area.

Examples of antecedence described earlier in this book include the Himalayas (p. 132), Papua New Guinea (p. 55) and various ranges in the western United States (p. 104).

Reversed drainage

The normal pattern of rivers is dendritic, with tributaries joining a main stream at acute angles that point downstream. The tributaries also have steeper gradients than the main stream. If the whole area is tilted back, the main stream can be reversed so that it flows in the opposite direction to its old course along the same stream bed. The tributaries, being steeper, are not affected by the tilting and continue to flow in the same direction.

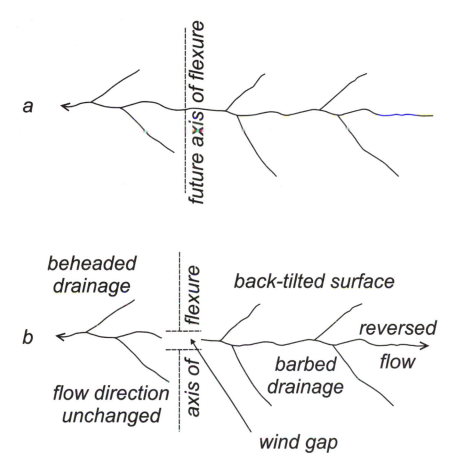

Figure 11.17 The formation of barbed drainage.

    a.   The top figure shows a prior condition of a simple dendritic stream flowing over an incipient axis of flexure.

    b.   After backtilting of the headwaters, or possibly upwarp of the flexure axis, the main channel upstream is reversed, but the steeper tributaries continue in their old direction, producing barbed drainage.

The drainage pattern is then said to be barbed: the tributaries still join the main stream at acute angles, but the angles point upstream relative to the main stream. If a warp occurs across a river, the downstream continues to flow in the same direction, with the same pattern. Only the flow has been reduced. The upstream part has reversal of the main stream, but the tributaries flow the same way so barbed drainage is produced (Figure 11.17). At the turnover point there will be remains of a broad, wide valley. This commonly has alluvium, but being very flat it is often poorly drained and may have swamps, even on what is now a watershed.

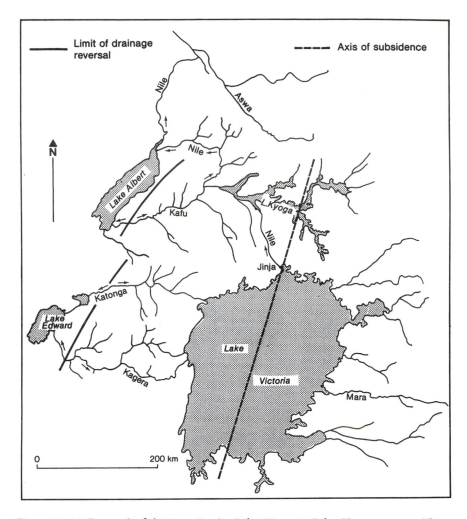

Figure 11.18 Reversal of drainage in the Lake Victoria–Lake Kyoga system. The original drainage was from east to west, with the Mara continuous with the Kagera. The drainage lines of the Katonga and Kagera are continuous between Lake Victoria and Lake Edward, as is the Kafu from Lake Kyoga to Lake Albert. Much of the drainage of these rivers is now reversed as shown by the arrows. The middle course of the Mara–Kagera river was drowned when downwarping created the Lake Victoria basin. Lake Victoria overflowed at the lowest point on its watershed at Jinja to form the stretch of the Nile between Lakes Victoria and Kyoga. Lake Kyoga was formed by similar backtilting of the Kafu River. It flowed up a northern tributary and overflowed into the Albert rift valley, forming the stretch of the Nile between Kyoga and Albert. From Lake Albert the Nile flows north along a continuation of the rift until it joins the Aswa which follows a major Precambrian mylonite band. Note the barbed drainage pattern of the reversed rivers.

The Kafu River in Uganda provides a good example (Figure 11.18). It used to flow from east to west, but has been reversed by relative uplift of the shoulder of the Lake Albert Rift Valley and, now flows to the east, with barbed tributaries maintaining their old direction. A small stretch of the Kafu beyond the axis of uplift retains the old direction, and there is a broad swampy valley between the old course and the headwaters of the reversed course. The original course of the valley must be older than the rift valley faulting.

Another good example is provided by the Clarence River in northern New South Wales. The most obvious feature is the barbed drainage on the southern side (Figure 11.19). (The northern side is ignored here because the drainage pattern is complicated by volcanic activity.) The headwaters (known as the Orara River) flow normally with simple dendritic drainage, and beyond the Great Divide the Condamine, which appears to be a former continuation of the Clarence, also has simple dendritic drainage. But between the Great Divide and the Orara junction all tributaries are barbed, suggesting reversal. This fits in with coastward tilting associated with the tectonics of the continental margin. Between the Condamine and the Clarence the old valley course can be traced, with gravels, right across the Great Divide of eastern Australia (sometimes incorrectly called the Great Dividing Range).

A different approach to reversal is presented by Grabert (1971). He believed that the Brazilian Shield, with its thick weathering cover, had been exposed since the Triassic or earlier, and from it drainage flowed towards surrounding seas. At that time the west-flowing rivers from the Shield would have flowed to the Andean depositional trough or geosyncline. But the Tertiary uplift of the Andes reversed the pattern and caused the Amazon to flow eastwards (along a structural low) and the Parana to flow southwards subparallel to the strike of the Andes. A modern and more detailed account of the evolution of major rivers in South America is provided by Potter (1998).

The cause of reversal is interesting. It is commonly assumed that uplift of a tectonic barrier can cause reversal, but Ollier (1992) pointed out that large rivers usually keep pace with uplift making antecedent gorges. To reverse rivers it is probably necessary to backtilt them. Uplift and back-tilting are not just relative terms that can refer to the same movement. The long profile of the river must actually be tilted. If this is so, it is the sagging of the Lake Victoria–Lake Kyoga axis that causes reversal of rivers to the west, not uplift of the Rift Valley shoulder. In the Clarence drainage it is lowering of land towards the east rather than uplift of the Great Divide that caused drainage reversal.

Radial drainage

Radial drainage is common on volcanic cones, which is generally an unremarkable feature. Radial drainage is also present on some very young gneiss-mantled domes, as described on p. 186.

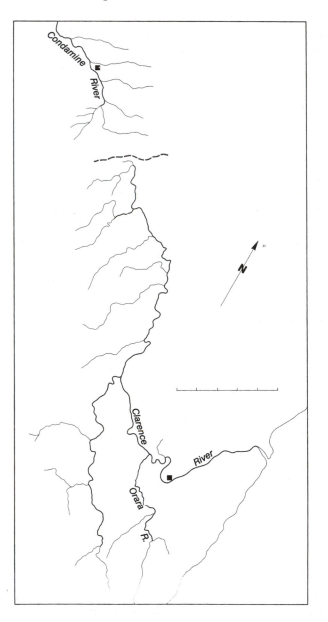

Figure 11.19 Barbed tributaries of the Clarence River, New South Wales. Only
            southern tributaries are shown as the northern side is complicated
            by volcanic derangement of drainage. Beyond the Great Divide the
            Condamine River was a continuation of the drainage before reversal
            of the Clarence. The Orara in the south, and the parallel tributary,
            the Mitchell, still has its original northerly direction. The course of
            the Clarence between the normal drainage of the Orara and the
            reversed Clarence flows along a tectonic trough to the sea.

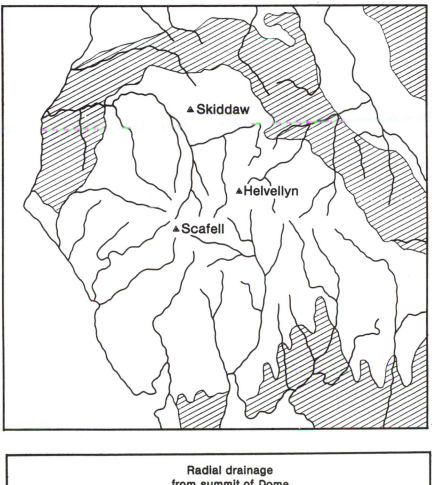

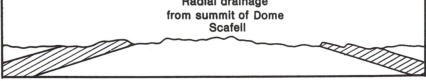

Figure 11.20 The dome of the Lake District, northern England.
Top. Radial drainage, and very simplified geology. The shaded area represents carboniferous limestone.
Bottom. Cross-section of the dome.

Figure 11.21 The original drainage pattern of Wales, as interpreted by Brown (1960). It is a partial radial system, centred on a high point in northern Wales.

Domes caused by broad uplift may also develop radial drainage. A classic example is the Lake District of northern England (Figure 11.20). Most authorities suggest that there was a Cretaceous cover, or at least a Cretaceous erosion surface before domal uplift, and the present drainage is superimposed from that dome. Certainly the radial drainage pattern has nothing to do with the complex bedrock structures of the region. Presumably a totally new dome arose from the sea, on which a radial drainage could be initiated, and then superimposed on the complex rocks beneath.

A partial radial pattern is discernible in the reconstructed ancient drainage of Wales (Figure 11.21), as deciphered by Brown (1960). Perhaps this had a Cretaceous cover too, for there is apparently no trace of earlier drainage.

Some geophysicists and geologists, such as Cox (1989) and Cope (1994), present scenarios in which a dome is made (by a mantle plume) and a radial drainage develops related entirely to this new uplift. In most instances there was an earlier drainage pattern before the doming, and normally some of the old pattern will be inherited. Totally new drainage patterns are unlikely to be found on domes imposed on pre-existing landforms.

Effects of faulting

Faulting produces fairly obvious effects on drainage. Figure 11.22 shows the effects of faults on west-flowing streams. At (1) we have beheaded streams that continue to flow in their original direction. East of the fault the drainage is blocked and may form a fault-angle lake (2) and flow out along the fault-angle depression (3). The former continuation of the river is now a beheaded river (4). A west-facing fault will give rise to a fault-angle lake only if the downthrown block is also backtilted, with reversal of drainage (5). It is possible for river erosion to keep pace with the uplift of a tilt block or horst if uplift is sufficiently slow, and as in the same situation with upwarps the result is antecedent drainage: the early formed river maintains its course across the later tectonic topography. The antecedent river usually flows in a gorge (6).

In New South Wales, Lake George and its associated features appear to provide a simple and elegant example of drainage strongly modified by faulting. The Lake George Fault cuts across drainage lines, which must therefore be older than the fault (Figure 11.23). The old Taylors Creek used to flow right across the area from east to west, but is now dammed back to make a lake in the fault-angle depression. To the north a number of creeks that formerly flowed north west are diverted by the fault, and after barbed junctions they flow along the base of the fault scarp into Lake George. To the south of Lake George the fault crosses the Molonglo River. This has a much bigger catchment than the northern rivers and was able to maintain its course across the rising fault block, cutting the Molonglo Gorge, an example of antecedent drainage. It is unusual to get three different responses to faulting – *diversion*, *damming* and *antecedence* – in such a small area.

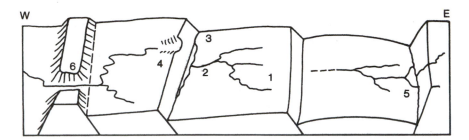

Figure 11.22 Some effects of faulting on drainage systems. For explanation see
text. West is to the left.

The Molonglo Gorge is by no means mountainous, but it illustrates principles that can be found in much larger areas of fault block mountains,
such as the drainage across the Owen Stanley Range at Milne Bay (p. 55).

Summary of drainage pattern relationships

In this chapter we have seen that the simplest pattern of drainage is the
dendritic pattern. This can be modified by differential erosion picking out
rock structure, and can be modified by warping or faulting. Tectonic changes
produce a whole range of minor landforms and associated drainage modifications. If drainage pattern were the only indication of tectonics any
hypothesis of modification would be tentative, but when long profiles, terrace
history, overflows, wind gaps, alluvial areas and so on all build up a consistent story we can use drainage pattern data with confidence to deduce
tectonic movements.

Major drainage patterns are on the same time scale as global tectonics,
and often pre-date the formation of rift valleys, mountain ranges or continental margins.

Mountain building and climate change

The formation of mountains affects climate at both a local or regional scale,
and on the global scale.

Regional climatic effects

The main local effects are changing temperatures by simple elevation,
inducement of orographic precipitation, production of rain shadows, and
effects on wind circulation.

The Sierra Nevada is a tilted fault-block mountain range, about 100 km
by 600 km, high on the eastern fault scarp edge and sloping gently to the
west. Axelrod (1962) used biogeographical evidence to determine the time

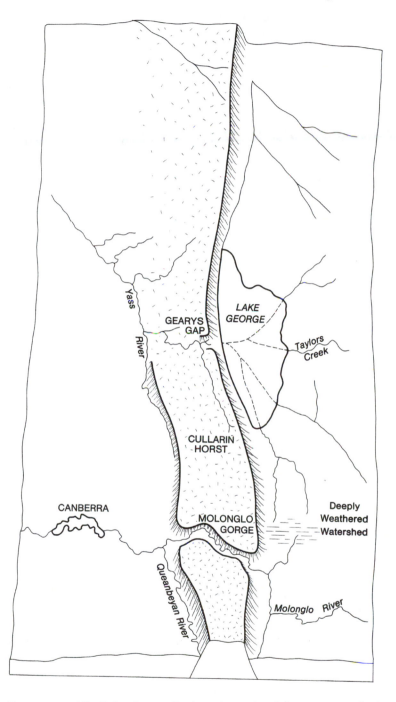

Figure 11.23 The Lake George Fault and associated features, near Canberra, Australia. For interpretation see text.

of uplift of the Sierra Nevada. At present the Owens Valley, east of the Sierra Nevada, has a very different climate from the lowlands to the west of the range, a climate modified by the existence of the high mountains between. In the early Pleistocene a similar vegetation, a pine–fir ecotone, was established right across the whole region, a situation that could only exist if the present climatic barriers were absent, so the major uplift is well into the Pleistocene.

Aspect is important in climatic and erosional effects on large volcanoes. The north-eastern slopes of the larger Hawaiian islands may be incised by deep canyons because of the high rainfall, whereas leeward slopes remain relatively undissected. The rate of erosion on a volcano may be reduced by the growth of another volcano to the windward, as happened when the Koolau dome cut off the trade winds from the Waianae dome (Ollier, 1988).

Simple uplift alone causes cooling. Partridge (1998) wrote: 'Uplift of 1000 m [in southern Africa] is equivalent, in its effects on surface temperatures, to the cooling experienced during an Ice Age in higher latitudes.' He thinks the close coincidence of South African uplift with a global period of cooling and increasing aridity between 2.8 and 2.6 Ma would have amplified its effects. Among the landscape responses to this were the creation for the first time of the Kalahari Desert, expansion of grasslands, and the turnover in antelope and other faunas that occurred in this area in the Late Pliocene.

The Quaternary uplift of the Tibet Plateau and the Himalayas introduced a powerful new geographical factor into the pattern of climate. In the Early and Middle Pleistocene when the average elevation of the Himalayan range was about 4400 m, the evidence from interglacial deposits shows that the north side was as warm as the south side at similar elevations. However, the uplift of the range to its present 6000 m average elevation made the Himalayas a much more effective climatic barrier, preventing warm, moist air from entering the Tibetan Plateau. Upheaval of the plateau essentially created the monsoons of east and south Asia (Manabe and Terpstra, 1974). Once monsoon conditions existed, exclusion of the Indian monsoon made the north side of the plateau colder and drier and this restricted the growth of glaciers.

Liu Tunsheng and Zhongli Ding (1998) described the palaeomonsoon as determined from the loess record. In the past 2.6 million years the palaeomonsoon record can be divided into 166 events. The monsoonal rainfall belt has experienced a wide, repeated advance–retreat change during the glacial–interglacial cycles of the Pleistocene. Both temporal and spatial changes of the monsoon system in the Quaternary can be linked closely to global ice-volume relationships. It therefore seems that although uplift of the Tibet Plateau may be vitally concerned with the onset of the monsoon, minor changes are controlled by orbital variations.

During the Pleistocene, climatic changes in high latitudes are characterised by repeated growth and decays of vast ice sheets. During the last glacial maximum the total ice volume was almost tripled and ice-covered area increased by over 2.5 times relative to the present. The additional ice cover in the last glacial maximum was mostly in the northern hemisphere, with less than 3 per cent in the southern hemisphere. In the northern hemisphere the additional growth of ice occurred mainly in North America and north and west Eurasia,. Expansion of sea ice occurred particularly in the Arctic Ocean. Although there are advocates for extensive glaciation in Tibet, most workers think ice cover was relatively small, which is a result of the dryness, and the cold dry climate was highly conducive to the formation of the observed huge volumes of loess.

Liu Tunsheng and Zhongli Ding (1998) also stress the important feature that the monsoon links the low pressure cell over the Tibet Plateau (India Low) to both the Pacific High and the Australia High, leading to inter-hemispheric temperature exchange, and so plays a part in global climate changes:

> Because the winter monsoon originates in the high latitudes of the Northern Hemisphere and can transfer climatic signals to low latitudes and even across the equator, we speculate that it may have played a part in the inter-hemispheric connection of climatic changes.

Global climatic effects

Global climate changes on many time scales. At present we are living in an interglacial within an ice age that extends through the Pleistocene and into the Upper Cenozoic. The Lower Cenozoic and the Mesozoic were generally warm periods. Many factors control climate. Astronomical causes (the Milankovich cycles) seem to account for much of the glacial–interglacial change. But what caused the ice age in the first place? Perhaps mountain building is responsible. Morner (1992), after summarising much earlier work, wrote, 'an important neotectonic–paleoclimatic linkage is hereby advocated'. Zhu (1997) suggested a coupled climatic–tectonic system.

Ruddiman and colleagues, especially Raymo and Kutzbach, have been major proponents of the mountain building control of climate, through a series of papers and books (see Ruddiman, 1997). Originally they suggested that uplift of the Tibet Plateau and the Andes led to the onset of global cooling. They maintain that uplift of Tibet and the Andes started about 40 million years ago and increased in the Tertiary, with most uplift in the last 10 million years. They go on to argue that the increase in mountain uplift caused a change in climate, and they propose a mechanism to bring it about. We think their basic argument – that mountain uplift caused global cooling – is correct, but we have problems with their proposed mechanism and their time scale.

Very simply, their proposal is that increased uplift leads to greater erosion and weathering. Weathering of rocks is by carbonation and removes carbon dioxide from the atmosphere. The reduction in carbon dioxide causes a negative greenhouse effect and leads to global cooling, and eventually to glaciation. Bicarbonates are taken to sea and precipitated by organisms to make limestone.

There appear to be a number of misconceptions and oversimplifications in this interrelationship of erosion, weathering and carbon dioxide.

First, erosion and weathering are two different processes that do not always occur together. Erosion can proceed, as it often does in mountainous regions, with little chemical alteration of the rock or mineral fragments that are loosened and transported. Physical weathering may be a necessary precursor to erosion, but not chemical weathering. Indeed, immature, so-called labile sediments with large amounts of unweathered mineral and lithic fragments are generally assumed by sedimentologists to be derived from the rapid erosion of mountains. In mountain regions physical weathering is much greater than chemical weathering. As expressed by Bickford, 1973: 'The abundance of rock fragments indicates that the sediments have not undergone extensive transport or been subjected to extensive weathering.'

Second, weathering is by no means restricted to carbonation, that is the interaction with CO_2, as expressed by Raymo and Ruddiman (1992) in the following simple formula:

chemical weathering
$$CaSiO + CO_2 \rightarrow CaCO_3 + SiO_2$$

This is the simple formula found in many elementary texts, but it is an oversimplification. In most situations *hydrolysis* is by far the most important process of silicate weathering (Ollier, 1984; Trescases, 1992). Although carbonation is important in the soil zone, the influence of CO_2 diminishes rapidly with depth as the CO_2 is used up. Outside areas where evapotranspiration exceeds precipitation, carbonate does not accumulate in the regolith. In brief, weathering profiles do not lock up carbon dioxide.

Bicarbonate is a useful associate of cations during the transport of weathering products to the sea, but carbon dioxide in rivers has a short residence time compared with landscape evolution. Whether carbonates are precipitated in the sea depends on many factors, but it is likely that the CO_2 is recycled through the atmosphere again.

An uplifted plateau may retain the regolith that was formed on it when it was a low level plain. It cannot be assumed that regoliths found on plateaus such as the Rocky Mountain Plateau, the East Brazil Plateau, or the Deccan Plateau were formed after the plateaus had reached their present altitude. The relationship between regolith and altitude is something to be investigated, not something to assume. In south-east Australian high plateaus, for example, dated lava flows over deep weathering profiles show they are of pre-Oligocene age.

A second problem is the effect on climate of plateaus and mountain ranges. Uplift alone will influence climate in several ways. When mountains are high enough they may start to accumulate ice which, in turn, will have many further effects on other aspects of climate and geomorphology. Mountain ranges will influence wind circulation and create orographic rain (tending to remove regolith) and rain shadows (helping to preserve regolith).

Uplift may cause more widespread, even global effects, depending on the location of the plateau. Uplift of the Antarctic Plateau may result in such massive ice accumulation that the whole world's climate and sea level is affected, but a similar plateau near the Equator may have less effect.

Ruddiman and Kutzbach (1991) may be working on a time scale that is too long, even if the principle is correct. They made a case for plateau uplift being a major cause of climatic change over the past 40 million years and especially the last 15 million years. They placed great emphasis on the plateaus of North America and Tibet, although of course there is much other high ground that could also contribute to the effect, such as the Andes and especially the Antarctic Plateau. They record the changing vegetation of the Tibetan region. At present the area supports grass and scrub; temperate forests prevailed from 5 to 10 million years ago, and tropical and sub-tropical forest before 30 million years ago. The changing vegetation sequence might imply a long term uplift of the Tibet Plateau, but some of the change to drier conditions might be caused by a rain shadow effect of the even higher Himalayas. This scheme seems to lack the fine detail of the Chinese workers described in Chapter 7, which also has a shorter time scale of 2 to 3 million years.

Although Raymo et al. (1988) suggest that mountain uplift would cause cooling by increasing rates of carbonation weathering and falling atmospheric levels of CO_2 (greenhouse cooling), the real world situation seems to be the exact opposite. The greatest deep weathering profiles all around the world seem to be of Mesozoic or Early Tertiary age. These ages do indeed correspond to a time of enhanced limestone deposition. But this was a period of broad plains with very little mountain building on a global scale. Since then uplifted plateaus have been associated with stripping of the old regolith to make etchplains (Summerfield and Thomas, 1987). In mountainous areas physical weathering is dominant. In general there appears to have been a reduction in chemical weathering since the Mid- or Early Tertiary, and deep weathering at present is confined to what are now the tropics.

Other factors that may affect the equation between tectonics, weathering, erosion and climate are the products of volcanoes. Some may be recycled, but the surprising amount of new gases (such as helium) in mid-oceanic eruptions suggests that much of this gas has been generated recently. Carbon dioxide is second only to water as the major greenhouse gas, and so volcanoes can affect climate by at least three processes: production of the dust veil and sulphate aerosols, both of which cause cooling, and an increase in greenhouse gas, which causes heating. Vast outpourings of basalt at about the Cretaceous–Tertiary boundary may be related to climate change, though it seems that warm temperatures were widespread before the eruptions. Eruption of a single volcano such as Paracutin can cause a lowering of world temperature by perhaps half a degree. Larger eruptions such as Krakatoa (1883) or Tamboro (1815) can cause cooling of more than a degree.

If, as suggested here, the uplift of Tibet, the Andes and many other highlands occurred mainly in the Pliocene and Pleistocene, the correspondence between uplift and climatic change is much greater than Ruddiman and his colleagues proposed. The orographic time scale matches the glaciation time scale much better than the 40–20 Ma of Ruddiman and colleagues. The correspondence of Tibetan uplift with the onset of the monsoon climate and the deposition of loess is the best example of detailed correlation.

A summary of the main features of the tectonic uplift–climate change argument is provided by the many contributors to a book on the topic edited by Ruddiman (1997). This focuses on the last 50 million years.

Ruddiman himself outlines the nature of the proposed connection and also outlines the objections to it. Several physical effects of uplift affect climate, including lapse-rate cooling on high elevation surfaces, diversion of jet stream meanders, and the creation and intensification of monsoon circulation. He then considers the biological effects of uplift on climate via CO_2. This presents his own favoured scenario which is the uplift–weathering hypothesis. This assumes that massively increased chemical weathering localised in regions of uplifted terrain is the critical factor that controls CO_2, and climate.

As support for the Ruddiman hypothesis, some workers have invoked the oceanwide $^{87}Sr/^{86}Sr$ isotopic ratio in calcareous marine sediments as an index of global chemical weathering. They argued that the abrupt increase in this ratio over the last 40 million years reflects increased chemical weathering of continental rocks. Objections came from alternative hypotheses ranging from input from mid-ocean ridges to anomalously enriched rocks in the Himalayas. We suggest that the timing is wrong – most uplift took place in the last 5 million years, rather than the least 40 million, so any change in strontium ratio requires a different explanation.

Just how uplift of mountains all over the globe would affect climate is not exactly clear. The most obvious effect of uplifting a mass of land into a plateau is that the plateau will become cooler. But to develop a large ice cap it is necessary to have large precipitation as well as low temperatures. In the climate of today it is probably impossible to generate an ice cap such as we see in Antarctica and Greenland because there is not enough precipitation. Our major ice caps have to be inherited from a previous era which combined high precipitation with low temperatures in high regions. The Tibet Plateau, being exceptionally high but mainly dry, would be the ideal place to generate loess, and this is what happens. There is one school that believes in widespread glaciation of the Tibet Plateau, but as well as lacking field support it would seem that theory is against it. There was never enough precipitation to make a vast ice cap in Tibet. On the other hand, it did have a great influence on the monsoon climate.

It was the growth of the world's biggest ice cap, in Antarctica, that domi-nated the world climate including the onset and continuation of the Late Cenozoic Ice Age. The uplift of the Antarctic Plateau should therefore be of paramount importance, exceeding the effects of uplifts in Tibet, the Andes, or North America. Ice accumulation actually began in Antarctica in the Oligocene at least, which does not support the concept of altitude-induced glaciation. In contrast, glaciation in the northern hemisphere is limited to the last 3 million years, and uplift may have had a significant impact on the global refrigeration as opposed to the Antarctic refrigeration. Sibrava (1997) notes that pronounced environmental changes started about 5.5 Ma in the period close to the Gauss-Matuyama palaeomagnetic boundary, and writes, 'This time line is regarded by some authors as the Plio-Pleistocene boundary.'

Despite the worldwide appearance of highlands in the Late Cenozoic, it was probably the Antarctic Plateau that gave rise to Late Cenozoic cooling and the Ice Age. In warmer times precipitation could have been sufficient to grow an ice cap on the Antarctic Plateau, which grew until sea ice formed. Only when sea ice has grown quite large does it have a significant effect on ocean temperatures, which works to sustain an ice age. The great spread of cooling via cold ocean currents eventually brought cooling into the northern hemisphere. Whereas Antarctic ice accumulation goes back to the Oligocene, the Greenland ice cap is only 3 million years old, and much of the glaciation of the northern hemisphere is younger still. Both Greenland and Antarctica have the basin shape typical of many continents with passive margins, which, combined with their latitude, made them very suitable collecting grounds for ice, which enhanced the onset of global glaciation.

In recent years climate modellers have been highly concerned with the factors of astronomical controls (Milankovic cycles) and greenhouse gases, but now a renewed emphasis is being placed on the effects of mountain building.

The tectonic impact may not just be to start a cold climate, but also to affect changes at a smaller scale. Zhu (1997), after work throughout the Tibet Plateau and the East Asian continent, came to three conclusions:

1   Active neotectonics is usually accompanied by changes to drastic cold climatic conditions on the Tibetan Plateau and in East Asia.
2   There are some quasi-cycles of 0.8 Ma (huge cycle), 0.4 Ma (main cycle) and smaller cycles and subcycles in Quaternary events of Tibet and East Asia.
3   The quasi-cycles may be global, and the periodic changes are called 'climatic–tectonic cycles'.

He goes further and says the interrelationship between climate and neotec-tonics is not a simple one of cause and effect. The cyclicity of the climatic–tectonic cycle may be conditioned by planetary movements of the Earth, especially changes in eccentricity:

When eccentricity increases, the Earth moves farther away from the Sun. This will cause the cooling of climate on the one hand and an increase in rotational inertia and deep geological processes on the other. This, in turn, brings about activation of tectonic movement. When the eccentricity decreases, the climate becomes warm and the tectonics stable.

12 Problems of mountain tectonics

Classification and distribution of mountains

The classification and distribution of mountains are related topics, so we deal with them together. Some hypotheses of mountain building would relate certain kinds of mountains to particular sites on the Earth, which offers some sort of test for the hypothesis. But which mountains are to be included?

Mountains are scattered all over the world. Local people often call even small hills 'mountains' if the surrounding country is notably flat. Maps prepared by non-geologists, such as climbers or naturalists, often do not conform to the image of plates that are emphasised by some geologists when they explain their theories.

The traditional classification of mountains, still used in elementary books, is:

fold mountains;
block faulted mountains;
mountains made by erosion (mountains of circumdenudation);
volcanoes.

This does not work. We have demonstrated (we hope) that fold mountains, in the sense of folding of rock and simultaneous creation of mountains, do not occur in the real world. And many block mountains consist of folded rocks, though the folding occurred much earlier than the mountain formation. Volcanoes remain a distinct class of mountains. All mountains are affected to some extent by erosion.

In plate tectonic theory mountains are formed at plate boundaries, so the basic classification of mountains is:

mountains at collision sites;
volcanoes, at both spreading sites and collision sites.

Subduction is usually thought to make fold mountains in the traditional style, and since even the mountains on such sites, such as the Coast Ranges of North America or the European Alps do not fit the description, this is a faulty hypothesis.

At a more elaborate level it is suggested that subduction leads to uplift of a plateau, such as the Tibet Plateau. This is a faulty hypothesis because the subduction occurs on a much longer time scale than the uplift of the plateau.

Plate tectonics has no real explanation for mountains on passive margins. Uplift is real, but there is no possibility of subduction. Ad hoc explanations have to be proposed, such as thermal effects or diapirs.

Volcanoes, in general, fit well into plate theory, and many volcanoes fall on either spreading or collision boundaries. But there are many volcanoes that do not fit into plate theory. Some passive margins (for example eastern Australia) have volcanic activity, but many do not. Some sea mounts fall on mid-ocean sites and hot-spot traces, but many others do not. Tristan da Cunha and Gough Island in the South Atlantic are both hundreds of kilometres from any plate edge, and have experienced eruptions millions of years apart. The distribution of sea mounts in the Pacific does not fit any plate theory.

We should make it clear that we have no objection to plate tectonics in general, for it explains many things. But we do object to the simplistic explanation of mountains and their distribution.

A descriptive classification of mountains might result in a list such as this:

broad swells (Urals, Scandinavia);
uplifted plateaus (Colorado, Kimberley);
uplifted plateaus with spreading (Iran, Tibet);
uplifted plateaus with spreading, and central graben (Andes);
plateau-marginal ranges (Himalaya, Zagros);
plateau uplift with gravity slides (New Guinea);
tilt blocks (Sierra Nevada, Ruwenzori);
uplifted plateaus with gravity disruption (Dolomites);
passive margin warps (East Australian Highlands, Southern Africa);
simple volcanic mountains (Hawaii);
Volcanic mountains associated with other mountain building factors (Andes, Cascades);
symmetrical mountain belts.

We believe it is premature to establish a new classification of mountains, but it is necessary to beware of over-simple classification, because it leads to over-simple thinking, and obscures the wide variety of mountains in the real world.

Putting different kinds of mountains on a map proves to be a very difficult exercise, because the classification of different mountains is not obvious or agreed, and because several different factors often occur at the same place.

For instance, the Andes are examples of uplifted plateaus with an inter-montane graben. They also represent symmetrical mountains. Along much of their length the Andes are close to the coast, which fits the subduction model, but some advocates of plate tectonics want to subduct Brazil under the Andes too.

This symmetry seen in the Andes can be traced as far as Mexico, and perhaps extended into the Rockies. But in North America there is a major extensional tract (the Basin and Range) between the Rockies and the Coast Range.

Passive margin mountains do have a fairly simple distribution, and have been mapped for a long time (e.g. Holtedahl, 1953; Jessen, 1943), but even these have their complications as explained in Chapter 10. Some have volcanoes (Eastern Australian Highlands), others do not; some have faults, others do not. In this instance there is great disagreement at present between the advocates of coastal downwarp, and those who propose uplift along an offshore fault.

It is remarkable that mountains in many parts of the globe, all characterised by rapid uplift after a period of rapid planation, should occur in so many different tectonic styles. These include the simple plateau (Colorado), Inter-montane plateau (Tibet), tilt block (Sierra Nevada), uplift of a broad swell (Alps, Apennines, Papua New Guinea), symmetrical uplift with a central graben (Andes), and isostatically induced uplift (Himalayas).

Why should a near-global pulse of mountain building take so many different forms?

Folds, fold belts and gravity structures

In this book we have seen that mountains are not made by folding, but by uplift of a plain, sometimes after folding. Nevertheless, because of the common assumption that folding, collision and mountain building go together, we need to discuss some general principles of folding.

Field evidence

Perhaps the first thing that non-geologists need to realise is that the 'truth' shown in geological cross-sections is of variable value.

The cross-section shown in Figure 12.1 helps to explain this. A geologist crossing the ground surface may note the dip of the strata and plot it on his map and section. Then the data must be interpreted, and can give rise to such diverse pictures as shown in these two versions: one has a simple gravity slide with folding of the upper mass; one has a series of deep-seated thrust faults. Other information should be used to determine which is the 'truth', but this is not always possible.

Sometimes regional mapping can help to determine which is which. Real geologists are taught to think in three dimensions, and a geological map

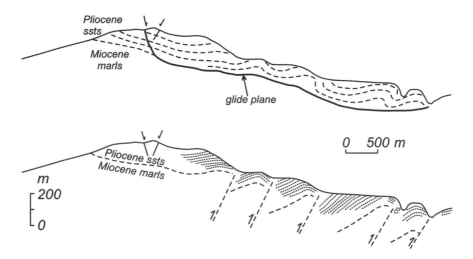

Figure 12.1 Two interpretations of the same outcropping pattern of strata.
Top. As a giant landslide.
Bottom. As the result of deep seated tectonic forces and thrust faults.

tells much more than a section. In some places a constraint can come from drilling, and a series of boreholes can constrain a section very well. This does not always deter those who want a different, theory-driven section, as shown in Figure 8.7 where abundant drilling shows gravity sliding, but plate tectonic interpreters simply ignore the raw data.

A variety of geophysical methods can help in constructing sections, but much depends on interpretation. The rift valley faults are clearly normal faults to a field geologist, and are now generally regarded as such, but because of Bullard's interpretation of a gravity survey they were thought for several decades to be thrust faults. A vast number of seismic surveys show an array of normal faults in deep sea trenches, but this does not prevent geophysicists proposing compression and subduction in the same areas.

In general the deeper the section, the greater the imagination needed to interpret and conduct it. This was very evident in the paper by Lees (1952), who despite being an oil geologist used to deep drilling, thought that sections *should* be carried to great depths. He was perhaps the last serious scientist to propose the shrinking earth as a mountain building mechanism.

Compression, crustal shortening and mountain building

The actual ground surface is seldom folded beyond a very gentle stage. Broad warping of the surface undoubtedly occurs, but this is quite different from the tight, small folds in rocks that are often observed in real geological exposures or reconstructed geological sections. There is usually no direct evidence that folding was accompanied by mountain building. Some folding could occur under the sea while sediments were still accumulating. Some folding could result from gravitational effects after the sedimentary pile was uplifted (see Chapter 8).

But the old idea that crustal shortening and folding go together and create mountains has not been dispelled, and most observed vertical uplift is explained as the result of lateral compression. For instance, Munoz and Charrier wrote in 1996, 'the uplift of mountain ranges can generally be related to crustal shortening processes'. And Okaya *et al.* (1997) wrote, 'The central Andes are a prime example of compressional deformation and mountain building.'

This crustal-shortening concept (included in plate-tectonic explanations of mountains) comes in many forms but with three features in common:

1   folding and mountain building are simultaneous processes;
2   lateral compression causes both;
3   the actual landforms are ignored, especially unfolded planation surfaces.

So ingrained is the concept of crustal shortening in mountain building that the logic has sometimes been reversed. Ruddiman (1997) edited a book on the influence of young mountain building on climate, and in one chapter Jordan *et al.* attempted to deduce the age of mountain uplift in the Andes from measures of crustal shortening.

In plate tectonic theory folding is caused by compression where plates collide. At the crudest a pile of sediments is simply squashed. Sediments may be subducted and folded as they go down. Or they may be scraped off the descending plate and thrust on to the upper plate as 'fold mountains'.

The idea of lateral compression to form fold mountains arose initially from early ideas of a shrinking Earth. A cooling, shrinking Earth would cause folds and mountains, like the wrinkles on the skin of a shrivelled apple. Although few now believe the Earth to be shrinking, and despite the certain knowledge that the strength of rocks is insufficient to permit folds to be created by lateral compression, it is proving very difficult for geologists to rid themselves of these outmoded concepts. The notion of vice-like compression of sediments has a vice-like grip on the minds of geologists.

The cause of folding

Applied force and body force

It cannot be over-stressed that it is not possible to fold rocks by a lateral push. Ideas gained by looking at folded tablecloths or at squashed clay in a squeeze-box do not relate to the folds we see in mountains. The compressive strength of rock is very much lower than the force that would be needed to push a slab of strata a few kilometres thick and hundreds of kilometres wide. If a sufficiently large force were somehow applied to such a slab, it would not move, but would simply be crushed at the point where the stress was applied.

Gravity applies a body force that acts on every grain, every molecule, in the rock mass, and affects the whole mass in a way in which an applied force cannot. A landslide is a small example of movement under a body force, and causes folding at the toe of the landslide. Folds are not brought about by lateral shortening of the Earth's crust: they are superficial features and the folds relate only to the few kilometres of folded rock. The rock mass slid under gravity, and was folded as a result of its sliding motion. Compression and shortening is confined to the folded beds. We can theoretically unravel the folds and see how big the sheet of rocks was before folding, but this measures the shortening of specific strata. It is not a measure of *crustal* shortening.

Lateral compression versus gravity tectonics

The idea of lateral compression to make folds is a naive, almost instinctive one. Simple analogies such as creating folds in a tablecloth by pushing the ends together make it obvious. But as shown in Figure 1.5 an arc or 'fold' can be created without shortening. In the tablecloth example it should be noted that although the cloth is folded, the table top is not. This separation between the folded and unfolded is often found in nature as a decollement surface.

The Jura folds are often likened to the folds that can be generated in a tablecloth when it is pushed across a table. The table doesn't fold, but the tablecloth does. This homely analogy is very misleading, for there is no geological equivalent of the moving hand that does the pushing. It is simply not possible to push a sheet of rocks sideways like a tablecloth. If the table were tilted and the cloth slid down into folds the analogy would be better. Some claim that subduction where plates collide provides the push, but subduction is driven by sea floor spreading, and the force has to be transmitted from the spreading site to the subduction site across thousands of kilometres of thin, unfolded crust.

The squeeze-box is a common little device to demonstrate folding of soft materials such as clay and sand. The layers are undoubtedly folded by lateral compression, but the base of the apparatus is not folded, and the decollement surface at the base of the sediments may be the most significant feature of the experiment. Of course, in the real world it is hard to find the equivalent of the block and screw that did the compressing.

Thin skinned tectonics and the wedge

It has long been recognised that some mountains are underlain by a decollement, so with the basic idea of compression it is necessary to push a thin mass of sediments over a stationary basement. This type of structure is very widespread, including the examples described in Chapter 8, and the Sawtooth Mountains of Montana (Figure 12.2). The lateral dimensions of this thin

Figure 12.2 Sawtooth Mountains, Montana (after Deiss, 1943).

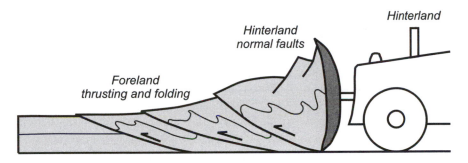

Figure 12.3 The 'wedge' explanation of thin skinned tectonics. A bulldozer pushes a wedge of sand, and thrusts proximal sand over distal sand. The base of the wedge is a decollement. One would not expect greatest deformation far in front of the bulldozer (which happens in nappes) but close to it. It seems improbable that normal faults would form close to the bulldozer in this situation.

skinned tectonics are very great, often over 100 km, so the applied force has to be enormous to push this great mountainous pile uphill.

Today, one common 'explanation' of thin skinned tectonics is the thrusting of a wedge (Figure 12.3), in which a bulldozer pushing a mass of sand simulates movement of a nappe. Distal parts of the sand are over-ridden by sand pushed from behind, moving over thrust faults or nappes.

But one would expect deformation to be greatest at the bulldozer end, where a force is applied, not at the distal part. Deformation should decrease and even disappear with distance from the applied force. This does not fit with the observed 'breakers at the nappe front' described later in this chapter.

A simple analogue is shown in Figure 12.4. If a stationary car has a force applied at the back (such as another car running into it) then the damage will be at the rear. A car that has moved under gravity (because the handbrake failed) and collided with a tree is crumpled at the front. This is exactly like the crumpling at the front of a nappe – the nappe was sliding under gravity until it hit something stable – the foreland.

Stamp (1950), writing of Palaeozoic folding in Britain, speculated on the effect of a massive thrust from the south east crushing the strata against a

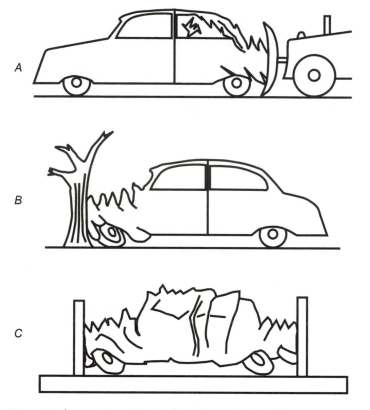

Figure 12.4   a.   A car that has been hit in the back by another car. The deformation is greatest at the site of applied force.
b.   A car that has rolled downhill under gravity (a body force) until it hit a tree. The deformation is greatest where it met an obstacle.
c.   A car in a squeeze box is damaged at both ends. The geological equivalent is compression (e.g. subduction) on opposite sides of a plate.

resistant mass or 'stable block.' He wrote, 'Would the folds increase in intensity towards the direction from which the pressure came (i.e. towards the south-east), or towards the 'stable block'? The former seems more likely as the simple experiment of pushing a mass of some kind against a pile of sheets of paper will show. But the folds in Scotland *decrease* in intensity towards the south-east.'

Gravity spreading

Simple gravity sliding down a plane would produce a break at the back of the sedimentary pile. Such gaps are seldom found, so Elliot and Johnson (1978) proposed a more realistic theory of gravity spreading.

The idea of gravitational spreading had been conceived much earlier by Jeffreys (1931), who calculated the inevitability of gravity structures from the strength of materials and wrote that, 'Whenever an elevation over 3 km has been produced, either fracture or flow must set in; whichever it is, its general effect will be the spreading out of the elevated rocks over the surrounding country.'

Symmetry, gravity spreading, or double subduction

In many mountain areas a symmetrical disposition is observed. One explanation is the spreading of an uplifted block. More commonly, such symmetry is explained by subduction from opposite sides.

Valley bulges and valley anticlines

It seems that the unloading due to river erosion can be sufficient to create anticlines along valleys. These range in size from less than a metre to hundreds of metres, and in the case of the Himalayas and the Colorado Plateau they can easily be confused with tectonic folds. These have been described in Chapter 8.

Criteria for recognising gravity structures

In this book we have stressed the possibility that many structures are made by gravity sliding or spreading rather than compression of stable blocks or plates. Here we list the criteria in support of such interpretations. We have seen that mountains are not made by folding, but by uplift of a plain, sometimes after *previous* folding of strata, and also that some gravity phenomena occur *after* uplift. The criteria that follow are arranged in two groups: pre-planation and post-uplift gravity structures.

The criteria for pre-planation gravity sliding

1   The total thickness of the nappes is often about 10 kilometres. This is much greater than the terrestrial relief, so is not controlled by the topography. Much of the nappe would have been below sea level at the time of sliding.
2   Most folding and faulting occurs at the front of the nappe. This is referred to in German as 'Branden der Deckenstirne' or 'Breakers at the nappe front'. The nappes of the Helvetide Alps of Switzerland have a sub-horizontal structure except along the front, where they are crumpled into narrow folds. Figure 4. 5 illustrates the idea very well too. Where the reaction between moving force and resistance is greatest is where crumpling would be expected: this is at the front with gravity sliding, at the back with push.

The nappe-front breakers are not restricted to the frontal part of external nappes, but normally occur in the frontal part of every nappe or nappe lobe.

3   The youngest beds travel furthest. In the Helvetide nappes, for example, the furthest travelled nappes in the north are mainly formed by Cretaceous rocks; then comes the Jura nappe formed mainly by Jurassic rocks; and last the Verucano nappe which is mainly Permian.

This follows from gravity tectonics, where the highest units have greatest potential energy. It is hard to see on a tangential pressure model how a differential push could move the higher parts of a series several tens of kilometres further than the lower parts. How could such a force continue to be applied once the nappe had moved away, if the pushing part is itself rooted in the crust?

4   Isolated fold sheets or intercutaneous nappes ('chevauchement intercutanes') occur. This refers to a situation where a series of folded and overthrust strata are found within a mass of seemingly tranquil, apparently unmoved strata. This is common in the Alps, and also in many other places. The intercutaneous nappes mark an internal zone of great disturbance that took place inside a thick series of sediments without disturbing either underlying or overlying strata. It is hard to see how this could be obtained by lateral compression, but it could easily result from a gravity slide during deposition of the sedimentary pile.

5   There is convergence of forces from more than two opposed directions. With one thrust it can be argued that the block is subducted; with two opposed thrusts it can be argued that a thrust has an opposite reaction, but if thrusts converge from three directions, such as north, south and west, the only reasonable explanation is that nappes are converging on a single stable block. The nappes around the Pelvoux Massif are the prime example.

6   Divergence of thrusts in plan of more than 180° suggests gravity spreading. Subduction would ideally make a plane, or a very gentle arc on the Earth's curved surface. But if the curve of a nappe front is very marked, it would require subduction from many directions, some opposed, which is impossible. The Apennines and the Carpathians are good examples of such arcs of thrusting.

The criteria for post-uplift gravity tectonics

1   There is preservation of a planation surface on top of the uplifted block.
2   Faults follow a straight line for long distances (many tens of kilometres) even in rugged topography, though in local detail they appear as thrusts. A continuous low angle fault would have a very irregular outcrop in rugged country.
3   Low angle thrust fault planes can be traced into steeper faults.

4 The overlying slab sometimes overlies soft material such as alluvium. If underthrusting did occur, it seems incredible that such material would not be scraped off. Complications arise if a younger movement takes place along an old fault. The great nappes of the Himalayas are prime examples of pre-planation thrusts, but there has been renewed movement on some nappes in more recent times. Maltman (1993) illustrates (his Figure 12) high-grade Precambrian Nanga Parbat gneiss thrust over unlithified glaciofluvial sediments, which is clearly a post-glacial, Quaternary movement.

5 Uplifted blocks are bounded by divergent faults. The alternative, of compression from opposite sides, requires exceptional tectonic forces.

6 Anticlinal valleys (valley bulges) occur. A river, however it originates, is not related to structures that lie even a few metres below its course, let alone several kilometres. So if a river follows an anticline, it is almost certainly because the anticline appeared *after* the river. If a dozen or more rivers follow anticlines, as in the Himalayas, the causal relationship seems definite. Such anticlines have their greatest amplitude near river level, and die out with depth.

7 Minor folds that appear along the foothills of mountains are likely to result from pressure of the mountains themselves, like uplifted zones in the toe of a landslide.

8 Deformation is greatest near the uplift and decreases with distance from the uplift. Figure 12.5 shows a cross section of the pre-Andean folds and faults in eastern Peru, with intense folding and faulting near the Andean complex and broad folds to the north east. This is the opposite of the 'breakers at the front' situation in pre-planation gravity slides.

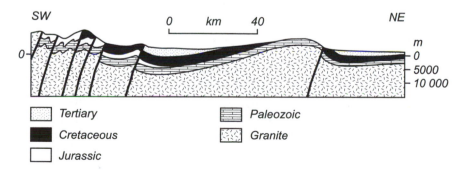

Figure 12.5 A generalised section across the Ucayali basin, eastern Peru (after Ham and Herrera, 1963). Maximum deformation is close to the direction of thrusting, not at the distal end of thrusts.

Gravity and isostasy

At sea level the value of gravity is dependent only on the latitude, and is less at latitudes near the equator than at latitudes near the poles. This value for gravity at a particular point on the spheroid is called the theoretical value for that point, and can be calculated simply by knowing the latitude.

Subtracting the theoretical value of gravity from an observed value of gravity at a point gives a difference called the 'gravity anomaly'. If this value is corrected to remove the effects of the elevation of the gravity instrument above the spheroid, the result is a 'free-air gravity anomaly'. The free-air anomaly is almost always positive because it does not take into account the gravitational pull of the rock that lies between the instrument and the spheroid. Such rocks between the instrument and the spheroid will exert an attraction depending on their density. The measured value of gravity, corrected both for elevation and for the gravitational attraction of the rocks between the instrument and the spheroid, minus the theoretical value, equals the 'Bouguer gravity anomaly'. Bouguer gravity anomalies are generally strongly negative over mountains and plateaus, and zero or positive over oceans.

The word 'isostasy' is used to describe the principle of flotation as applied to continents and oceans.

There are two theories concerning the way in which isostasy acts to buoy up or compensate mountain masses. Pratt's theory assumes that different parts of the lithosphere have different density and float on a uniformly dense substratum. The less dense crustal blocks float higher, forming mountains, and the more dense blocks form basins and lowlands. This seems to be essentially the case for the difference between continents and oceans. Airy's theory assumes that the various parts of the lithosphere have approximately the same density but have different thicknesses. High mountains therefore not only project upwards, but have roots extending into the denser substratum. Thick lithosphere should form mountains, and thin lithosphere should form lowlands. Both theories assume the presence of a dense fluid or plastic layer in which the lithospheric blocks float – the layer that we now call the asthenosphere. Both theories account for the deficiency of mass under high mountains, but Airy's theory is now known to be a better explanation of mountains within continental regions.

The geophysical root of the Alps is postulated to explain the negative Bouguer anomaly over much of the region. The negative anomaly corresponds to a thickening of the crust, but not of the upper, sialic layer. In northern Europe there is no trace of a similar root beneath the Hercynian uplands.

Compelling evidence for the general existence of isostasy in the southwestern United States is shown by the very good correlation of generalised topography and the raw Bouguer anomalies, as shown by Gilluly (1970) in the western United States. The greatest negative anomalies coincide precisely with the topographic highs of the San Juan Mountains and the Mount Elbert mass in Colorado; topographic and anomaly gradients are concordant around the southern and western sides of the Colorado and Mogollon Plateaus; and even the Colorado River and Death Valley are reflected.

The laws of flotation act on the continents just as they would on a raft or on an iceberg. If we melt the top of an iceberg, the berg rises further out of the water. If we increase the load on a raft, it sinks further into the water. If we erode our continents they will rise anew. If we load our continents with a heap of sediments or an ice cap they will be depressed.

The adjustment time will not be the same for a continent as for a raft or iceberg. The raft sinks as soon as a sailor climbs aboard, and rises as soon as he jumps off. Loading and unloading continents is a slower process, depending on the viscosity of the earth materials.

Perhaps the best documented information on crustal movements in response to load comes from those areas which had ice caps in the last ice age and are now 'rebounding' after the ice has melted. In the Baltic and in Canada contours on the amount and rate of uplift can be drawn, and the land is still rising (at about 1 cm per year in Stockholm). Greenland is a saucer shaped country with a rim of mountains and a deep central area presumably depressed by the Greenland ice cap. If the icecap melts we must not expect the saucer-shape to remain, for the central area will start to rise as the load of the overlying ice is removed.

Isostatic rebound can also follow removal of the weight of a large lake such as Lake Bonneville, Utah. The rebound of the Bonneville Basin after unloading of its water load indicates a mantle viscosity nearly an order of magnitude smaller than that deduced from the rebound of Fennoscandia; the difference might be explained by the very much smaller area involved, so a thinner zone of the mantle flowed.

For the vertical movements there have to be compensating horizontal movements of material in the underlying materials. If a boat sinks deeper upon loading, water under the boat moves out of the way. If Greenland sinks under the weight of ice, some compensating sub-crustal material under Greenland must move sideways out of the way. If an iceberg rises as it melts, more sea water moves in to occupy the space beneath the iceberg. If a mountain range rises, the space under the range must be filled by the inflow of material below.

Gravity anomalies may result from two quite different causes: the rocks may not be in isostatic equilibrium, so they are tending to slowly rise or sink towards a new equilibrium; or the rocks may simply be less or more dense than the usual.

Most gravity anomalies seem to tell us whether the underlying earth materials are denser or lighter than the average. Continents are generally lighter, being siallic, and ocean floors are denser, made of basalt sima. Mountain ranges and high plateaus have negative gravity anomalies (lighter) as they have a great thickness of lighter materials. The Red Sea has a positive anomaly because it is floored by dense sima. The rift valleys of Africa have negative anomalies because they are floored by thousands of metres of light sediment.

Deep sea trenches such as those off Chile and Peru, or in front of the island arcs of Indonesia and the West Pacific, have marked negative anomalies. The simplest explanation for this is that they are full of light sediment, like the rift valleys, but various dynamic explanations have also been proposed.

Dynamic explanations of gravity anomalies provide quite different solutions to earth problems, so gravity data alone cannot give firm answers to tectonic problems. For instance, when the negative anomaly over rift valleys was first reported, it was thought that to produce this anomaly the fallen block in the middle of the rift must be held down against a tendency to rise and so it was proposed that the rift valleys were actually bounded by thrust faults. There is abundant geological evidence against such an interpretation, and the much simpler hypothesis that the negative anomaly results from the thickness of accumulated sediment is preferred. Deep sea trenches can be explained dynamically too. The first hypothesis was that some sort of down-fold (a so-called tectogen) was holding down sediments. Nowadays, underthrusting at a subduction zone is the generally preferred explanation. In some instances the simple effect of the sediment mass may provide sufficient explanation (Worzel, 1976).

Isostatic effects may be marked at continental margins. Assume a continental edge is being eroded, and the resultant sediments deposited in the sea. The weight of the sediments depresses the crust, creating space for more sediments. The mountain mass, being reduced in thickness as a result of erosion, rises isostatically. Transfer of subcrustal material takes place by some sort of flow mechanism. Once initiated this sort of process might be self sustaining, with the sedimentary belt getting thicker and thicker, and the mountainous edge of the continent being repeatedly uplifted. A significant point of this idea is that it might be checked by comparing the denudation chronology of the mountains, worked out by geomorphic methods, against the sedimentary history worked out by stratigraphic methods. Thus periods of planation and reduced erosion should correspond with periods of reduced sedimentation in marine basins, while periods of marked uplift and erosion should correspond to increased deposition, and formation of less mature sediments. The situation seems to prevail on some passive margins. In geological time there are records of many long and narrow sedimentary basins, known as geosynclines. These seem to take several geological periods to fill, while the message in the present book is that mountains are uplifted in a few million years – a considerable mismatch of time scales.

Gravity can have other effects where materials of different density rise as plugs or diapirs. The most obvious examples are salt domes. Some granite plutons rise through the surrounding sediments in a similar way, pushing aside and doming up the neighbouring rock. Gneiss-mantled domes are another but more complex example.

Isostasy and mountain building

Of interest to geomorphologists, because mountain building and erosion are brought together, is the idea of isostatic uplift compensating for erosion. This has been applied to continental margins and also to major valleys of the Himalayas, the Colorado Plateau and elsewhere.

In brief, erosion of large valleys removes a load from the crust, which is compensated by underflow of sima at depth, which raises the topography again. Suppose a plateau at 2000 m is deeply eroded along the plateau edge. Isostatic uplift will raise the edge to a height above 2000 m, and the ridge tops will be at an elevation considerably higher than the plateau. This mechanism is especially probable in mountain ranges bounding plateaus, such as the Himalayas and Kunlun Mountains bounding the Tibetan Plateau; the Zagros and Elburz Mountains bounding the Iranian Plateau; or the Western and Eastern Cordillera bounding the Altiplano of the Andes.

Of course, rivers crossing the mountains must flow downhill, and have flowed from the plateau across the mountainous area since before the uplift of the mountains. This is called 'antecedent drainage', and the rivers such as the Arun that cross the Himalayas from the Tibet Plateau are classic examples. The point is that although isostatic compensation can account for the uplift of the bordering mountains, the original plateau uplift is a precursor for the out-flowing drainage. In debates about the effect of isostasy it is not enough to have gross morphology and geophysical evidence: drainage pattern is essential.

Molnar and England (1990) reconsidered the relationship of isostasy to uplift and climate. They suggest simply that erosion is a driving force of uplift, and since erosion is climatically controlled to some degree that climate may be a driving force of uplift. 'An alternative cause of these phenomena [deep incision and high mountain altitudes] is late Cenozoic global climate change.' They do not discuss drainage patterns or the relationship of mountain ranges to plateau uplift

As described in Chapter 4, the Swiss Alps of today were produced by strong uplift in the Pleistocene, accompanied by severe glaciation. Molnar and England (1990) suggest that although the relief is 4000 m, the mean elevation of the Alps is only 2000 m, and the inferred uplift of 2000 m could simply be the isostatic response to the removal of material by the incision of deep valleys.

If an erosional plain is graded to sea level, it may or may not be in isostatic equilibrium. This depends on the prior history. If a considerable overburden has been removed, there could be isostatic uplift. The effect may be enhanced by loading on the bordering seafloor by the deposition of material eroded from the plain area. The uplifted area would, of course, be subject to further erosion, and the ultimate landscape depends on the balance of different tectonic and erosional processes. This is roughly the situation in passive continental margins, discussed in Chapter 10. There is still controversy over the nature of uplift in such situations.

But most other mountains and plateaus tend to have very distinct edges, suggesting uplift of distinct blocks, and to raise such blocks by isostasy alone seems improbable. Antecedent rivers draining high plateaus show that although erosion can enhance uplift in some areas (by isostatic compensation of interfluves as well as valleys), it cannot be the prime driving force of plateau uplift, which is the real basis of mountain building. You cannot reach the threshold for isostasy if you start with a plain – and since we have planation surfaces in almost all our mountains it is necessary to start with a plain.

If somehow a plain came into existence that was in isostatic equilibrium, there is no way in which erosion alone could cause uplift. There has to be some force from within the Earth to push up the land and induce valley erosion, or there has to be a past geological history that leaves a plain out of equilibrium to cause uplift.

Isostasy and volcanoes

One of the fundamental differences between non-volcanic mountains and volcanoes is that the former are part of the Earth's crust but the latter may be regarded as a heavy load placed upon the crust. In this respect a large volcanic cone resembles an ice sheet, a large lake, or a huge sedimentary pile. As with these other loads, volcanoes are subject to isostatic forces and settle under their own weight.

Large volcanoes contribute, by their own mass, to tectonic adjustment around them. The ocean crust is rather thin, and when a submarine volcano erupts it is presumed that it draws its magma from a considerable area around the vent. When activity ceases, a heavy volcano is resting on a fairly limited area of thin crust, and the volcano then starts to sink. It may depress the surrounding crust to form an annulus of lowland around the base. Quite often the volcanoes are truncated by marine erosion to give them a flat top and many flat topped volcanoes have now subsided to depths well below the zone of wave activity. They are known as 'guyots'. A volcano that subsides without flattening is called a 'seamount'. Some volcanoes in tropical areas provide a base for the growth of coral reefs, and the routine sinking of their base allows coral growth to attain great thickness (thousands of metres) and accounts for the succession of reef types from fringing reef through atoll to barrier reef.

Symmetry of mountains

Some mountain ranges are symmetrical, others markedly asymmetrical, and some mountains are ranged symmetrically around a central region. This symmetry is presumably telling us something about the origin of the mountains, but there is much disagreement about what the message is.

Median plateaus and outfacing ranges

At the largest scale there are very large plateaus, bounded by mountain ranges. The largest is the Tibet Plateau, with the Himalayas to the south and the Kunlun Mountains to the north. The Himalayas are thrust to the south, and the Kunlun to the north, so there is symmetrical divergence of thrusting. The symmetry is restricted to this assemblage of two ranges and a plateau: south of the Himalayas is the plateau of India, and north of the Kunlun comes the Tarim Depression, then the Tian Shan horst, and a series of block faulted ranges stretching into central Asia.

Another example is the Iranian Plateau with the Zagros Mountains one side, and the Elburz Mountains on the other (Figure 12.6). Such plateaus between two ranges may be termed 'median plateaus' (Holmes, 1965). German workers knew them long ago and call them 'Zwichengebergen'.

Divergence in the Andes

In the Andes, many authors have depicted the symmetry of fan-like divergence of faults, including Mégard (1987) and Litherland *et al.* (1993), and the official Geological Map of Ecuador. Andean symmetry goes beyond the median plateau stage because there is a symmetrical graben, the Inter Andean Depression, running along the centre of the Andes, separating the westward thrusted Western Cordillera from the eastward thrusted Eastern Cordillera

The symmetry of the Andes seems to run the whole length of the chain. A cover picture in *Geology* (July 1996) highlighted the apparent symmetry of the Andes as seen on a computer generated relief map of South America. A caption suggested that an explanation might be found in a variation on the Wilson cycle (opening and closing of an ocean), as outlined in an earlier paper (Russo and Silver, 1996). But as explained in Chapter 6 the Andean Cordilleras have a planation surface on top, which is Miocene and uplift is mainly Plio-Pleistocene. If formation of the Andes is so rapid, there is no time for a Wilson cycle to operate.

Two other suggestions have been made for the symmetry – subduction and spreading.

At present the vast majority of writers assume the Andes to be caused by subduction of the Pacific Plate beneath the South American Plate. Gravity

Figure 12.6 A median plateau. The Iranian Plateau is bordered by the divergent Zagros Mountains and the Elburz Mountains. The outward thrusting from the median plateau suggests either subduction on both sides of the plateau as the opposed forelands approach, or gravity spreading from the Iranian Plateau when it stood higher, or both.

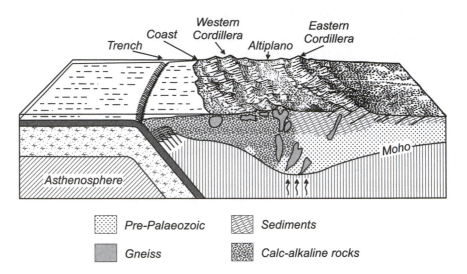

Figure 12.7 Cross-section of the Andes showing symmetrical features despite one-sided subduction (after James, 1973). Variations on this figure appear in many textbooks.

spreading can make a symmetrical group of structures, but to make a symmetrical Andean structure by subduction may require underthrusting from both sides, which is not part of the general paradigm.

The early advocates of subduction realised that there was symmetry in the Andes, and managed to show the symmetry on their diagrams (Figures 12.7 and 12.8), usually without any discussion or explanation. But how can subduction of seafloor on the Pacific side produce apparently symmetrical structures on the opposite, continental side of the Andes?

Other authors, sufficiently impressed by the apparent symmetry, boldly subduct Brazil under the Andes to match the Pacific subduction. Reutter *et al.* (1988) and Kennan *et al.* (1997) describe or depict low angle faults which they believe show subduction from east as well as west. Kennan *et al.* wrote, 'Thin-skinned shortening in the SubAndes accommodated c.140 km of underthrusting of the Brazilian shield beneath the Cordillera Oriental.' Reutter *et al.*, write of underthrusting of the (eastern) foreland under the mobilised crust of the central part of the orogen.

The Andean symmetry may be carried further through Central America and Mexico, where the Central Plateau is bounded by the Sierra Madre Occidental and the Sierra Madre Oriental.

Symmetry in Europe

The European Alps reveal a different kind of symmetry. The northern part has northward thrusts and the southern Alps has southern thrusts, but there

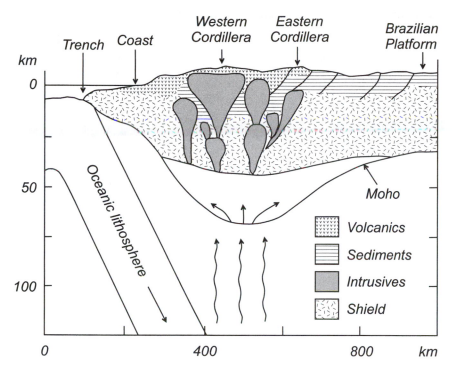

Figure 12.8 Cross-section of the Andes after Brown and Hennessy, 1978.

is no plateau in the middle. Instead there is a belt of metamorphic rock, not directly attached to the thrusts on either side, called the 'root zone'.

The Pyrenees between Spain and France are somewhat similar, with northward thrusts on the north side and southward thrusts on the south side, but there is no root zone between.

Other mountain ranges around the Mediterranean, such as the Apennines, are highly asymmetrical, but when all the ranges are plotted it is possible to see yet another symmetry. The Apennines are symmetrically matched by the Southern Alps across the Po Plain, and the Dinaric Alps across the Adriatic. Unlike the ranges bordering a median plateau, these ranges are convergent. The convergence causes neither folding nor uplift. Patterns can be drawn all around the Mediterranean to display various kinds of symmetry. Unfortunately different authors produce different maps and matches.

Symmetry in the Pacific

In the Pacific a symmetry can be detected in the island arcs, and was described by Krebs (1975) in a paper that relates arc–trench systems to mountain systems and to global vertical tectonics instead of the horizontal movements of

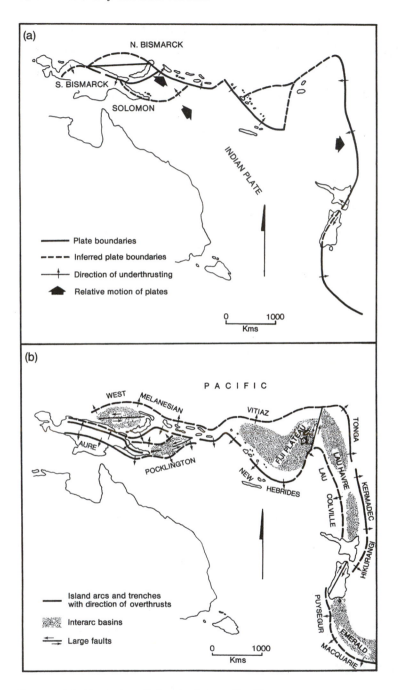

Figure 12.9 Tectonics of the south-west Pacific
   a.   Plate tectonics version
   b.   Two-sided symmetrical system (after Krebs, 1975).

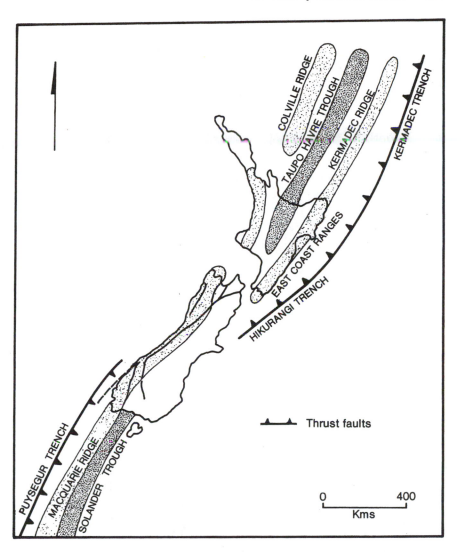

Figure 12.10 New Zealand interpreted as a symmetrical system (after Krebs, 1975).

plate tectonics. Krebs takes into account the remnant or inactive trenches and arcs as well as the active ones. A complex of bilaterally symmetrical structures is revealed. On the axis of the system there is an elongate interarc basin, bordered on both sides by convergent dipping thrust planes or planes of seismic activity. Usually only one side is active at present, and so gives rise to the apparently asymmetrical behaviour of island arcs. The axis is therefore bounded by either active or inactive arcs and trenches. The key areas for understanding the system are the interarc basins which represent the top of diapir-like upwelling material from the asthenosphere. These subcrustal

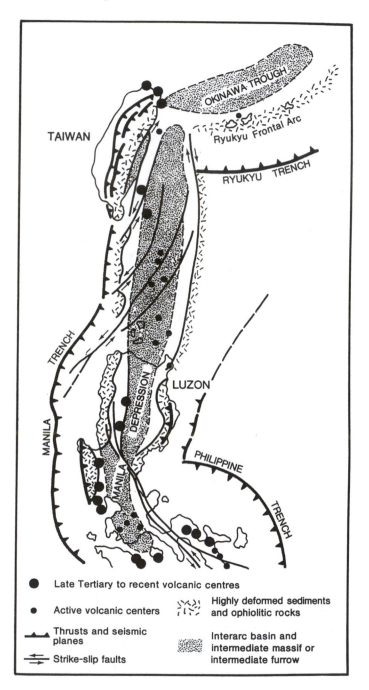

Figure 12.11 The Philippines interpreted as a symmetrical system (after Krebs, 1975).

asthenoliths are characterised by crustal thinning, extension, high heat flow, deep earthquakes, positive gravity anomalies, extrusion of mantle-derived tholeiite basalts, and intrusion of ultramafic massifs. Basically vertical forces cause the asthenolith to rise, and then it spreads horizontally under gravity to produce a host of secondary effects. Krebs certainly gives enough criteria to put his model to the test, but in the plethora of plate tectonic explanations for Pacific structure his model seems to have been lost.

The interpreted bilateral symmetry of the southwest Pacific is shown in Figure 12.9, and the more detailed map of New Zealand (Figure 12.10) shows how bilateral symmetry can be found in that country which is generally regarded as very asymmetrical.

Figure 12.11 shows Krebs' version of the double arc of the Philippines. The plate tectonic explanation of the double-arc system is the arc–arc collision. The New Guinea area is complicated (on Krebs' or any other model) and Figures 12.12 and 12.13 show how it may be interpreted on the bilateral hypothesis. There is convergence on the Solomon Basin, with diverging Benioff zones at the New Britain Trench and the Woodlark Trench. The Woodlark Basin and the Bismarck Basin are regarded as typical interarc basins with upwelling asthenoliths. The Bismarck and Woodlark Basins are about 3 km deep with thin sedimentary cover, high heat flow, crustal extension linear magnetic anomalies, spreading centres, tholeiitic basalts, deep seismic shocks and large strike-slip faults; while the Solomon Basin is 5 km deep and has thicker sediments and an upper crust 15–20 km thick.

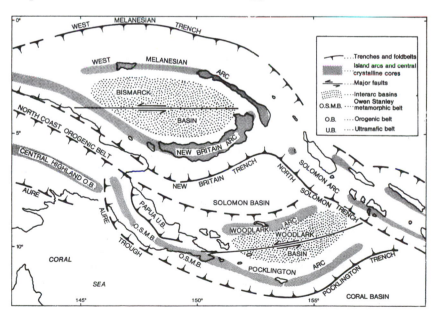

Figure 12.12 Papua New Guinea interpreted as a symmetrical system (after Krebs, 1975).

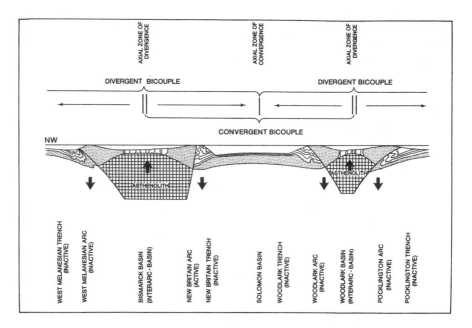

Figure 12.13 Cross-section of Papua New Guinea to show the tectonic components of the symmetrical system (after Krebs, 1975).

The Mediterranean may also be of the bilateral spreading asthenolith type, and other workers have found anomalous structure, with dome-shaped low velocity channels that might correspond to an asthenolith (Figure 12.14). A spreading origin of the Mediterranean as deduced by Glangeaud (1957) is shown in Figure 12.15.

Lack of symmetry

It was the lack of symmetry that worried some earlier workers. If, as they assumed, the folding and thrusting was caused by compression, perhaps between two rigid blocks, why was the folding not symmetrical? Stamp (1950, p. 101) was concerned about the asymmetry of Palaeozoic structures in Wales. 'Some geologists claim that such structures [asymmetrical folds and thrusts] can be produced by equal pressure in two directions, or rather that unilateral thrust cannot exist in the earth's crust parallel to the surface. It seems more likely, however, that given equal pressure in two directions an ordinary [symmetrical] anticlinorium or synclinorium would be produced.'

Several explanation of asymmetry are possible.

1   Original folding was symmetrical, but we are only seeing one half of the resultant structure at present. In the case of Wales, with thrusting

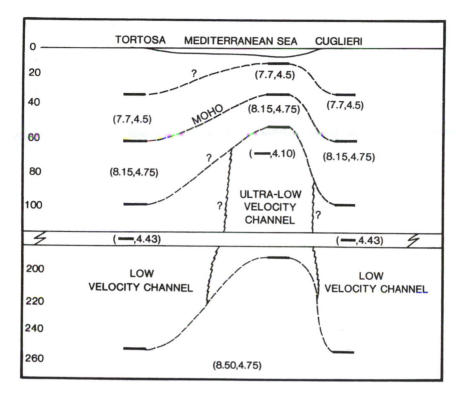

Figure 12.14 Structure of the upper mantle under the western Mediterranean, showing a possible asthenolith (after Berry and Knopoff, 1967).

from the west, perhaps the matching part is in Appalachia, with thrusting from the east. The advent of seafloor spreading and the reality of continental drift made this idea more plausible.

2   Two thrust areas may be separated by a more rigid median block. The median plateaus described earlier might be such blocks.

3   Perhaps in some instances there really is a mechanism for causing asymmetrical thrusting and folding.

Conclusion

In conclusion we might recognise that symmetry is real in many mountain regions, but takes different forms.

Three main explanations have been offered.

1   Symmetrical subduction causing opposing convergent thrust belts.

2   A rising central region (perhaps driven by a rising asthenolith), gives rise to spreading, with divergent thrusts. The geometry of the thrusts in explanations 1 and 2 is the same.

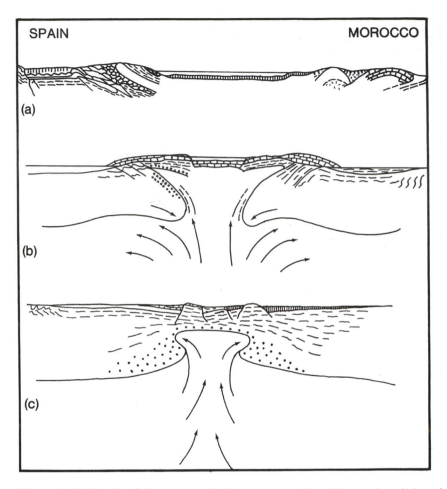

SPAIN   MOROCCO

(a)

(b)

(c)

Figure 12.15 Origin of the western Mediterranean, with a rising asthenolith and
divergent spreading structures (after Glangeaud, 1957).

3   As illustrated by many diagrams of the Andes, a one-sided thrust caused
     by subduction really can somehow create a symmetrical mountain range.

Plate tectonics and mountains

Plate tectonics is the ruling theory, even the current dogma, in Earth
Sciences, and is supposed to explain, among many other things, how mountains are formed. Since we take a very different view in this book, we must
examine plate tectonics theory and its explanation of mountain building,
and how it differs from the explanations given in this book. A vast literature is available to explain plate tectonics, so we have only a very brief
summary here, quoting advocates of plate tectonics (Bickford, 1973):

According to plate tectonic theory, mountains grow at consumption zones . . . Orogenesis, or mountain building, takes place in five stages. First, the initiation of plate motion produces a consumption zone and a trench forms. During the next stage, melting along the descending lithospheric plate initiates volcanism and an island arc, or a mountain arc arises that is composed almost entirely of volcanic rocks but cored, or underlaid, by plutonic rocks. Weathering and erosion of the island or mountain arc and of the nearby continent produce sediments that settle into the shallow ocean basins in the third stage. In and adjacent to the trench and the mountain or island arc, great thicknesses of sediment accumulate . . . The abundance of rock fragments indicates that the sediments have not undergone extensive transport or been subjected to extensive weathering . . . In the fourth phase, continuing plate movement compresses and deforms the accumulated sediments . . . Finally, as the sedimentary and volcanic rocks of the arc continue to be depressed, the entire rock mass thickens and deforms plastically. The growing mass reaches greater elevations. . . . Igneous material derived from melting of the downthrust sediments and from heating along the Benioff Zone intrudes the deformed and metamorphosed sediments. All of these processes are accompanied by vertical uplift, and as the rock mass rises, erosion accelerates – and a mountain belt is formed.

From the descriptions of real mountains presented earlier, the reader can see that this is a very rare scenario, and many of the processes described are on the wrong time scale to match real mountain building.

Plate tectonics and mountain building

In relating plate tectonic boundaries to geomorphology it is clear that there is some correlation between topographic features and plate boundaries, but this is far from perfect. A great many mountains, plateaus and other landscape features have no apparent relationship to plate tectonic situations. Some of the postulated explanations of mountains by plate tectonic theory are rather tenuous and require special events such as reversals of subduction direction, changes of subduction position, and similar complications. Some mountains may also be related to subduction zones of the past, so we may have palaeo-plate tectonics to add to plate tectonics in our search for the explanation for the origin of mountains. The elements of plate tectonics are vital to understanding the world, but alone will not enable us to account for the topography of the Earth.

The Wilson Cycle

The Wilson cycle, named after J. Tuzo Wilson who suggested the idea, refers to a successive recurrence of plate tectonic spreading and convergence with a period generally in the 100 million year range. This may have some relevance to long term geology, but its only application to mountain building is in the, perhaps outrageous, suggestion of Russo and Silver (1996) that the symmetry of the Andes could be accounted for by two phases of mountain uplift separated by a Wilson cycle. The Wilson cycle is in fact well beyond the time scale of the origin of any mountains at the Earth's surface today.

Difficulties with and objections to plate tectonics

1 The total length of spreading sites is three times longer than that of subduction sites.

2 Plate tectonic theory does not explain why subduction is located almost entirely around the Pacific, while spreading is present in all oceans.

3 The spreading sites are not static, but move away from continents. The circum-Antarctic spreading site is the best example. It was once just bounding Antarctica, but has moved away in all directions to its present position. Spreading is also symmetrical around most of Africa. As it moved towards the equator, the circum-Antarctic ridge also grew longer. Plate tectonics has not provided any mechanism for spreading sites to grow longer. As a geometrical consequence of the mobility of spreading sites, subduction sites are also mobile.

4 The North America plate rides indiscriminately over the North Pacific (and other) plates with no regard to spreading sites, plate margins, or transform faults.

5 The chemical and petrological work allegedly achieved by subduction is quite remarkable. Subducted sediments are presumed to mix, melt and contaminate basalt to produce granite batholiths and andesitic magmas that are common in collision sites (andesite and granite do *not* have the same composition). The proportions added from varied sources should be quite variable, and the possibilities of reaction numerous.

6 After subduction the descending slab is supposed to return to the mid-ocean ridge as part of the convection cell. Mid-Ocean Ridge Basalt has a very consistent and rather odd composition. How can MORB, with such a complicated history, be so uniform in composition? Also eruptions at the mid-ocean ridge erupt helium, which is so light that it escapes from the Earth and is not recycled, and juvenile (new) water.

7 Most of the world's great rivers drain to passive margins and most sediment is deposited there (Potter, 1978). How do sediments deposited on passive margins ever get back into the rock cycle or the plate tectonic cycle?

8 Island arcs in the western Pacific are explained as the result of subduction of the Pacific plate. The collision might be expected to cause compression, but instead of compression we find further seafloor

spreading on the other side of the arc, the back-arc basin. 'Subduction roll-over' is the special pleading in this case, but it is hard to apply in three dimensions.

9 Subduction at an island arc. Island arcs are conical surfaces intersecting the Earth's surface. If the direction of plate movement is constant, as seems to be the case (Figure 12.16) how can it give rise to a conical surface? Alternatively, if the subduction is perpendicular to the arc, as suggested by most cross-sections, all the subducted streams must be meeting at the point of the cone, which gives a space problem as material piles up. But we do not find uplift in such places, but more sea-floor spreading.

10 Subduction around curved mountain ranges. Subduction is invoked to explain curved mountain ranges such as the Apennines (p. 66) and the Carpathians (p. 86). But if subduction is perpendicular to the mountain range, the subducted slabs must be converging at some place within the arc, which should cause accumulation of material and presumably uplift, but this area is always a relative lowland.

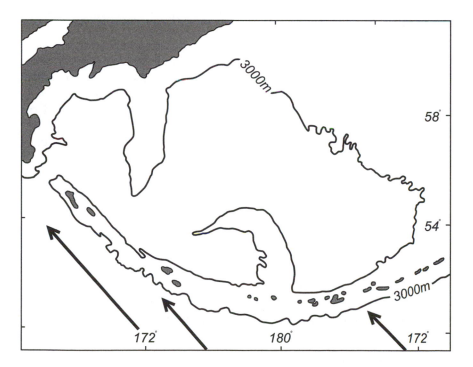

Figure 12.16 Sketch of the Aleutian island arc. It has the same sort of curvature as other island arcs, though the direction of seafloor motion and possible subduction is not perpendicular to the arc, and in the east is even parallel to the arc.

11 A rock mass cannot move simultaneously in opposite directions. But the Po Plain appears to be subducted under both the Southern Alps to the north, and the northern Apennines to the south (Chapter 4). The Pelvoux Massif appears to be subducted to the north, south and west (Chapter 8).

12 If subduction is the cause of mountain building, why did mountain uplift occur mainly in the last 5 million years, while subduction is supposedly a continuous process. that worked over the past 50 to 200 Ma in different parts of the world?

13 Subduction fails to explain why there is a period of still-stand, when land was extensively planated before the period of mountain uplift on a global scale.

14 The symmetry of many mountain ranges, discussed in the last section, presents a further problem. Some of advocates of subduction have the same process of subduction causing underthrust on the near side and overthrust on the distal side, with remarkably similar results. Others subduct Brazil, Russia or other continental masses – which is a huge leap from subduction as originally conceived, and for which there is scant evidence.

15 Stöcklin (1989) pointed out that subduction and spreading had to be equal at the same time, and objected to the plate tectonic concept of subduction of the Indian Plate under Tibet because of the lack of geological evidence for the existence of the vast Late Palaeozoic Tethys Ocean supposed to have been available for Mesozoic subduction. He concludes, rather, that the excess of crustal expansion in the Indian Ocean over crustal shortening in the orogenic belt is evidence for expansion of the Earth.

16 The real problem with subduction is that it can do everything. Plate collision may be invoked 'to explain uplift (making mountains), or subsidence (making deep trenches). It may make folds by compression, but makes backarc basins by tension. The fact that the subduction hypothesis can account for both uplift and subsidence, compression and tension, means that it has too many degrees of freedom. It can account for opposite effects and it is not testable' (Ollier and Pain, 1988).

17 Plate tectonics as a general principle has been enormously helpful in many aspects of geology, but its practitioners have neglected the ground surface, and have often been uncritical in their time scales. The geomorphology of mountains and their recent origin make plate tectonics an improbable mechanism for mountain building.

Plate tectonics and orthodoxy

It should be possible to apply plate tectonics where it fits, and use other mechanisms where appropriate.

Foose (1973) was convinced of vertical and gravity tectonics in his own region (the Middle Rocky Mountains) but was willing to accept horizontal compression in neighbouring areas. He believes that the validity of a tectonic interpretation for any specific region need not obviate the validity of a totally different interpretation in another crustal region. More significantly, he wrote, 'Final answers to global-scale tectonic problems will be derived not by blurring the facts concerning tectonic features and forces in smaller crustal units but by sharpening them.'

But most of the time it seems that the orthodox plate tectonic practitioners will not tolerate other explanations. We have had our own work rejected by referees because it was not couched in the proper language.

As Zeil (1979) noted (see Chapter 6), 'In the process [of applying plate tectonics to mountain building] a number of geophysical and geological data were inadequately considered or were suppressed in favour of the grandiose tectonic picture.' To this we could add that most geomorphology is omitted or suppressed.

Some authors express deeper disquiet about the plate tectonic orthodoxy. Van Bemmelen (1975) wrote: 'This basic assumption [plate tectonics] suppresses alternative approaches to the dynamic history of our planet.' This is particularly so in the study of mountain building. Mountains have been forced in the Procrustean bed of plate tectonics where the great bulk of geological and geomorphic information shows they do not fit in time, space or process.

Writing on the topic of plate tectonics in geologic education, Lowman (1992) noted there is a dismaying intellectual climate in geology. It has become a bland mixture of descriptive research and interpretive papers in which the interpretation is a facile cookbook application of plate tectonics concepts used as confidently as trigonometric functions. The theory of plate tectonics has given Earth Sciences for the first time a master plan of compelling simplicity. But it can stifle research on the great problems of tectonics by convincing students that these problems have already been solved. Lownman advises that geology teachers 'should broaden their treatments, and remember that the theory of plate tectonics was not delivered on Mount Sinai carved in stone'.

The analogue with religious belief was also noted by Baars (1972): 'The concept of the New Global Tectonics may be likened to a new religion, since hard facts are lacking; if one is not a 'believer' one is considered an 'atheist' with regard to the many theories and interpretations of the 'clergy' – the oceanographers and geophysicists.' It will be clear to readers that we are not totally converted to the religion. You can believe what you like, but please don't send missionaries!

Significance of still-stands and the great speed of planation

Before most mountains were uplifted there was a period of tectonic still-stand or at least quietness when planation surfaces were eroded.

Plains of terrestrial erosion are made by river erosion, which cuts down to sea level (base level), accompanied by valley widening and landscape flattening. This process takes time. If there is renewed uplift, the erosional process starts all over again. We do not have much idea of how long it takes to form an erosion surface of low relief. Estimates are usually on the longer time scale, like those of Lester King. He envisaged a series of planation surfaces initiated in the Cretaceous, Eocene, and Miocene. In other words, he had whole geological periods to play with. It is generally thought, and supported by reported rates of erosion (e.g. Young and Saunders, 1986), that erosion rates become slower as relief is reduced, and that to form a low relief planation surface takes many millions of years.

The remarkable thing is that plains of great perfection are ever made, despite all the obvious possibilities of complication. But they are real, and planation surfaces were widespread before the uplift of the many mountains of Plio-Pleistocene age. In turn, this implies that there was a period of tectonic quiet in many areas immediately before the Plio-Pleistocene uplift when planation could operate with little tectonic disruption.

In Tibet there appears to be no problem. Li (1995) points out that the development of planation surfaces requires a lengthy period of tectonic stability, perhaps for 10 to 20 million years. He concludes that the Peak Surface formed between 38 and 20 million years ago, and the Main Surface between 18 and 3 million years ago. But in many other places that length of time is not available, and planation must have occurred much faster.

There is nothing very special about the climate in the Late Miocene–Early Pliocene period when there often occurred planation that suggests an increased erosion rate, and in any case the mountains discussed are in a wide range of latitudinal and climatic situations. At present, the cause of the observed high rate of planation remains a mystery.

It is even more difficult to make a planation surface if the land is rising tectonically, yet the planation surfaces are there. This suggests tectonic quiet in many different places. It is virtually a global tectonic quiet period. Why should this be? And why should a period of tectonic quiet be followed so rapidly by a period of great uplift? Furthermore, in many regions the planation surfaces cut structures that indicate high tectonic activity just before the planation.

Coltorti and Pieruccini (2000) were aware of these problems in discussing their suggestions that the Apennine planation surface is a wave cut platform created across the whole region in the lower Pliocene. They wrote, 'We realise that the development of a wave-cut platform, so wide that it was able to flatten the whole Peninsula in a very short time span has problems in regard to present day models of landscape modelling'; and 'A further question to be solved is the presence of a period of tectonic quiescence because it is even harder to explain such wide platforms in a context of active mountain building.'

One of the biggest obstacles to our hypothesis of widespread Plio-Pleistocene mountain building is that the period of time available for the preceding planation is too short. Only further investigation can clarify this point.

Significance of widespread Plio-Pleistocene uplift

Our compilation of mountains throughout the world shows that a major phase of uplift occurred in the Plio-Pleistocene In previous chapters we have put the pros and cons and debated the issues, but a very brief compilation is shown in Table 12.1. We do not believe this is an artefact of our sampling, as workers all over the world have come to the same conclusion. If it true, it has several important implications and puts constraints on possible mechanisms and theories.

1   Synchronism of mountain building. There has been much debate among tectonicians, stratigraphers and other geologists about whether mountain building occurs at particular times in geological history, or if there are 'orogenies' in different parts of the world at different places.

The conclusion reached here suggests that at least the latest orogeny was roughly synchronous over a large area of the world. However, there is no doubt that some uplift took place at different times. Passive margin uplift is frequently older, possibly going back to the Early Tertiary, and some may be older still. Some, like the Scandinavian warp, seems to have a resurgent action on the same site even since the Mesozoic. Nevertheless there does seem to have been a major pulse of uplift in the Pliocene extending into the Pleistocene and even to the present. In some instances this uplift was preceded or accompanied by folding (Apennines, New Guinea), but in many other places it was not. Uplift of many mountains was synchronous: orogeny (in the sense of folding of rocks) was not. The folding of the Caledonian rocks of Scandinavia and Scotland is much older than the uplift that made the mountains. The structures in the Carpathians are much older than the Late Pliocene uplift.

2   Uplift occurred over a relatively short and distinct time. Some earth process switched on and created mountains after a period with little or no significant uplift. This is a deviation from uniformitarianism.

3   The mountain building period is generally relatively short. It does not appear to be on the same time scale as granite intrusion which takes tens of millions of years, or plate tectonics which is continuous. The same rapid uplift occurs in areas where hypotheses such as mantle plumes do not seem appropriate. We do not yet know what causes this short, sharp period of uplift, but at least the abandonment of naive mountain building hypotheses might lead to further realistic explanations.

Table 12.1 Summary of planation and uplift in mountain areas

EUROPE	
SWISS ALPS	
Planation	Pliocene
Uplift	Quaterrnary
Eastern Alps uplift	Late Miocene
THE JURA	
Uplift	Pleistocene
THE APENNINES	
Emerged from the sea	Latest Pliocene
Uplift	Early–Middle Pleistocene
THE PYRENEES	
2000 m of uplift since the beginning of	Pliocene
THE CENTRAL CORDILLERAS OF SPAIN	
Uplift (major)	Plio-Pleistocene
BAETIC CORDILLERA	
Planation surface	Upper Miocene
Uplift	Pliocene
WESTERN CARPATHIANS	
Summit level	Upper Miocene
Uplift	Pliocene
SOUTHERN CARPATHIANS	
Younger planation	Pliocene
Uplift and basalt volcanics	Pliocene
THE CAUCASUS	
Planation not younger than	Upper Pliocene
THE URAL MOUNTAINS	
Uplift start	Pliocene–Pleistocene boundary
Uplift intense movement	Middle Pleistocene
THE SUDETEN	
Uplift (main)	Pliocene–Early Quaternary
ASIA	
THE TIBETAN PLATEAU	
Planation (main)	Pliocene
Uplift in phases:	A 3.4 Ma
	B 2.5 Ma
	C 1.8 Ma
THE HIMALAYAS	
Uplift start	3.4 Ma
Uplift main	Quaternary
THE KUNLUN MOUNTAINS	
Uplift	Late Tertiary–Quaternary
THE TIEN SHAN	
Uplift	Quaternary
TURFAN DEPRESSION	
Subsidence	Quaternary

Table 12.1 continued

THE ALTAI MOUNTAINS	
Peneplain	Tertiary
Uplift	Tertiary
THE TRANSBAIKAL MOUNTAINS	
Planation surface	Pre-Mid Tertiary
Uplift	Mid- and Late Tertiary
THE KARAKORAM	
Uplift	Late Neogene–present
SHANXI MOUNTAINS AND GRABEN	
Planation	Miocene
Faulting began	Late Pliocene
Uplift (maximum)	Middle Pleistocene
JAPANESE MOUNTAINS	
Planation	Late Tertiary
Uplift	Pliocene–Early Pleistocene
TAIWAN	
Uplift	Plio-Pleistocene
NORTH AMERICA	
SIERRA NEVADA	
Uplift	Post-Pliocene
Start of western tilting	5 Ma
BASIN AND RANGE FAULT BLOCK LANDSCAPE	
Planation surface	between 7 and 4 Ma
Start of normal faulting	4 Ma
THE COLORADO AND OTHER PLATEAUS	
Uplift (main)	Late Pliocene–Recent
THE BIGHORN MOUNTAINS	
Planation surface	Middle Tertiary
Uplift	End Tertiary and Pleistocene
THE ROCKY MOUNTAINS	
Major uplift within the past	5 million years
THE COAST RANGES	
Uplift	Late Pliocene
CANADIAN CORDILLERA	
Planation surface	Late Miocene
Uplift	Pliocene
THE CASCADE RANGE	
Peneplain	Pliocene
Uplift	4–5 Ma
SOUTH AMERICA	
COLOMBIA	
Planation surface	Plio-Pleistocene

Table 12.1 continued

CHILE	
Uplift	Pliocene and Pleistocene
BOLIVIA	
Planation surface	Pliocene
Uplift	Plio-Pleistocene
ECUADOR	
Planation	Upper Miocene
Uplift	Plio-Pleistocene
OTHER REGIONS	
ETHIOPIAN RIFT	
Uplift	2.9 and 2.4 Ma
WESTERN RIFT	
Uplift	3–2 Ma
RUWENZORI	
Uplift	within the last 3 million years
NEW GUINEA MOUNTAINS	
Erosion surface	Plio-Pleistocene
Uplift	2 Ma
NEW ZEALAND MOUNTAINS	
Planation surface	Late Tertiary
Uplift (major)	Pliocene

In relation to the long-standing debate about whether there are distinct mountain building periods or whether the process might be continuous through Earth history, the answer seems to be definitely in favour of the former. We are seeing the results of a distinct and remarkably young mountain building period. This is a deviation from strict uniformitarianism.

This rapid mountain building period has apparently been an accepted idea in Russia for quite a while. Pavlides (1989) first brought the idea to Western notice, and he attributes it to Obruchev (1964) whose work we have not seen. The mountain building period is known as the 'Neotectonic Period', and was defined by Obruchev as 'Young tectonic movements occurring in the late Tertiary and early half of the Quaternary'. Crustal movements of this period are said to be of first order magnitude in the Earth's history and characterised by epeirogeny. Apparently some Russian workers maintained that the term should be applied only to vertical crustal movement of old cratonic provinces, but this seems to be needlessly restrictive, and Hoshino (1998) considers that the Neotectonic Period also affected the mobile belts – it is a global crustal phenomenon.

For the past 30 years plate tectonics has dominated geology, and the greatest single theme has been subduction, which allegedly formed mountains and also the structures within them. Subduction may have a role in other studies, but because it is a continuous process that has allegedly operated for hundreds of millions of years, it is a most improbable mechanism to make mountains in a few million years, complete with their erosion surfaces. *In areas of Plio-Pleistocene mountains there is no time for subduction to be an effective mechanism.*

Most of the Plio-Pleistocene mountains are parts of what is generally referred to as the Alpine orogeny. The Andes, the Tibet Plateau and Himalayas, and the European Alps themselves are all classic examples. A few other mountain areas not regarded as 'Alpine' also have planation surfaces uplifted in the Pleistocene, such as the Urals, Ruwenzori, the ranges of central Spain, and those of Central Asia. The 'Alpine Orogeny' seems to be a real thing, in the sense of mountain building as well as 'the formation of folds and rock structures' as described in Chapter 1. What has become clear in this book is that most of the rock structures such as folds and faults were formed before the vertical uplift that actually formed the mountains.

The short mountain building period is not on the same time scale as granite intrusion which takes tens of millions of years, or plate tectonics which is continuous. The same rapid uplift occurs in areas where hypotheses such as mantle plumes do not seem appropriate. In the plate tectonic explanation of mountains there is a seamless run-on from the processes that allegedly fold rocks to the formation of mountains. But even some of the exponents of plate tectonics have been describing the obvious extensional tectonics of very late Cenozoic times in contrast to the compressive regime of earlier times. We are coming to an understanding of a major tectonic change in many areas, from a compressive to an extension system, linked somehow by vertical uplift and mountain building. To make progress will require an integration of structural and tectonic geology with geomorphology.

The 6 million year period of mountain building which seems to emerge from our collation, or the Neotectonic Period of some Russian workers, appears to have been recognised, at least partially, by earlier workers. It may be the equivalent of the Mediterranean Movement (Aubouin, 1965), the Antillean Revolution (Schuchert, 1935), and the Pasadenian Orogeny (Stille, 1955).

Mountain building mechanisms

Mountains are made by erosion of plateaus, which are themselves made by vertical uplift of plains. What causes the vertical uplift? Over 20 mechanisms have been suggested (Table 12.2) so how can a selection be made?

Most tectonicians, convinced by the plate-tectonic paradigm, believe that the vertical movement must be derived somehow from horizontal movement, so have collision and subduction as the driving force. Burg and Ford (1997, p. 13) wrote: 'The inference from both observation and modelling is that there *must* be subduction involved in orogens formed at plate

Table 12.2 Twenty causes of tectonic uplift

---

1 Thermal expansion due to a mantle plume or hot-spot.
2 Thermal expansion due to overriding and subduction of a hot mid-ocean ridge or spreading centre.
3 Thermal expansion due to shear heating along a lithosphere–asthenosphere interface.
4 Expansion accompanying partial melting (the increase in volume on fusing basalt is about 8 per cent).
5 Hydraulic reactions such as serpentinisation (10 per cent expansion).
6 Introduction of volatiles due to deep-seated dehydration of hydrous minerals.
7 Expansion due to depletion of 'fertile' mantle in garnet and iron resulting from basalt genesis.
8 Crustal thickening due to horizontal transfer of mass in the lower crust.
9 Deep-seated solid state reactions such as eclogite-basalt transformation.
10 Subduction at a very shallow angle, perhaps horizontal.
11 Simple subduction, or continent–continent subduction.
12 Cessation of subduction and resulting thermal equilibrium of static slab.
13 Isolation of a plateau by listric normal faulting in the surrounding area.
14 A piece of cooling lithosphere detaches from the crust and is replaced by a counterflow of asthenosphere, which warms the crustal rocks and causes uplift (due to thermal expansion) and volcanism.
15 Intrusion of magma into the lower crust.
16 Intrusion of sills.
17 Isostatic uplift after scarp retreat.
18 Isostatic uplift after regional erosion.
19 Underplating, the addition of unspecified lighter material to the base of the crust.
20 Isochemical phase change with volume increase in the lower crust or upper mantle.

---

boundaries' (our italics). This is a vast over-statement. Nobody has observed subduction, and there are plenty of models of mountain building without it. After all, mountains were studied long before plate tectonics was ever thought of. Similarly, Okaya *et al.* (1997) claim that in Bolivia, 'The Altiplano rose uniformly by lower crustal thickening.' There is absolutely nothing to demonstrate this proposed cause.

The complications of assumed subduction are often bizarre. Are the Apennines really being underthrust from all directions from north through east to south? Is Brazil being subducted under Bolivia? Furthermore, convergence of plates does not necessarily produce collision or high topography, as already described for the Po Valley. Eocene marine sediments as old as or younger than the alleged collision between India and Asia show that the Indus–Tsangpo collision zone was below sea level (Burg and Chen, 1984). The Caribbean collision site today is under the sea (Dercourt *et al.* 1993) as is the site of the collision between Timor and Australia (Karig *et al.* 1987).

In fact, subduction seems to be one explanation of mountain building that can be eliminated, because it is on the wrong time scale. The Himalayas rose in the past 3.4 Ma, but the alleged subduction of India under Asia occurred over 55 million years. The Andes rose in the past few million years but Pacific subduction has allegedly been active for 200 million years.

The many mechanisms of mountain building proposed in the past may be summarised as:

1   Mechanisms that cause vertical uplift
    Granite intrusion
    Mantle plumes and asthenoliths
    Hot spots
    Rise of a subducted plate under a continental edge

2   Mechanisms of crustal shortening
    Collision of landmasses (pre-plate tectonics)
    The bow-wave of continent grinding into ocean (Wegener)
    Convergence of opposed sides of a geosyncline
    Collision of plates (subduction, obduction, flakes)
    Tectogene and eversion of sediments
    Crustal thickening at collision site, with isostatic rise

Our own view of mountain building may be summarised as follows:

1 Rock structures under plains, plateaus or mountains may not be the cause of the plain, plateau or mountain.
2 Plains are made by erosion, and vertical uplift of plains creates plateaus.
3 Mountains are usually plateaus or eroded plateaus.
4 Some structures, especially monoclines and vertical faults, may be associated with uplift.
5 There are no fold mountains in the sense that folding and mountain formation occurred simultaneously and by the same force.
6 A plateau may spread laterally after uplift, with the formation of thrust faults and post-uplift folds.
7 Deep incision of a plateau can cause isostatic response, with formation of new structures including anticlines along major valleys and even major mountain ranges.
8 Major drainage patterns are on the same time scale as global tectonics, and often pre-date the formation of rift valleys, mountain ranges or continental margins.
9 Theories of mountain building must explain both the period of tectonic quiet that permitted erosion of a planation surface, and the usually young and rapid uplift that made a plateau.
10 Subduction is a continuous and long lived process that does not readily explain either the tectonic quiet, or the young and rapid uplift of most mountains.

The sequence of problems

In this chapter we have examined some specific problems in the study of mountains, but problems do not come singly. To study any mountain region a series of questions must be asked:

What is the rock type?
What is the rock structure?
What is the age of the rock?
What is the age of the structure?
What is the form and distribution of the mountains?
What is the drainage system and how could it have evolved?
What is the chronology of landform development?
What is the regional and global tectonic setting?

Some of these questions are fairly simple, but others present problems. Ideally all these components should be studied, but often conclusions are based on partial information, such as seismicity or petrology, and from this inadequate basis a big leap is made to the origin of the mountains.

We might be accused of bias ourselves, as we have stressed the geomorphic aspects and partly neglected other lines of evidence. But we have done this deliberately, because this is the part most frequently left out. We still have to convince geologists and geophysicists that geomorphology is not a Quaternary addition that can be ignored. It is on the same time scale as global tectonics and mountain building.

We believe that the ideal paper on mountain building would treat all aspects of the mountains, and point out where different data sources disagree, and where there is conflict in interpretation.

Perhaps the biggest problem in the study of mountains is that the contribution of geomorphology will continue to be ignored!

13 Science and the origin of mountains

As explained in this book, mountains are made by uplift of originally low-lying continental areas. If the uplifted area remains undissected it is a plateau; if it is deeply dissected it will be a typical mountain chain with isolated peaks rather than a continuous high surface. In this way theories of mountain building are really theories of plateau formation

Our overall conclusion, that mountains are made by erosion of uplifted plateaus, has many precursors. We might well have written the following:

> In the older classifications of mountains 'fold mountains' figured prominently, while the term 'mountains of circumdenudation' was reserved for residuals of high standing plateaux of horizontally bedded rocks that had become deeply dissected by erosion.
>
> It has long been recognized that there is little correspondence between the fold structures and the actual forms of mountains formed of folded rocks.
>
> Not many years ago all the great mountain ranges of the world which are built of folded rocks were believed to have originated as 'fold mountains' and to be now in process of reduction by erosion for the first time.
>
> In many parts of the world evidence has come to light which proves that mountain ranges are really dissected plateaux, *though composed of folded rocks*; that is to say, they are *two-cycle*, or perhaps *multi-cycle*, mountains, the region having been worn down by erosion to small relief at least once, and possibly more than once, prior to a nearly even uplift, which was followed by deep dissection of the plateau so formed.
>
> Instead of even doming taking place to upheave a plateau that is later destroyed by erosion, the land surface is sometimes broken into blocks, some of which are uplifted.

In fact these words were written by Charles Cotton in 1918. He seemed to think that the old ideas would soon be gone, but his optimism was not justified. Eighty years on we are still trying to get the simple message across, and we have no more reason to be optimistic than Cotton had. Indeed, the dead weight of orthodoxy and the preference for models over ground truth that prevails today suggests that we have less reason for optimism, not more.

Theories and bandwagons

Ideas about mountain building have been subject to fads throughout the history of Earth Science. The shrinking Earth, geosynclines, and latterly plate tectonics have all provided 'answers', usually flawed by the scientific fallacy of a single cause, and biased by selective evidence and the rule of dogma.

Most theories of mountain building are really theories of folding of rocks, and of the effects of subduction at continental margins. The fact that many mountains are not made of folded rock, or not located on suitable continental margins, is generally ignored. Some mountains are at the edges of the plates delineated in plate tectonic theory, but many are not. Interaction at plate edges may give rise to mountains, but there is a great variety in the styles of mountains even within this situation. The Andes of Peru, for example, seem to have a very different geological and tectonic history from the Coast Ranges of North America. Plate tectonic theories of mountain formation relate in some way to the hypothesis of subduction, but the many mountain chains in positions where subduction is impossible, such as the Drakensberg of South Africa or the Trans Baikal Mountains in the middle of Asia, show that subduction is not necessary for mountain formation. Subduction, like folding of rock in geosynclines, has nothing to do with mountain building.

Some mountains may relate to the changes in continental configuration in plan. Thus the opening of the Bay of Biscay may be related to compressional thickening under the Pyrenees; the opening of the Arabian Sea may relate to the Baluchistan orocline. But it is clear that many mountains are not in any such situation so this is merely an occasional accompaniment to mountain building. Some mountains may relate to collision of continental blocks, like the Himalayan Mountains where India collided with Asia, or the Urals where Asia collided with Europe, but again there are far more mountain ranges where collision cannot be invoked. And, as explained earlier, the mountain formation does not seem to be on the same time scale as the alleged collision.

All we know for certain is that mountains exist because they have been pushed up, generally after a period of planation at low altitude. Vertical tectonics caused the uplift and numerous theories attempt to explain vertical tectonics. Some, especially the plate tectonic explanation, translate the vertical movement into horizontal movement – the collision of plates causes vertical uplift in the area of compression. But there are many other possibilities as listed in Table 12.2.

The primacy of physics and the neglect of landscape evidence

Physicists (sometimes in the guise of geophysicists) are often very willing to tell geologists how things work. Lyttleton, for instance, wrote a book called *The Earth and its Mountains* (1982). In this he wrote, 'one generalization on which all geologists agree is that mountain building involves a reduction in the surface of the globe, a shortening of distance between points on its surface, and that all mountain building is the product of a single mechanism – sqeezing by horizontal compression.' His idea was that internal melting led to an unstable Earth, which eventually caused collapse that suddenly reduced the outer radius by 70 km, intensely fracturing the outer shell and making mountains. The Earth was then stable for a while, but internal temperature rose until another violent episode broke the exterior. Catastrophic periods of mountain building occurred at intervals of 100 million years, 'enough for some 200 mountain ranges of volume comparable with the Rocky Mountains'. We hope readers of this book will realise that all we know of mountains, their structure and their age, makes nonsense of this story, but at the time of its publication it was reviewed favourably by some, especially for its 'mathematical rigour'. No amount of mathematics will help if the basic facts are wrong.

Physics does not have a good record in geology. Lord Kelvin (using the best physics of his day) grossly underestimated the age of the Earth despite massive geological evidence for a longer time scale. Sir Harold Jeffreys dismissed continental drift as 'quantitatively insufficient and qualitatively inapplicable. It is an explanation which explains nothing which we wish to explain.' He probably held back continental drift (and plate tectonics) for 50 years.

Of course physics is necessary as the basic science. It was the discovery of radioactivity that showed Kelvin's error. It was work on palaeomagnetism that convinced physicists and geologists of the reality of continental drift. Jeffreys' work on rock mechanics, quoted in Chapter 10, seems still to be sound, though generally neglected by physicists today. Physics has much to offer, but it should not automatically take precedence over geological facts. It seems a good idea for geologists to derive their own theories without waiting for permission from the physicists.

Causes

Some people feel very uncomfortable if they do not have a mechanism to explain things. Even if it were agreed that mountains result from vertical uplift, we still want to know the cause or mechanism of uplift. Maybe we shall never know, but the lack of knowledge of the cause does not affect the knowledge that the mountains were uplifted. Perhaps the best safeguard is to have lots of possible mechanisms, but not feel compelled to pick just one as *the* causal mechanism. One of the greatest, and commonest, errors in the history of science is the fallacy of single cause. It is unlikely that we shall ever find one single cause of mountain building that accounts for vertical tectonics in all mountains.

The great professor of Geology at Cambridge, O.T. Jones, spent his life studying geosynclines and wondering why they sink. He concluded that they sink 'because they have that sinking feeling'. Likewise, perhaps, mountainous areas feel uplifted! The first thing to accept is that geosynclines did sink; we might find a mechanism later. Similarly mountains were uplifted. We must accept that even if we do not know, or disagree on, the mechanism.

The ultimate cause has a fatal fascination. Indeed it is the first thing that most people want to know. But it is the thing we are striving to find out. If we first get the geometry right, then the timing, we might work out the kinematics, and if we know that we might, just possibly, venture on the driving force.

Tectonicians and geophysicists can, it seems, cheerfully subduct a trench or throw up a mountain range without considering the effects on the landforms at the ground surface. Is the ground surface of an appropriate age? Have the rivers been diverted? Does the amount of uplift seem to be of the right order for the proposed mechanism?

In many discussions of mountains there is a mass of data on geophysics and structure, but little on the actual landscape. We should use all available information, including that of geomorphology. Unfortunately those wrestling with these problems usually ignore the directly observable evidence of the ground surface. The greatest weight is given to obscure geophysical evidence, while the most obvious and readily available evidence, the topography, is ignored. As Petriovsky wrote (1985), 'The study of the relief of the Earth is much easier and cheaper than the study of the Earth's depths and uses direct observation.' The idea is echoed by Kalvoda (1992), 'many papers on the global tectonics of lithospheric plates are based only on geological and geophysical data, and the direct evidence of surface phenomena is neglected or only conceived intuitively and understood from a physiographic point of view.'

Geophysicists are not entirely to blame for this situation. It is unfortunate that western geomorpholists have been very slow in accepting tectonic ideas into their concepts, and even when they do they often slavishly adopt orthodox plate tectonic ideas (e.g. Morisawa, 1975; Summerfield, 1991).

Orthodoxy

Another problem arises from orthodoxy. Anyone who disagrees with the ruling theory is regarded as an ignorant fool by the majority, and authoritarian orthodoxy even goes so far as the suppression of publications that do not fit the orthodox scenario (nowadays plate tectonics). There was a time when plate tectonics (in the older version of continental drift) was itself unorthodox. Chamberlin (1928) said geologists might well ask if theirs could still be regarded as a science when it is 'possible for such a theory as this [drift] to run wild'. Even the idea of nappes having moved for scores of kilometres was once unorthodox, and 'Its mere possibility was then as firmly denied, as is now the possibility of continental drift' (Van Waterschoot Van der Gracht, 1928).

Orthodoxy has its drawbacks. As Ollier wrote (1985): 'It seems very reasonable to suppose that we will make advances in the next twenty-five years just as we have in the past twenty-five or in any quarter of a century that we like to think of. Some of today's ideas will certainly be replaced. So if you are completely orthodox today, you will almost certainly be wrong in the future. It is as well to be kind to the oddballs, for this year's orthodoxy may be next year's ruling theory.'

Modelling and ground truth

The latest obstacle to the flow of reason is an increasing disregard for ground truth, or what used to be called field evidence. Ollier (1999) expressed it as follows:

> Gansser (1991) wrote 'During the classical exploration in the 19th and early 20th centuries the ratio between facts and theories was 1:0.5. Plate tectonics changed it to 1:3 and with geophysics, geochemistry and structural analysis the ratio became 1:5.' Nowadays, with the dominance of modelling it is perhaps 1:10. It would be nice to reverse this sorry state of affairs. It would be good if scientists would start from their own factual information in the study of mountains, rather than follow simplistic theories derived from other sources.

The study of mountains makes a fascinating chapter in the history of science. We believe that for the past few decades it has been hampered by rigid orthodoxy and lack of field observations. The time is ripe for renewed interest in the origin of mountains, and their relationships to many other aspects of science ranging from structural geology to climatology and biogeography. The message of this book is that there is still a lot we do not know, and a lot of fascinating work to do.

References

Allmendinger, R.W. and Jordan, T.E. 1997. The Central Andes. In van der Pluijm, B.A. and Marshak, S. *Earth Structure: an Introduction to Structural Geology and Tectonics.* WCB/ McGraw-Hill, Dubuque, Iowa, 430–434.

Allmendinger, R.W., Jordan, T.E., Kay, S.M. and Isacks, B.L. 1997. The evolution of the Altiplano-Puna Plateau of the Central Andes. *Ann. Rev. Earth Planet. Sci.*, 25, 139–174.

Andriessen, P.A.M., Helmens, K.F., Hooghiemstra, H., Riezbos, P.A. and Van Der Hammen, T. 1994. Pliocene–Quaternary chronology of the sediments of the high plain of Bogotá, Eastern Cordillera, Colombia. *Quat. Sc. Rev.*, 12, 483–503.

Atwood, W.W. 1940. *The Physiographic Provinces of North America.* Ginn, Boston.

Aubouin, J. 1965. *Geosynclines.* Elsevier, Amsterdam.

Augustinus, P.C. 1992. Outlet glacier trough size – drainage area relationships, Fiordland, New Zealand. *Geomorphology*, 4, 336–347.

Axelrod, D.I. 1956. Mio-Pliocene floras from west-central Nevada. *Univ. California Publications in Geological Sciences*, 33.

Axelrod, D.I. 1962. Post-Pliocene uplift of the Sierra Nevada, California. *Bull. Geol. Soc. Am.*, 73, 183–98.

Baars, D.L. 1972. *The Colorado Plateau.* Univ. New Mexico Press, Albuquerque.

Baker, B.H. and Mitchell, J.G. 1976. Volcanic stratigraphy and geochronology of the Kedong–Olorgesailie area and the evolution of the South Kenya rift valley. *J. Geol. Soc. Lond.*, 132, 467–484.

Baker, B.H., Mohr, P.A. and Williams, L.A.J. 1972. Geology of the eastern rift system of Africa. *Geol. Soc. Am. Spec. Paper* 136.

Barrett, P.J., Hambrey, M.J. and Robinson, P.R. 1991. Cenozoic glacial and tectonic history from CSIRO-1, McMurdo Sound. In Thomson, M.R.A., Crame, J.A. and Thomson, J.W. eds. *Geological Evolution of Antarctica.* Cambridge University Press, Cambridge, 651–661.

Bartolini, C. 1980. Su alcune superfici sommitali dell'Appennino Settentrionale (prov. di Lucca e di Pistoia). *Geogr. Fis. Dinam. Quat.*, 3, 42–60.

Bartolini, C. 1986. The neotectonic map of Italy and adjoining seas. *Mem. Soc. Geol. It.* 31, 53–57.

Bashenina, N.V. 1984. Ural Mountains. In Embleton, C. ed. *Geomorphology of Europe.* Macmillan, London, 404–411.

Bates, R.L. and Jackson, J.A. 1987. *Glossary of Geology.* 3rd. edn. American Geological Institute, Alexandria, Virginia.

Behrendt, J.C. and Cooper, A. 1991. Evidence of rapid Cenozoic uplift of the shoulder escarpment of the Cenozoic West Antarctic rift system and a speculation of possible climatic forcing, *Geology*, 19, 315–319.

Berry, M.J. and Knopoff, L. 1967. Structure of the upper mantle under the western Mediterranean basin. *J. Geophys. Res.*, 72, 3613–3626.

Bickford, M.E. 1973. *Geology Today*. CRM Books, Del Mar, California.

Bieber, D.W. 1983. Gravimetric evidence for thrusting and hydrocarbon potential of the east flank of the Front Range, Colorado. In Lowell, J. D. ed. *Rocky Mountain Foreland Basins and Uplifts*. Rocky Mountain Association of Geologists, Denver, 245–255.

Bird, P., 1978. Initiation of intracontinental subduction in the Himalaya. *J. Geophys. Res.*, 83, 4975–4987.

Bird, J.M. and Dewey, J.F. 1970. Lithosphere plate-continental margin tectonics and the evolution of the Appalachian orogen. *Geological Society of America Bulletin*, 81, 1031–1059.

Bishop, P. 1986. Horizontal stability of the Australian continental drainage divide in South Central N.S.W. during the Cainozoic, *Australian Journal of Earth Sciences*, 33(3), 295–307.

Bishop, P. 1988. The eastern highlands of Australia: the evolution of an intraplate highland belt. *Progress in Physical Geography*, 12(2), 159–182.

Bohannon, R.G. and Geist, E. 1998. Upper crustal structure and Neogene tectonic development of the Californian continental borderland. *Bull. Geol. Soc. Am.*, 110, 779–800.

Borisevich, D.V. 1992. Neotectonics of the Urals. *Geotectonics*, 26, 41–47.

Borradaile, G.J. ed. 1975. *Progress in Geodynamics*. North-Holland, Amsterdam.

Bowman, L. 1916. *The Andes of southern Peru*. Henry Holt & Co., London.

Bremer, H. 1985. Randschwellen: a link between plate tectonics and climatic geomorphology, *Zeitschrift für Geomorphologie Supplement Band*, 54, 11–21.

Bretz, J.H. 1962. Dynamic equilibrium and the Ozark land forms. *Am. J. Sci.*, 260, 427–438.

Briceño, H.O. and Schubert, C. 1990. Geomorphology of the Gran Sabana, Guayana Shield, southeastern Venuzuela. *Geomorphology*, 3, 125–141.

Bridges, E.M. 1990. *World Geomorphology*. Cambridge University Press, Cambridge.

Brookfield, M.E. 1998. The evolution of the great river systems of southern Asia during the Cenozoic India–Asia collision: rivers draining southwards. *Geomorphology*, 22, 285–312.

Brooks, C.K. 1985. Vertical crustal movements in the Tertiary of central East Greenland: a continental margin at a hot-spot. *Z. Geomorph. Supp. Bd.*54, 101–117.

Brown, E.H. 1960. *The Relief and Drainage of Wales*. University of Wales Press, Cardiff.

Brown, G.C. and Hennessy, J. 1978. The initiation and thermal diversity of granite magmatism. *Phil. Trans. R. Soc. Lond.*, 288A, 631–643.

Buchan, J. 1924. *The Last Secrets*. Riverside Press, Cambridge.

Büdel, J. 1959. Die 'Doppelten Einebnungsflächen' in den feuchten Tropen. *Z. Geomorph.* 1, 201–226.

Büdel, J. 1982. Translated by L. Fisher and D. Busche. *Climatic Geomorphology*. Princeton University Press, Princeton, NJ.

Bullard, E., Everett, J.E. and Smith, A.G. 1965. The fit of the continents around the Atlantic. *Phil. Trans. R. Soc. A.*, 258, 41–51.

Burchfiel, B.C., 1983. The Continental Crust. *Scientific American*, September, 86–98.

Burg, J. P. and Chen, G.M. 1984. Tectonics and structural zonation of southern Tibet, China. *Nature*, 311, 219–223.

Burg, J.P. and Ford, M. eds. 1997. *Orogeny through Time*. Geol. Soc. Lond. Spec. Publ. No. 121.

Burk, C.A. and Drake, C.L. eds. 1974. *The Geology of Continental Margins*. Springer-Verlag, Berlin.

Burke, K. and Lucas, L. 1989. Thrusting on the Tibetan Plateau within the last 5 Ma. In Sengör, A.M.C. ed. *Tectonic Evolution of the Tethyan Region*. Kluwer Academic, Amsterdam, 507–512.

Carey, S.W. 1958. *The Tectonic Approach to Continental Drift. Continental Drift: A Symposium*. Geol. Dept., Univ. Tasmania, Hobart.

Carey, S.W. 1976. *The Expanding Earth*. Elsevier, Amsterdam.

Carey, S.W. 1988. *Theories of the Earth and Universe*. Stanford University Press.

Cassano, E., Anelli, L., Fichera, R. and Cappelli, V. 1986. *Pianura Padana. Interpretazione integrata di dati geofisici e geologici*. AGIP.

Castellarin, A., Cantelli, L., Pesce, A.M., Mercier, J.L., Picotti, V., Pini, G.A., Grosser, G. and Selli, L. 1992. Alpine compressional tectonics in the Southern Alps. Relationship with the N. Apennines. *Annales Tectonicae*, 1, 62–94.

Cervenny, P.F., Naeser, N.D., Zeitler, P.K., Naeser, C.W. and Johnson, N.M. 1988. History of uplift and relief of the Himalaya during the past 18 million years. Evidence from fission-track ages of detrital zircons from sandstones of the Siwalik Group. In Kleinspehn, K.L. and Paulo, C. eds. *New Perspectives in Basin Analysis*. Springer-Verlag, New York, 43–61.

Chai, B.H.T. 1972. Structure and tectonic evolution of Taiwan. *Am. J. Sci.*, 272, 389–422.

Chamberlin, T.E. 1928. Some of the objections to Wegener's theory. In Waterschoot van der Gracht, W.A.J.M. van ed. *Theory of Continental Drift: a Symposium on the origin and movement of land masses both inter-continental and intra-continental, as proposed by Alfred Wegener*. American Association of Petroleum Geologists, Tulsa, 83–7.

Chang, C. 1996. *Geology and Tectonics of Qinghai-Xizang Plateau*. Science Press, Beijing.

Chatterjee, S. and Hotton, N. eds. 1992. *New Concepts in Global Tectonics*. Texas Tech. Univ. Press, Lubbock.

Chester, D.K. and Duncan, A.M. 1982. The interaction of volcanic activity in Quaternary times upon the evolution of the Alcantara and Simeto Rivers, Mount Etna, Sicily. *Catena*, 9, 319–342.

Childs, O.E. and Beebe, B.W. 1963. *Backbone of the Americas*. Amer. Ass. Pet. Geol. Mem. 2, Tulsa.

Chinzei, K. 1966. Younger Tertiary geology of the Mabechi River Valley, Northeast Honshu, Japan. *J. Fac. Sci. Univ. Tokyo*, II, 16, 161–208.

Choubert, G. and Faure-Muret, A. 1974. Moroccan Rif. In Spencer, A.M. ed. *Mesozoic–Cenozoic Orogenic Belts*. Sc. Acad. Press, Edinburgh, 37–46.

Choukroune, P. and Séguret, M. 1973. Tectonics of the Pyrenees: role of compression and gravity. In De Jong, K.A. and Scholten, R. 1973. *Gravity and Tectonics*. Wiley, NewYork, 141–156.

Ciabatti, M., Curzi, P.V. and Ricci, L.F. 1986. Sedimentazione Quaternaria nell'Adriatico centrale. *Atti Riunione Gruppo Sedimentologia CNR, Ancona*, 125–139.

Clapperton C.M. 1993. *Quaternary Geology and Geomorphology of South America*. Elsevier Science, Amsterdam.

Clark, A.H., Mayer, A.E.S., Mortimer, C. and others. 1967. Implications of the isotopic ages of ignimbrite flows, southern Atacama desert, Chile. *Nature*, 215, 723–724.

Clarke, J.D.A. 1994. Evolution of the Lefroy and Cowan palaeodrainage channels, Western Australia. *Australian Journal of Earth Sciences*, 41(1), 55–68.

Cobbing, E.J., Pitcher, W.S., Wilson, J.J, Baldock, J.W., Taylor, W.P., McCourt, W. and Snelling, N.J. 1981. The geology of the Western Cordillera of northern Peru. *Overseas Mem., Institute of Geol. Sc., Nat. Env. Res. Council*, 5, 124–125.

Colbert, E.H. 1973. Continental drift and distribution of fossil reptiles. In Tarling, D.H. and Runcorn, S.K. eds. *Implications of Continental Drift to Earth Sciences*. Academic Press, London, 393–412.

Coleman, M. and Hodges, K. 1995. Evidence for Tibetan uplift before 14 Myr ago from a new minimum age for east west extension. *Nature*, 374, 449–52.

Coltorti, M. and Ollier, C.D. 1999. The significance of high planation surfaces in the Andes of Ecuador. In Whalley, B., Smith, B. and Warke, P.A. eds. *Uplift, Erosion and Stability*. Geol. Soc. London Spec. Publ., 162, 239–253.

Coltorti, M. and Pieruccini, P. 2000. The planation surface across the Italian Peninsula: a late Lower Pliocene plain of marine erosion as a key tool in neotectonic studies. *J. Geodynamics*, 29(3–5), 323–328.

Cope, J.C.W. 1994. A latest Cretaceous hotspot and the southeasterly tilt of Britain. *J. Geol. Soc. Lond.*, 151, 905–908.

Copeland, P. and Hanson, T.M. 1990. Episodic rapid uplift of the Himalayas revealed by 40Ar/39Ar analysis of detrital K-feldspar and muscovite, Bengal Fan. *Geology*, 18, 354–357.

Costa, C.H., Giaccardi, A.D. and Diaz, F.G. 1999. Palaeo-landsurfaces and neotectonic analysis in the southern Sierras Pampeanas, Argentina. In Whalley, B., Smith, B. and Warke, P.A. eds. *Uplift, Erosion and Stability*. Geol. Soc. London Spec. Publ., 162, 229–238.

Cotton, C.A. 1918. Mountains. *New Zealand Journal of Science and Technology*, 1, 280–285. Reprinted in Cotton, C.A. 1955. *New Zealand Geomorphology*. New Zealand University Press, Wellington.

Coward, M.P., Dewey, J.F. and Hancock, P.L. eds. 1987. *Continental Extensional Tectonics*. Geol. Soc. London Spec. Publ. 28.

Cox, K.G. 1972. The Karroo volcanic cycle. *J. Geol. Soc. Lond.*, 128, 311–36.

Cox, K.G. 1989. The role of mantle plumes in the development of continental drainage patterns. *Nature*, 342, 873–876.

Crawford, R.A. 1974. The Indus suture line, the Himalaya, Tibet and Gondwanaland. *Geol. Mag.*, 111, 369–383.

Crawford, R.A. 1979. Gondwanaland and the Pakistan region. In Farah, A. and De Jong, K.A. eds. *Geodynamics of Pakistan*. Geological Survey of Pakistan, 380–382.

Crough, S.T. and Thompson, G.A. 1977. Upper mantle origin of Sierra Nevada uplift. *Geology*, 5, 396–399.

Cucci, L. and Cinti, F.R. 1998. Regional uplift and local tectonic deformation recorded by the Quaternary marine terraces on the Ionian coast of northern Calabria (southern Italy). *Tectonophysics*, 292, 67–83.

D'Angelo, E.P. and Le Bert, L.A. 1968. Relacion entre estructura y volcanismo cuaternario Andino en Chile. *Pan-American Symposium on the Upper Mantle.* Mexico, March 1961, 39–47.

Davis, G.H. 1975. Gravity-induced folding off a gneiss dome complex, Rincon Mountain, Arizona. *Bull. Geol. Soc. Am.,* 86, 979–990.

Davis, G.H. 1984. *Structural Geology of Rocks and Regions.* Wiley, New York.

Davis, W.M. 1896. Plains of marine and sub-aerial denudation, *Bull. Geol. Soc. Am.,* 7, 377–398.

Davis, W.M. 1899. The geographical cycle. *Geogr. J.,* 14, 481–504.

Davis, W.M. 1904. A flat-topped range in the Tian-Shan. *Appalachia,* 10, 277–289.

De Jong, K.A. and Scholten, R. 1973. *Gravity and Tectonics.* Wiley, New York.

De Sitter, L.U. 1952. Pliocene uplift of Tertiary mountain chains. *Am. J. Sci.,* 250, 297–307.

De Wit, M.C.J. 1988. Aspects of the geomorphology of the north-western Cape, South Africa. In Dardis, G.F. and Moon, B.P. eds. *Geomorphological Studies of Southern Africa.* A.A. Balkerna, Rotterdam, 57–69.

Deiss, C.A. 1943. Structure of central part of Sawtooth Range, Montana. *Bull. Geol. Soc. Am.,* 54, 1123–1167.

Demangeot, J. 1965. *Géomorphologie des Abruzzes adriatiques.* C.N.R.S., Paris.

Demek, J. 1984. Carpathian Mountains. In Embleton, C. ed. *Geomorphology of Europe.* Macmillan, London, 355–373.

Denham, D. and Windsor, C.R. 1991. The crustal stress pattern in the Australian continent. *Exploration Geophysics,* 22, 101–105.

Dercourt, J., Ricou, L-E. and Vrielynck, B. 1993. *Atlas Tethys Palaeoenvironmental Maps.* Gauthier-Villars.

Derry, L.A. and France-Lanord, C. 1997. Himalayan weathering and erosion fluxes: climate and tectonic controls. In Ruddiman, W.F. ed. *Tectonic Uplift and Climate Change.* Plenum Press, New York, 289–312.

Dewey, J.F., Pitman, W.C., Ryan, W.B. and Bonning, J., 1973. Plate tectonics and the evolution of the Alpine system. *Bull. Geol. Soc. Am.,* 84, 3137–3180.

Di Girolamo, P. and Morra, V. 1988. Il magnetismo mesozoico-quaternario della Campania: petrologia e significato geodinamico. *Mem. Soc. Geol. Italiano,* 41, 165–179.

Dietz, R.S. 1972. Geosynclines, mountains and continent building. *Scientific American,* 226, 30–33.

Dingle, R.V. 1977. The anatomy of a large submarine slump on a sheared continental margin (SE Africa). *J. Geol. Soc. Lond.,* 134, 293–310.

Dohrenwend, J.C., Wells, S.G., McFadden, L.D. and Turrin, B.D. 1987. Pediment dome evolution in the eastern Mohave Desert, California. In Gardiner, V. ed. *International Geomorphology 1986 Part II.* Wiley, Chichester, 1047–1062.

Doré, A.G. 1992. The base Tertiary surface of southern Norway and the northern North Sea. *Norsk. Geol. Tidsskr.,* 72, 259–265.

Downie, C. 1964. Glaciations of Mount Kilimanjaro, Tanganyika. *Bull. Geol. Soc. Am.,* 75, 1–16.

Dresch, J. 1958. Problemes morphologiques des Andes Centrales. *Ann. Geogr.,* 67, 130–151.

Duff, P. McL.D. ed. 1992. *Holmes' Principles of Physical Geology.* Chapman & Hall, London.

Dumitrashko, N.S. 1984. Caucasus mountains and Armenian highlands. In Embleton, C. ed. *Geomorphology of Europe*. Macmillan, London, 393–403.

Eardley, A.J. 1963. Relation of uplifts to thrusts in Rocky Mountains. *Am. Assoc. Petrol. Geol. Mem.*, 2, 209–219.

Eaton, G.P. 1987. Topography and origin of the southern Rocky Mountains and Alvarado Ridge. In Coward *et al.* eds 355–369.

Elliot, D. and Johnson, M.R.W. 1978. Discussion on structures found in thrust belts, *Journal of the Geological Society of London*, 135, 259–260.

Elston, W.E. 1978. Rifting and volcanism in the New Mexico segment of the Basin and Range province, Southwestern USA. In Neumann, E.R. and Ramberg, I.B. eds. *Petrology and Geochemistry of Continental Rifts*. Reidel, Dortrecht, 79–86.

Embleton, C. ed. 1984. *Geomorphology of Europe*. Macmillan, London.

Engelen, G.B. 1963. *Gravity Tectonics in the Northwestern Dolomites (N. Italy)*. Geologica Ultraiectina. No. 13. Rijksuniversiteit te Utrecht.

Ernst, W.G. 1973. Interpretative synthesis of metamorphism in the Alps. *Geological Society of America Bulletin*, 84(6), 2053–2078.

Escher, A. and Watt, W.S. eds 1976. *Geology of Greenland*. Geological Survey Greenland, Odense.

Evamy, D.D., Haremboure, J., Kamerling, P., Knaap, W.A., Molloy, F.A. and Rowlands, P.H. 1979. Hydrocarbon habitat of Tertiary Niger Delta. *Am. Assoc. Petrol. Geol. Bull.*, 62, 1–39.

Evanoff, E. 1990. Early Oligocene paleovalleys in southern and central Wyoming: evidence of high local relief on the late Eocene unconformity. *Geology*, 18, 443–446.

Fairbridge, R.W. ed. 1975. *The Encyclopedia of World Regional Geology Part I: Western Hemisphere*. Dowden, Hutchinson & Ross, Stroudsberg.

Falvey, D.A. and Mutter, J.C. 1981. Regional plate tectonics and the evolution of Australia's passive continental margins. *BMR Journal of Australian Geology and Geophysics*, 6, 1–29.

Fenneman, N.M. 1931. *Physiography of Western United States*. McGraw Hill, New York.

Fenneman, N.M. 1938. *Physiography of Eastern United States*. McGraw Hill, New York.

Findlay, A.L. 1974. The structure of foothills south of the Kubor Range, Papua New Guinea. *Aust. Petrol. Exp. Assoc. J.*, 14, 14–20.

Földvary, G.Z. 1988. *Geology of the Carpathian Region*. World Scientific, Singapore.

Foose, R.M. 1973. Vertical tectonism and gravity in the Big Horn Basin and surrounding ranges of the Middle Rocky Mountains. In De Jong, K.A. and Scholten, R. eds. *Gravity and Tectonics*. Wiley, NewYork, 443–455.

Frisch, W., Kuhlemann, J., Dunkl, I. and Brugel, A. 1998. Palinspastic reconstruction and topographic evolution of the Eastern Alps during Late Tertiary tectonic extrusion. *Tectonophysics*, 297, 1–15.

Fuller, R.E. and Waters, A.C. 1929. The nature and origin of the horst and graben structure of southern Oregon. *J. Geol.*, 37, 204–239.

Gabrielse, H. 1975. Canada – Cordilleran Region, Interior and Western Belts. In Fairbridge, R.W. ed. *The Encyclopedia of World Regional Geology Part I: Western Hemisphere*. Dowden, Hutchinson and Ross, Stroudsberg, 179–187.

Gale, S.J. and Hoare, P.G. 1997. The glacial history of the northwest Picos de Europa of northern Spain. *Z. Geomorph.*, 41, 81–96.

Gansser, A. 1973. Facts and theories on the Andes. *J. Geol. Soc. London*, 129, 93–131.

Gansser, A. 1982. The morphogenetic phase of mountain building in Hsü, K. ed. *Mountain Building Processes*, 221–228.

Gansser, A., 1991. Facts and theories on the Himalayas. *Eclogie Geol. Helv.*, 84, 33–59.

Gao, M.K. 1998. Late Cenozoic continental dynamics of East Asia. In *Proceedings of International Symposium on New Concepts in Global Tectonics*, Tsukuba, Japan, 1998. Organizing Committee of International Symposium on New Concepts in Global Tectonics, 41–46.

Gautier, A. 1965. Relative dating of peneplains and sediments in the Lake Albert Rift area. *Am. J. Sci.*, 263, 537–547.

Geikie, J. 1912. *Mountains: Their Origin, Growth and Decay*. Oliver and Boyd, Edinburgh.

Gilbert, G.K. 1890. *Lake Bonneville*. US Geol. Surv. Monograph 1.

Gilchrist, A.R. and Summerfield, M.A. 1990. Differential denudation and flexural isostasy in formation of rifted-margin upwarps. *Nature*, 346, 739–742.

Gilluly, J. 1970. Crustal deformation in the western United States. In Johnson, H. and Smith, B.L. eds. *The Magatectonics of Continents and Oceans*. Rutgers Univ. Press, New Brunswick, 47–73.

Gjessing, J. 1967. Norway's paleic surface. *Norsk. Geogr. Tidsskr.*, 21, 69–132.

Glangeaud, L. 1957. Essai de classification géodynamique des chaines et des phenomènes orogeniques. *Rev. Geog. Phys. Geol. Dynam.*, 1, 214.

Godard, A. ed. 1982. Les Bourrelets Marginaux des hautes latitudes. *Bull. Assoc. Géogr. Franç.*, 489, 239–269.

Grabau, A.W. 1927. Summary of Cenozoic and Psychozoic. *Bull. Geol. Soc. China*, 6.

Grabert, H. 1971. Die Prae-Andie Drainage des Amazonas Stromsystems, *Muenster Forsch. Geol. Palaeontol.*, 20, 51–60.

Graf, W.L. ed. 1987. *Geomorphic Systems of North America*. Boulder, Colorado. Geol. Soc. America, Centennial Special Volume 2.

Graf, W.L., Hereford, R., Laity, J. and Young, R.A. 1987. Colorado Plateau. In Graf, W.L. ed. *Geomorphic Systems of North America*. Boulder, Colorado. Geol. Soc. America, Centennial Special Volume 2.

Grecula, P., Hovorka, D. and Putis, M. 1997. *Geological Evolution of the Western Carpathians*. Geocomplex, Bratislava.

Griffiths, J.R. 1971. Reconstruction of the South-West Pacific margin of Gondwanaland. *Nature*, 234, 203–207.

Grove, A.T. 1986. Geomorphology of the African Rift System. In Frostick, L.E. *et al.* eds. *Sedimentation in the African Rifts*. Spec. Publ. Geol. Soc. Lond., 25, 9–18.

Gupta, S. 1997. Himalayan drainage pattern and the origin of fluvial megafans in the Ganges foreland basin. *Geology*, 25, 511–514.

Hack, J.T. 1960. Interpretation of erosional topography in humid temperate regions. *Am. J. Sci.*, 258, 80–97.

Ham, C.K. and Herrera, L.J. 1963. Fold of subandean fault system in tectonics of eastern Peru and Ecuador. In Childs, O.E. and Beebe, B.W. *Backbone of the Americas*. Amer. Ass. Pet. Geol. Mem. 2, Tulsa, 47–61.

Hamilton, W. 1987. Crustal extension in the Basin and Range Province, southwestern United States. In Coward, M.P., Dewey, J.F. and Hancock, P.L. *Continental Extensional Tectonics*. Geological Society Special Publication, 28, 155–156.

Hampton, C. 1997. Structural evolution of the East African rift system. In Schluter, T. ed. *Geology of East Africa*. Gebruder Borntraeger, Berlin, 265–301.

Harrison, J.V. and Falcon, N.L. 1934. Collapse structures. *Geol. Mag.*, 71, 529–539.

Harrison, J.V. and Falcon, N.L. 1936. Gravity collapse structures and mountain ranges, as exemplified in South-western Persia. *Q. J. Geol. Soc. Lond.*, 92, 91–102.

Heim, A. 1927. (trans. E. Montag) The summit-level of the Alps. *Proc. Liverpool Geol. Soc.*, 1928, 15, 90–109.

Heritsch, F. 1929. (trans. P.G.H. Boswell) *The Nappe Theory in the Alps*. Methuen, London.

Hills, E.S. 1956. A contribution to the morphotectonics of Australia, *J. Geol. Soc. Aust.*, 3, 1–15.

Hills, E.S. 1961. Morphotectonics and the geomorphological sciences with special reference to Australia. *Q. J. Geol. Soc. Lond.*, 117, 77–89.

Hills, E.S. 1975. *The Physiography of Victoria*. Whitcombe and Tombs, Melbourne.

Hirn, A., Nercessian, A., Sapin, M., Jobert, G., Xu, M.Z.X., Gao, E.Y., Lu, D.Y. and Teng, J.W. 1984. Lhasa block and bordering sutures. *Nature*, 307, 25–27.

Hollingworth, S.E. and Rutland, R.W.R. 1968. Studies of Andean Uplift Part 1 – Post-Cretaceous evolution of the San Bartolo area, north Chile. *Geol. J.*, 6, 49–62.

Hollingworth, S.E., Taylor, J.H. and Kellaway, C.A. 1944. Large-scale superficial structures in the Northampton Ironstone Field. *Q. J. Geol. Soc. Lond.*, 100, 1–44.

Holmes, A. 1965. *Principles of Physical Geology*. Nelson, London.

Holtedahl, O. 1953. On the oblique uplift of some northern lands. *Norsk. Geogr. Tidsskr.* 14, 132–139.

Hooghiemstra, H. 1989. Quaternary and Upper Pliocene glaciation and forest development in the tropical Andes: evidence for a long high-resolution pollen record from the sedimentary basin of Bogotà, Colombia. *Palaeogeogr. Palaeoclimatolo. Palaeoecol.*, 72, 11–26.

Hoorn, C. 1995. Andean tectonics as a cause for changing drainage patterns in Miocene northern South America. *Geology*, 23, 237–240.

Hoshino, M. 1998. *The Expanding Earth: evidence, causes and effects*. Tokai Univ. Press, Tokyo.

Hsü, K. ed. 1982. *Mountain Building Processes*. Academic Press, London.

Hunt, C.B. 1967. *Physiography of the United States*. W.H. Freeman, San Francisco.

Hutchinson, *Pocket Dictionary of Geography* (1993). Helican, Oxford.

Illies, J.H. 1972. The Rhine Graben rift system – Plate tectonics and transform faulting. *Geophys. Surveys*, 1, 27–60.

Isacks, B.L. 1988. Uplift of the Central Andean Plateau and bending of the Bolivian Orocline. *J. Geophys. Res.*, 93, 3211–3231.

Ishida, T. and Ohta, Y. 1973. Ramechhap–Okhaldhunga Region. In Ohta, Y. and Akiba, C. eds. *Geology of the Nepal Himalayas*. Saikon, Sapporo.

Jablonski, N.G. ed. 1997. *The Changing Face of East Asia during the Tertiary and Quaternary*. Univ. Hong Kong Press.

Jackson, J.A. 1997. *Glossary of Geology*. 4th edn. American Geological Institute, Alexandria, Virginia.

Jacob, A F., 1983. Mountain front thrust, southeastern Front Range and northeastern Wet Mountains, Colorado. In Lowell, J. D. ed. *Rocky Mountain Foreland Basins and Uplifts*. Rocky Mountain Assoc. Geol., Denver, 229–244.

James, D.E. 1971. Plate tectonics model for the evolution of the central Andes. *Bull. Geol. Soc. Am.*, 82, 3325–3346.

James, D.E. 1973. The evolution of the Andes. *Scientific American*, 229, 60–69.

Jeffreys, H. 1931. On the mechanics of mountains. *Geol. Mag.*, 68, 433–442.

Jenkins, D.A.L. 1974. Detachment tectonics in western Papua New Guinea. *Bull. Geol. Soc. Am.*, 85, 533–548.

Jensen, L.N. and Schmidt, B.J. 1993. Neogene uplift and erosion in the northeastern North Sea: magnitude and consequences for hydrocarbon exploration in the Farsund Basin. In Spencer, A.M. ed. *Generation, Accumulation and Production of Europe's Hydrocarbons*. European Assoc. Petrol. Geol., Special Publ. 3, 79–88.

Jessen, O. 1943. *Die Randschwellen der Kontinente*. Petermann Geogr. Mitt., Ergängungsh. 241.

Johnson, B.D., Powell, C.McA. and Veevers, J.J. 1976. Spreading history of the eastern Indian Ocean, and greater India's northward flight from Antarctica and Australia. *Geological Society of American Bulletin*, 87, 1560–1566.

Johnson, D.W. 1931. A theory of Appalachian geomorphic evolution. *Journal of Geology*, 39, 497–508.

Jordan, T.E., Reynolds, J.H. and Erikson, J.P. 1997. Variability in age and initial shortening and uplift in the central Andes, 16–33°30′S. In Ruddiman, W.F. ed. *Tectonic Uplift and Climate Change*. Plenum Press, New York, 41–61.

Kalvoda, J. 1992. *Geomorphological Record of the Quaternary Orogeny in the Himalaya and the Karakoram*. Elsevier, Amsterdam.

Karatson, D. 1999. Erosion of primary volcanic depressions in the Inner Carpathians volcanic chain. *Z. Geomorph. Suppl. Bd.*, 14, 49–62.

Karig, D.E. 1971. Origin and development of marginal basins in the western Pacific. *J. Geophys. Res.*, 76, 2542–2561.

Karig, D.E., Barber, A.J., Charlton, T.R., Klemperer, S. and Hussong, D.M., 1987. Nature and distribution of deformation across the Banda Arc–Australian collision zone at Timor. *Bull. Geol. Soc. Am.*, 98, 18–32.

Karner, G.A. and Weissell, J.K. 1984. Thermally induced uplift and lithospheric flexural readjustment of the eastern Australian highlands. *Geol. Soc. Aust. Abstracts*, 12, 293–294.

Kashfi, M.S. 1992. Geological evidence for a simple horizontal compression of the crust in the Zagros Crush Zone. In Chatterjee, S. and Hotton, N. eds. *New Concepts in Global Tectonics*. Texas Tech. Univ. Press, Lubbock, 119–129.

Katz, H.R. 1971. Continental margins in Chile – is tectonic style compressional or extensional? *Bull. Am. Assoc. Petrol. Geol.*, 55, 1753–1758.

Kear, D. 1957. Erosional stages of volcanic cones as indicators of age. *New Zealand J. Sci. & Tech.*, B38, 671–682.

Kennan, L., Lamb, S.H. and Hoke, L. 1997. High-altitude palaeosurfaces in the Bolivian Andes: evidence for late Cenozoic surface uplift. In Widdowson, M. ed. 1997, 307–323.

Khaiun, V.E. 1985. *Geology of the USSR*. Gebruder Borntraeger, Berlin.

Khaiun, V.E. 1994. *Geology of Northern Eurasia* (second part of the *Geology of the USSR*). Gebruder Borntraeger, Berlin.

Khobzi J. and Usselmann P. 1973. Problèmes de geomorphologie en Colombie. In Numéro Special sur Les Andes. *Revue de Geographie Physique et de Geologie Dynamique*, 15, 193–206.

King, B.C., Le Bas, M.J. and Sutherland, D.S. 1972. The history of the alkaline volcanoes and extrusive complexes of eastern Uganda and western Kenya. *J. Geol. Soc. Lond.*, 128, 173–205.

King, L.C. 1953. Canons of landscape evolution. *Bull. Geol. Soc. Am.*, 64, 721–51.

King, L.C. 1955. Pediplanation and isostacy: an example from South Africa. *Q. J. Geol. Soc. Lond.*, 111, 353–359.

King, L.C. 1962. *The Morphology of the Earth*. Oliver and Boyd, Edinburgh.

King, P.B. 1969. *Tectonic map of North America – a discussion to accompany the tectonic map of North America*. U.S. Geol. Surv. Prof. Paper 628.

Kleeman, J.D. 1984. The anatomy of a tin mineralising A-type granite. In Flood, P.G. and Runnegar, B. eds. *New England Geology*. Geol. Dept., Univ. New England, 327–334.

Korn, H. and Martin, H. 1959. Gravity tectonics in the Naukluft Mountains of South-West Africa. *Bull. Geol. Soc. Am.*, 70, 1047–1078.

Korsch, R.J. 1982. Mount Duval: geomorphology of a near-surface granite diapir. *Z.Geomorph.* 26, 151–162.

Krebs, W. 1975. Formation of southwest Pacific island arc-trench and mountain systems: plate or global vertical tectonics. *Bull. Am. Assoc. Petrol. Geol.*, 59, 1639–66.

Kroonenberg, S.B., Bakker, J.G.M. and Van der Wiel, M. 1990. Late Cenozoic uplift and paleogeography of the Colombian Andes: constraints on the development of the high-Andean biota. *Geol. en Mijnbouw*, 69, 279–290.

Krzyszkowski, D. and Biernat, J. 1999. Terraces of the Bystrzyca River, Middle Sudetes, and their deformation along the Sudetic Marginal Fault. In Krzyszkowski, D. The Late Cainozoic Evolution of the Sudeten and its Foreland. *Geologia Sudetica*, 31(2), 307–316.

Krzyszkowski, D. and Stachura, R. 1998. Neotectonically controlled fluvial features, Walbrzych Upland, Middle Sudeten Mts, southwestern Poland. *Geomorphology*, 22, 73–91.

Lamb, S., Hoke, L., Kennan, L. and Dewey, J. 1997. Cenozoic evolution of the Central Andes in Bolivia and northern Chile. In Burg, J.P. and Ford, M. eds. *Orogeny Through Time*. Geol. Soc. Lond. Spec. Publ. No. 121, 237–264.

Lamb, S. and Hoke, L. 1997. Origin of the high plateau in the Central Andes, Bolivia, South America. *Tectonics*, 16, 623–649.

Lambeck, K. and Stephenson, R. 1986. The post-Palaeozoic uplift history of south-eastern Australia. *Aust. J. Earth Sci.*, 13, 253–270.

Laubscher, H.P. 1971. The large-scale kinematics of the western Alps and the northern Apennines and its palinspastic implications. *Am. J. Sci.*, 271, 193–226.

Lees, G.M. 1952. Foreland folding. *Q. J. Geol. Soc. Lond.*, 108, 1–34.

Li, J. 1995. *Uplift of Qinghai-Xizang (Tibet) Plateau and Global Change*. Lanzhou University Press.

Li, Y., Yung, J., Xia, Z. and Mo, D. 1998. Tectonic geomorphology in the Shanxi graben system, northern China. *Geomorphology*, 23, 77–89.

Lidmar-Bergström, K. 1995. Relief and saprolites through time on the Baltic Shield. *Geomorphology*, 12, 45–61.

Lillie, A.R. 1980. *Strata and Structure in New Zealand*. Tohunga Press, Auckland.

Litherland, M., Zamora, A. and Eguez, A., 1993. *Mapa geologico del Ecuador, escala 1:1.000.000*. CODIGEM-BGS, Quito.

Liu T. and Zhongli D. 1998. Loess and palaeo-Monsoon. *Annual Review of Earth and Planetary Science* 1998. Annual Reviews Inc., Palo Alto.

Locardi, E. 1985. Neogene and Quaternary Mediterranean volcanism: the Tyrrhenian example. In Stanley, D.J. and Wezel, F.C. eds. *Geological Evolution of the Mediterranean Basin.* Springer-Verlag, New York, 273–291.

Locardi, E. 1988. The origin of the Apenninic arcs. *Tectonophysics*, 146, 105–123.

Loomis, F.B. 1937. *Physiography of the United States.* Doubleday Doran, Garden City, NY.

Lowell, J. D., ed. 1983. *Rocky Mountain Foreland Basins and Uplifts.* Rocky Mountain Association of Geologists, Denver.

Lowman, P.D. 1992. Plate tectonics and continental drift in geologic education. In Chatterjee, S. and Hotton, N. eds. *New Concepts in Global Tectonics.* Texas Tech. Univ. Press, Lubbock.

Lowry, W.D. and Baldwin, E.M. 1952. Late Cenozoic geology of the lower Columbia River Valley, Oregon and Washington. *Bull. Geol. Soc. Am.*, 63, 1–24.

Lucchita, I. 1979. Late Cenozoic uplift of the southwestern Colorado Plateau and adjacent lower Colorado River region. *Tectonophysics*, 61, 63–95.

Luongo, G. 1988. Tettonica globale dell'Italia meridionale: subduzione o bending? *Mem. Soc. Geol. Italiano*, 41, 159–163.

Lyttleton, R.A. 1982. *The Earth and its Mountains.* Wiley, Chichester.

Maack, R. 1969. Die Serra do Mar im Staate Parana. *Die Erde*, 100, 327–347.

Mackin, J.H. 1947. Altitude and local relief in the Bighorn area during the Cenozoic. *Wyoming Geol. Assoc. Field Conference in Bighorn Basin. Guidebook*, 103–120.

Madole, R.F., Bradley, W.C., Loewenherz, D.S., Ritter, D.F., Rutter, N.W. and Thorn, C.E. 1987. Rocky Mountains, in Graf, W.L. ed. *Geomorphic Systems of North America.* Geol. Soc. America, Centennial Special Volume 2. Boulder, Colorado.

Maltman, A. ed. 1993. *The Geological Deformation of Sediments.* Chapman & Hall, London.

Manabe, S. and Terpstra, T.B. 1974. The effects of mountains on the general circulation of the atmosphere as identified by numerical experiments. *J. Atmospheric Sci.*, 31, 3–42.

Mantovani, E., Albarello, D., Babbucci, D. and Tamburelli, C. 1992. *Recent Geodynamic Evolution of the Central Mediterranean Region.* Dept. of Earth Sciences, University of Siena.

Matheson, D.S. and Thomson, S. 1973. Geological implications of valley rebound. *Can. J. Earth Sci.* 10, 961–978.

Mathews, W.H. 1991. Physiographic evolution of the Canadian Cordillera. In Gabrielse, H. and Yorath, C.J. eds. *Geology of the Cordilleran Orogen in Canada.* Geological Survey of Canada, Geology of Canada no. 4, 403–418.

Maund, J.G., Rex, D.C., Le Roex, A.P. and Reid, D.L. 1988. Volcanism on Gough Island: a revised stratigraphy. *Geol. Mag.*, 125, 175–181.

Mazzanti, R. and Trevisan, L. 1978. Evoluzione della rete idrografica nell'Appennino Centro-Settentrionale. *Geogr. Fis. Dinam. Quat.*, 1, 55–62.

McKee, B. 1972. *Cascadia.* McGraw-Hill, New York.

McLaughlin, D.H., 1924. Geology and physiography of the Peruvian Cordillera. Departments of Junin and Lima. *Bull. Geol. Soc. Am.*, 35, 591–632.

Mégard, F. 1987. Structure and evolution of the Peruvian Andes. In Schaer, J.P. and Rogers, J. eds. *The Anatomy of Mountain Ranges.* Princeton, University Press, Princeton, NJ. 179–210.

Meyerhoff, H.A. and Olmsted, E.W. 1963. The origins of Appalachian drainage. *American Journal of Science*, 332, 21–41.

Migon, P. and Lach, J. 1999. Geomorphological evidence of neotectonics in the Kaczawa sector of the Sudetic marginal fault, southwestern Poland. In Krzyszkowski, D. The Late Cainozoic Evolution of the Sudeten and its Foreland. *Geologia Sudetica*, 32(2), 307–316.

Miller, E.L. and Gans, P.B. 1997. The North American Cordillera. In van der Pluijm, B.A. and Marshak, S. *Earth Structure: an Introduction to Structural Geology and Tectonics.* WCB/ McGraw-Hill, Dubuque, Iowa, 424–429.

Milnes, A.G. 1987. Tectonic evolution of the southern Andes, Tierra del Fuego: a summary. In Schaer, J.P. and Rogers, J. *The Anatomy of Mountain Ranges.* Princeton University Press, Princeton, NJ, 173–176.

Molnar, P. and England, P. 1990. Late Cenozoic uplift of mountain ranges and global climate change: chicken or egg? *Nature*, 346, 29–34.

Molnar, P. and Tapponnier, P. 1978. Active tectonics of Tibet. *J. Geophys. Res.*, 83, 5361–5375.

Morisawa, M. 1975. Tectonics and geomorphic models. In Melhorn, W.N. and Flemel, R.C. eds. *Theories of Landform Development*. George Allen & Unwin, London, 199–216.

Morner, N.A. 1992. Neotectonics and Palaeoclimate. In Morner, N.A., Owen, L.A., Stewart, I. and Finzi, C.V. eds. *Neotectonics Recent Advance* (Abstract Volume). Quaternary Research Association, Cambridge, 39.

Mortimer, C. 1973. The Cenozoic history of the southern Atacama Desert, Chile. *J. Geol. Soc. Lond.*, 129, 505–526.

Muhs, D.R., Thorson, R.M., Clague, J.J., Mathews, W.H., McDowell, P.F. and Kelsey, H.M. 1987. Pacific coast and mountain system. In Graf, W.L. ed. *Geomorphic Systems of North America*. Geol. Soc. America, Centennial Special Volume 2. Boulder, Colorado.

Muñoz, N. and Charrier, R. 1996. Uplift of the W. border of altiplano, N. Chile. *J. S. Am. Earth Sci.*, 9, 171–181.

Myers, J.S. 1975. Vertical crustal movement of the Andes in Peru. *Nature*, 254, 672–674.

Myers J.S., 1976. Erosional surfaces and ignimbritic eruption, measures of Andean uplift in northern Peru. *Geol. J.*, 11, 29–44.

Näslund, J.O. 1998. Palaeosurfaces, glacial landforms, and landscape development in Dronning Maud Land, East Antarctica. In Näslund, J.O. *Ice Sheet, Climate, and Landscape Interactions in Dronning Maud Land, Antarctica*. Dissertation 11, Department of Physical Geography, Stockholm University.

Näslund, J.O. 2000. Palaeosurfaces, glacial landforms, and landscape development in Dronning Maud Land, East Antarctica. *Palaeogeogr., Palaeoclim., Palaeoecol.*, in press.

National Geographic Society 1993. *Exploring Your World: The Adventure of Geography*. National Geographic Society, Washington, D.C.

Ni, J. and York, J.E. 1978. Late Cenozoic tectonics of the Tibetan Plateau. *J. Geophys. Res.* 83, 5377–5384.

Nitchman, S.P., Caskey, S.J. and Sawyer, T.L. 1990. Change in Great Basin tectonics at 3–4 Ma – a hypothesis. *Geol. Soc. America Abstracts, Cordilleran Section*, 33, no. 3, 72. (1990).

Oakeshott, G.B. 1971. *California's Changing Landscape*. McGraw-Hill, New York.

Oberlander, T. 1965. *The Zagros Streams: a New Interpretation of Transverse Drainage in an Orogenic Zone.* Syracuse Univ. Press, Syracuse, NY.

Obruchev, S.V. 1964. The principal subdivisions of the Proterozoic in the U.S.S.R. Report of the 22nd Session – International Geological Congress, India, 153–154.

Ohta, Y.and Akiba, C. eds. 1973. *Geology of the Nepal Himalayas.* Saikon, Sapporo.

Okaya, N., Tawackoli, S. and Giese, P. 1997. Area-balanced model of the late Cenozoic tectonics and evolution of the central Andean arc and back arc (lat 20 degrees – 22 degrees S). *Geology*, 25, 367–370.

Ollier, C.D. 1959. A two cycle theory of tropical pedology. *Journal of Soil Science*, 10, 137–148.

Ollier, C.D. 1960. The inselbergs of Uganda. *Z. Geomorph.*, 4, 43–52.

Ollier, C.D. 1977. Early landform evolution. In Jeans, D.N. ed. *Australia: a Geography.* Sydney University Press, 85–98.

Ollier, C.D. 1978. Tectonics and geomorphology of the Eastern Highlands. In J.L. Davies and M.A.J. Williams. Eds. *Landform Evolution in Australasia*, ANU Press, Canberra, 5–47.

Ollier, C.D. 1982. The Great Escarpment of Eastern Australia: tectonic and geomorphic significance. *Journal of the Geological Society of Australia*, 29, 13–23.

Ollier, C.D. 1984. *Weathering.* Second edition. Longman, Harlow, Essex.

Ollier, C.D. 1985. Orthodoxy. *Geology Today*, July 99–100.

Ollier, C.D. 1988. *Volcanoes.* Blackwell, Oxford.

Ollier, C.D. 1992. A hypothesis about antecedent and reversed drainage. *Geogr. Fisica e Dinamica Quaternaria*, 14, 243–246.

Ollier, C.D. 1993. Age of soils and landforms in Uganda. *Israeli J. Earth Sci.*, 41, 227–231.

Oliier, C.D. 1995. Tectonics and landscape evolution in southeast Australia, *Geomorphology*, 12(3), 37–44.

Ollier, C.D. 1999. Geomorphology and mountain building, *Geografia Fisica e Dynamica Quaternario, Supplementa III*, Tomo 3, 49–60.

Ollier, C.D. and Marker, M.E. 1985. The Great Escarpment of southern Africa, *Zeitschrift für Geomorphologie Supplement Band*, 54, 37–56.

Ollier, C.D. and Pain, C.F. 1980, Actively rising surficial gneiss domes in Papua New Guinea. *J. Geol. Soc. Aust.*, 27, 33–44.

Ollier, C.D. and Pain, C.F. 1981. Active gnesis domes in Papua New Guinea – new tectonic landforms. *Z. Geomorph.*, 25(2), 133–145.

Ollier, C.D. and Pain, C.F. 1988. Morphotectonics of Papua New Guinea, *Z. Geomorph.*, Supplementband, N.F., 69, 1–16.

Ollier, C.D. and Pain, C.F. 1994. Landscape evolution and tectonics in southeastern Australia, *AGSO Journal of Australian Geology and Geophysics*, 15(3), 335–345.

Ollier, C.D. and Pain, C.F. 1996. *Regolith, Soils and Landforms.* Wiley, Chichester.

Ollier, C.D. and Pain, C.F. 1997. Equating the basal unconformity with the palaeoplain: a model for passive margins. *Geomorphology*, 19, 1–15.

Ollier, C.D. and Powar, K.B. 1985. The Western Ghats and the morphotectonics of peninsular India, *Zeitschrift für Geomorphologie Supplement Band*, 54, 57–69.

Ollier, C.D. and Stevens, N.C. 1989. The Great Escarpment in Queensland. In R.W. Le Maitre, ed. *Pathways in Geology: Essays in Honour of Edwin Sherbon Hills.* Blackwell, Melbourne, 140–152.

Ollier, C.D. and Taylor, D. 1988. Major geomorphic features of the Kosciuszko–Bega region, *BMR J. Australian Geol. Geophys.*, 10, 357–362.

Ollier, C.D. and Wyborn, D. 1989. 'Geology of alpine Australia'. In Good, R. ed. *The Scientific Significance of the Australian Alps*, Australian Alps, National Parks Liaison Committee and Australian Academy of Science, Canberra, 35–53.

Ollier, C.D., Gaunt, G.F.M. and Jurkowski, I., 1988. The Kimberley Plateau, Western Australia: a Precambrian erosion surface, *Z. Geomorph.*, 32: 239–246.

Otuka, Y. 1933. The Japanese coastline. *Geogr. Review Japan*, 9, 819–843.

Oxburgh, E.R. 1972. Flake tectonics and continental collision. *Nature*, 239, 202–204.

Page, W.D. and James, M.E. 1981. The antiquity of erosion surfaces and late Cenozoic deposits near Medellin, Colombia: implication to tectonics and erosional rates. *Mem. primo Sem. Cuat. Col.*, *Revista CIAF*, 6, 421–454.

Pain, C.F. 1985. Morphotectonics of the continental margins of Australia. *Zeitschrift für Geomorphologie, Supplement Band*, 54, 23–35.

Pain, C.F. 1986. Geomorphology of metamorphic core complex mountains in Arizona and S.E. California. *Z. Geomorph.*, 30(2), 51–166.

Pain, C.F. and Ollier, C.D. 1981. Geomorphology of a Pliocene granite in Papua New Guinea. *Z. Geomorph*, 25(3), 249–258.

Pain, C.F. and Ollier, C.D. 1984. Drainage patterns and tectonics around Milne Bay, Papua New Guinea. *Rev. Geom. Dyn.*, 32, 113–120.

Parker, S.P. ed, 1984. *McGraw-Hill Dictionary of Earth Sciences*, McGraw-Hill, New York. 837p.

Parrish, R.R. 1983. Cenozoic thermal evolution and tectonics of the Coast Mountains of British Columbia: 1 Fission-track dating, apparent uplift rates, and pattern of uplift. *Tectonics*, 2, 601–631.

Partridge, T.C. 1997. Late Neogene uplift in eastern and southern Africa. In Ruddiman, W.F. ed. *Tectonic Uplift and Climate Change*. Plenum Press, New York, 63–86.

Partridge, T.C. 1998. Of diamonds, dinosaurs and diastrophism: 150 million years of landscape evolution in southern Africa. *South African J. Geol.*, 101, 167–184.

Partridge, T.C. and Maud, R.R. 1987. Geomorphic evolution of southern Africa since the Mesozoic. *South African Journal of Geology*, 90(2), 179–208.

Pavlides, S.B. 1989. Looking for a definition of Neotectonics. *Terra Nova*, 1, 233–235.

Pazzaglia, F.J. and Gardner, T.W. 1993. Fluvial terraces of the lower Susquehanna River. *Geomorphology*, 8, 83–113.

Peirce, H.W., Damon, P.E. and Shafiqullah, M. 1979. An Oligocene (?) Colorado Plateau edge in Arizona. *Tectonophysics*, 61, 1–24.

People's Republic of China Ministry of Geology and Mineral Resources, 1982. *Regional Geology of Xinjiang Uygur Autonomous Region*. Geological Memoirs 32. Geological Publishing House, Beijing.

Petriovsky, M.V. 1985. *Morphotectonics: Definition, Terms, Conditions and Urgent Problems*. Morphotectonics Working Group Meeting, Sofia, 1983. Bulgarian Acad. Sci., Sofia.

Peulvast, J.P. 1988. Pre-glacial landform evolution in two coastal high latitude mountains: Lofoten–Vesteralen (Norway) and Scoresby Sund area (Greenland). *Geografiska Annaler*, 70A, 351–360.

Pickford, M., Senut, B. and Hadoto, D. 1993. *Geology and Palaeobiology of the Albertine Rift Valley Uganda–Zaire*. Volume 1 Geology. CIFEG Occas. Publ. 24. Orleans.

Pierce, W.G. 1957. Heart Mountain and South Fork detachment thrusts of Wyoming. *Am. Assoc. Petrol. Geol. Bull.*, 41, 591–626.

Pinter, N. and Brandon, M.T. 1997. How erosion builds mountains, *Scientific American*, 276, 60–65.

Pitcher, W.S. and Bussell, M.A. 1977. Structural control of batholithic emplacement in Peru: a review. *J. Geol. Soc. Lond.*, 133, 249–256.

Plasienka, D., Grecula, P., Putis, M. and others. 1997. Evolution and structure of the Western Carpathians: an overview. In *Geological Evolution of the Western Carpathians*, ed. P. Grecula, D. Hovorka and M. Putis. Mineralia Slovaca Corporation, Geocomplex, Bratislava, Slovak Republic, 1–24.

Plasienka, D., Havrila, M., Michalik, J., Putis, M. and Rehakova, D. 1977. Nappe structure of the western part of the Central Carpathians. In Plasienka, D., Hok, J., Vozar, J. and Elecko, M. eds. *Alpine Evolution of the Western Carpathians and Related Areas*. Introductory articles to the excursion. Bratislava, 139–142.

Platt, J.P. 1997. The European Alps. In van der Pluijm, B.A. and Marshak, S. eds, *Earth Structure: an Introduction to Structural Geology and Tectonics*. WCB/McGraw-Hill, Dubuque, Iowa, 408–415.

Pluijm, B.A. and Marshak, S. 1997. *Earth Structure: an Introduction to Structural Geology and Tectonics*. WCB/ McGraw-Hill, Dubuque, Iowa.

Potter, D.B. and McGill, G.E. 1978. Valley anticlines of the Needles District, Canyonlands National Park, Utah. *Bull. Geol. Soc. Am.*, 89, 952–960.

Potter, P.E. 1978. Significance and origin of big rivers. *Journal of Geology*, 86, 13–33.

Potter, P.E. 1998. The Mesozoic and Cenozoic paleodrainage of South America: a natural history. *J. South American Earth Sci.*, 10, 331–344.

Price, R.A. and Mountjoy, E.W. 1970. Geologic structure of the Canadian Rocky Mountains between Bow and Athabasca rivers – progress report. In Wheeler, J.O. ed. *Geological Association of Canada, Special Paper 6*, 7–25.

Priest, G.R., Woller, N.J.M., Black, G.L. and Evans, S.H. 1983. Overview of the geology of the central Oregon Cascade Range. In Priest, G.R. and Vogt, B.F. eds. *Geology and Geothermal Resources of the Central Oregon Cascade Range, Oregon*. Dep. Geol. and Mineral Industries Special Paper 15, 3–28.

Radelli, L., 1967. Géologie des Andes Colombiennes. *Trav. du Lab. de la Fac. des Sc. de Grenoble*, mem. 6.

Ratschbacher, L., Frisch, W., Linzer, H.G. and Merle, O. 1991. Lateral extrusion in the Eastern Alps. *Tectonics*, 10, 257–271.

Raymo, M.E. and Ruddiman, W.F. 1992. Tectonic forcing of late Cenozoic climate. *Nature*, 359, 117–122.

Raymo, M.E., Ruddiman, W.F. and Froelich, P.N. 1988. Influence of late Cenozoic mountain building on ocean geochemical cycles. *Geology*, 16, 649–653.

Reeves, F. 1946. Origin and mechanics of thrust faults adjacent to the Bearpaw Mountains, Montana. *Bull. Geol. Soc. Am.*, 57, 1033–1047.

Reusch, H. 1901. Nogle bidrag till forstaaelsen af hvorledes Norges dale og fjelde er blevne til. *Norsk Geol. Unders.* 32, 124–263. *Aarbog* 1900.

Reutter, K-J., Giese, P., Gotze, H-J., Scheuber, E., Schwab, K., Schwarz, G. and Wigger, P. 1988. Structures and crustal development of the central Andes between 21° and 25°S. In Bahlburg, H., Breitkreuz, C. and Giese, P. eds. *The Southern Central Andes*. Springer-Verlag, Berlin, 231–261.

Riis, F. 1996. Quantification of Cenozoic vertical movements of Scandinavia by correlation of morphological surfaces with offshore data. *Global and Planetary Change*, 12, 331–357.

Ripper, I.D. and McCue, K.F., 1982. Seismicity of the Indo-Australian/ Solomon Sea plate boundary in the southeast Papua region. *Tectonophysics*, 87, 355–369.

Ripper, I.D. and McCue, K.F. 1983. The seismic zone of the Papuan Fold Belt. BMR *Journal of Australian Geology and Geophysics*, 8, 147–156.

Rodgers, J. 1997. Exotic nappes in external parts of orogenic belts. *Am. J. Sci.*, 297, 174–219.

Rondeel, H.E. and Simon, O.J. 1974. Betic Cordilleras. In *Mesozoic–Cenozoic Orogenic Belts; Data for Orogenic Studies; Alpine–Himalayan Orogens*. Special Publication – Geological Society of London, 4, 23–35.

Royden, L.H. and Burchfiel, B.C., 1997. The Tibetan Plateau and surrounding regions. In van der Pluijm, B.A. and Marshak, S. eds, *Earth Structure: an Introduction to Structural Geology and Tectonics*. WCB/MC Graw Hill, Dubuque, Iowa, 416–423.

Ruddiman, W.F. ed. 1997. *Tectonic Uplift and Climate Change*. Plenum Press, New York.

Ruddiman, W.F. and Kutzbach, J.E. 1991. Plateau uplift and climatic change. *Scientific American*, March, 42–50.

Ruiz, E. 1981. El Quaternario de la region Garzon-Gigante, Alto Magdalena (Colombia). *Mem. primo Sem. Cuat. Colombia, Revista CIAF*, 6, 505–523.

Russo, R.M. and Silver, P.G. 1996. Cordillera formation, mantle dynamics, and the Wilson Cycle. *Geology*, June, 24, 511–514.

Rutten, M.G., 1969. *The Geology of Western Europe*. Elsevier, Amsterdam.

Sala, M. 1984a. The Iberian Massif. In Embleton, C. ed. *Geomorphology of Europe*. Macmillan, London, 294–322.

Sala, M. 1984b. Baetic Cordillera and Guadalquivir Basin. In Embleton, C. ed. *Geomorphology of Europe*. Macmillan, London, 323–340.

Sala, M. 1984c. Pyrenees and Ebro Basin Complex. In Embleton, C. ed. *Geomorphology of Europe*. Macmillan, London, 269–293.

Sato, Y. and Denson, N.M. 1967. Volcanism and tectonism as reflected in the distribution of nonopaque heavy minerals in some Tertiary rocks of Wyoming and adjacent states. *U.S. Geol. Surv. Prof. Paper* 575C, C42–C54.

Schaer, J.P. and Rogers, J. 1987. *The Anatomy of Mountain Ranges*. Princeton University Press, Princeton, NJ.

Schipull, K. 1974. Geomorphologische Studien in zentralen Südnorwegen mit Beiträgen über Regelungs- und Steurungssysteme in der Geomorphologi. *Hamburger Geogr. Stud.*, 31.

Schroder, J.F. 1993. *Himalaya to the Sea*. Routledge, London.

Schubert, C. and Huber, O. 1990. *The Gran Sabana*. Langoven, Caracas.

Schuchert, C. 1935. *Historical Geology of the Antilles and Caribbean Region*. Hafner, New York.

Seeber, L. and Gornitz, V. 1983. River profiles along the Himalayan Arc as indicators of active tectonics. *Tectonophysics*, 92, 335–367.

Segerstrom, K. 1963. Valley widening and deepening processes in the high Andes of Chile. *Special Paper – Geological Society of America* 73.

Selby, M.J. 1985. *Earth's Changing Surface*. Clarendon Press, Oxford.

Sengör, A.M.C. ed. 1989. *Tectonic Evolution of the Tethyan Region*. Kluwer Academic, Amsterdam.

Shackleton, R.M. and Chang, C. 1988. Cenozoic uplift and deformation of the Tibetan Plateau. *Phil. Trans. Roy. Soc. Lond.*, 327, 365–377.

Sibrava, V. 1997. Quaternary environmental changes in Asia and the Pacific (Results of the IGCP Project 296). In Jablonski, N.G. ed. *The Changing Face of East Asia during the Tertiary and Quaternary*. University of Hong Kong Press, Hong Kong, 3–16.

Simmons, G.C. 1966. *Stream Anticlines in Central Kentucky*. US Geol. Surv. Prof. Paper 550–D, D9–D11.

Smith, A.G. 1982. Late Cenozoic uplift of stable continents in a reference frame fixed to South America. *Nature*, 296, 400–404.

Smith, A.G. and Drewry, D.J. 1984. Delayed phase change due to hot asthenosphere causes Transantarctic uplift? *Nature*, 309, 536–538.

Smith, D.G. ed, 1981. *The Cambridge Encyclopedia of Earth Sciences*, Cambridge University Press, Cambridge. 496p.

Smith, R.B. and Easton, G.P. eds. 1978. Cenozoic tectonics and regional geophysics of the western Cordillera. *Geol. Soc. America Memoir* 152, 1–31.

Soeters, R., 1981. Algunos datos sobre la edadde dos supérficies de erosion en la Cordillera Central de Colombia. *Mem. primo Sem. Cuat. Colombia*, Revista CIAF 6, 525–528.

Sonnenberg, F.P., 1963. Bolivia and the Andes, in Childs, O.E. and Beebe, B.W. (eds), Backbone of the Americas; Tectonic History From Pole to Pole, a Symposium. *American Association of Petroleum Geologists, Memoir* 2, 36–46.

Soons, J.M. and Selby, M.J. 1982. *Landforms of New Zealand*. Longman Paul, Auckland.

Spencer, E.W. 1965. *Geology: a Survey of Earth Science*. Crowell, New York.

Spry, M.J., Gibson, D.L. and Eggleton, R.A. 1999. Tertiary evolution of the coastal lowlands and the Clyde River palaeovalley in southeast New South Wales. *Aust. J. Earth Sci.*, 46, 173–180.

Stamp, L.D. 1950. *An Introduction to Stratigraphy*. 2nd edn. Murby, London.

Stevens, G.R. 1980. *New Zealand Adrift: the theory of continental drift in a New Zealand setting*. A.H. and A.W. Reed, Wellington.

Stewart, J.H. 1978. Basin-range structure in western North America: a review. In Smith, R.B. and Easton, G.P. eds. Cenozoic tectonics and regional geophysics of the western Cordillera. *Geol. Soc. America Memoir*, 152, 1–31.

Stille, H. 1936. The present tectonic state of the earth. *Bull. Am. Assoc. Petrol. Geol.*, 20, 849–80.

Stille, H. 1955. Recent deformation of the Earth's crust in the light of those of earlier epochs. *Geol. Soc. America Spec. Paper*, 62, 171–191.

Stöcklin, J. 1980. Geology of Nepal and its regional frame. *J. Geol. Soc. Lond.*, 137, 34–60.

Stöcklin, J. 1983. Himalayan orogeny and Earth expansion. In Carey, S.W. ed. *Expanding Earth Symposium, Sydney 1981*. Univ. Tasmania, 161–164.

Stöcklin, J. 1989. Tethys evolution in the Afghanistan–Pamir–Pakistan region. In Sengör, A.M.C. ed. *Tectonic Evolution of the Tethyan Region*. Kluwer Academic, Amsterdam, 241–264.

Suggate, R.P. 1982. The geological perspective. In Soons, J.M. and Selby, M.J. *Landforms of New Zealand*. Longman Paul, Auckland, 1–13.

Sugimura, A. and Uyeda, S. 1973. *Island Arcs, Japan and its Environs*. Elsevier, Amsterdam.

Summerfield, M.A. 1991. *Global Geomorphology: An Introduction to the Study of Landforms*. Longman Scientific and Technical, London.

Summerfield, M.A. and Thomas, M.F. 1987. Long-term landform development: editorial introduction. In Gardiner, V. ed. *International Geomorphology 1986*, Part II. Wiley, Chichester, 927–933.

Suslov, S.P. 1961. *Physical Geography of Asiatic Russia*. W.H. Freeman, San Francisco.

Sweeting, M.M. 1955. Landforms in north-west County Clare, Ireland. *Trans. Inst. Br. Geogr.*, 21, 33–49.

Tabbutt, K.D. 1990. Temporal constraints on the tectonic evolution of Sierra de Famatina, northwestern Argentina, using the fission track method to date tuffs interbedded in synorogenic clastic sedimentary strata. *J. Geol.*, 98, 557–566.

Tarling, D.H. and Runcorn, S.K. eds. 1973. *Implications of Continental Drift to Earth Sciences*. Academic Press, London.

Thomas, M.F. 1989. The role of etch processes in landform development II: etching and the formation of relief. *Z. Geomorph.*, 33, 257–274.

Thomas. M.F. 1994. *Geomorphology in the Tropics*. Wiley, Chichester.

Thomson, M.R.A., Crame, J.A. and Thomson, J.W. eds. 1991. *Geological Evolution of Antarctica*. Cambridge University Press, Cambridge.

Thornbury, W.D. 1965. *Regional Geomorphology of the United States*. Wiley, New York.

Thornbury, W.D. 1969. *Principles of Geomorphology*. Wiley, New York.

Tingey, R.J. 1985. Uplift in Antarctica, *Zeitschrift für Geomorphologie Supplement Band*, 54, 85–99.

Trescases, J.J. 1992. Chemical weathering. In Butt, C.R.M. and Zeegers, H. eds. *Regolith Exploration Geochemistry in Tropical and Subtropical Terrains*. Elsevier, Amsterdam, 25–40.

Trimble, D.E. 1980. Cenozoic tectonic history of the Great Plains contrasted with that of the southern Rocky Mountains: A synthesis. *The Mountain Geologist*, 17, 59–69.

Trumpy, R. 1980. *An Outline of the Geology of Switzerland*. Wepf, Basel.

Trunkó, L. 1996. *Geology of Hungary*. Gebruder Borntraeger, Berlin.

Twidale, C.R. 1994. Gondwanan (Late Jurassic and Cretaceous) palaeosurfaces of the Australian Craton. *Palaeogeog., Palaeoclim., Palaeoecol.*, 112, 157–186.

Twidale, C.R. and Harris, W.K. 1977. The age of Ayers Rock and the Olgas, central Australia. *Trans. R. Soc. S. Aust.*, 101, 45–50.

Uddin, A. and Lundberg, N. 1998. Cenozoic history of the Himalayan–Bengal system: Sand composition in the Bengal basin, Bangladesh. *Bull. Geol. Soc. Am.*, 110, 497–511.

Ufimtsev, G.F. 1990. Morphotectonics of the Baikal Rift Zone (U.S.S.R.). *Geogr. Fis. e Dinam. Quaternaria*, 13, 3–22.

Ufimtsev, G.F. 1994. The continental rejuvenated mountain belts. *Geogr. Fis. e Dinam. Quaternaria*, 17, 87–102.

Unruh, J.R. 1991. The uplift of the Sierra Nevada and implications for late Cenozoic epeirogeny in the western Cordillera. *Bull. Geol. Soc. Am.*, 103, 1395–1404.

van Bemmelen, R.W. 1975. Some basic problems in geonomy. In Borradaile, G.J., Ritsema, A.R., Rondeel, H.E. and Somin, O.J. eds. *Progress in Geodynamics*, North-Holland, Amsterdam, 9–20.

van den Berg, J., Klootwjk, C.T. and Wonders, T. 1975. in Borradaile, G.J., Ritsema, A.R., Rondeel, H.E. and Somin, O.J. eds. *Progress in Geodynamics*, North-Holland, Amsterdam, 165–175.

van der Beek, P.A., Braun, J. and Lambeck, K. 1999. Post-Palaeozoic uplift history of southeastern Australia revisited: results from a process-based model of landscape evolution. *Aust. J. Earth Sciences*, 46, 157–172.

van der Gracht, W.A.J.M. van Waterschoot, 1928. The problem of continental drift. In van der Gracht, W.A.J.M. van Waterschoot ed. *Theory of Continental Drift: a Symposium.* Am. Assoc. Petrol. Geol. Tulsa, 1–75.

Van der Hammen, T.H., Werner, J.H. and Van Dommelen, H., 1973. Palynological record of the upheaval of the Northern Andes: a study of the Pliocene and Lower Quaternary of the Colombian Eastern Cordillera and the early evolution of its High-Andean biota. *Palaeolgeogr. Palaeoclim., Palaeoecol.*, 16, 1–22.

Viers, G. 1967. La Quebrada de Humahuaca (province de Jujuy, Argentine) et les problèmes morphologiques des Andes sèches. *Annales de Geographie* 76, 411–433.

Wager, L.R., 1937. The Arun River drainage pattern and the rise of the Himalaya. *Geogr. J.*, 89, 239–50.

Walker, E.H. 1949. Andean uplift and erosion surfaces near Uncia, Bolivia. *Am. J. Sci.*, 247, 646.

Wayland, E.J. 1934. Peneplains and some other erosional platforms. *Ann. Rep. Bull. Uganda Geol. Surv. Dept.* 1933, 77–79.

Webb, J.A., Finlayson, B.L., Fabel, D. and Ellaway, M. 1991. The geomorphology of the Buchan karst – implications for the landscape history of the Southeastern Highlands of Australia. In Williams, M.A.J., De Dekker, P. and Kershaw, A.P. eds. *The Cainozoic in Australia: a re-appraisal of the evidence.* Geological Society of Australia Special Publication 18, 210–234.

Wegener, A. 1929. *The Origin of Continents and Oceans* translated from the fourth revised German edition by J.G. Skerl. Methuen, London.

Weidick, A. 1976. Glaciation and the Quaternary of Greenland. In Escher, A. and Watt, W.S. eds. *Geology of Greenland.* Geological Survey Greenland, Odense, 431–458.

Wellman, H.W. 1975. The obduction–subduction part of the Australian-Pacific plate boundary in New Zealand. *Int. Union Geodesy Geophys. Commission on Crustal Movements*, 10, 50.

Wellman, P. 1987. Eastern Highlands of Australia: their uplift and erosion. *BMR J. Aust. Geol. Geophys.*, 10, 277–286.

Wellman, P. 1988. Tectonic and denudational uplift of Australian and Antarctic highlands, *Zeitschrift für Geomorphologie*, 32(1), 17–29.

Wellman, P. and McDougall, I. 1974. Cainozoic igneous activity in eastern Australia. *Tectonophysics*, 23, 49–65.

Wernicke, B. 1981. Low-angle normal faults in the Basin and Range Province: Nappe tectonics in an extending orogen. *Nature*, 291, 645–648.

Wezel, F.C. 1982. The Tyrrhenian Sea: a rifted krikogenic-swell basin, *Mem. Soc. Geol. It.*, 24, 531–568.

Wezel, F.C. 1988. A young Jura-type fold belt within the Central Indian Ocean. *Boll. Oceanologia Teorica e Applicata*, 6, 75–90.

Widdowson, M. ed. 1997. *Palaeosurfaces: Recognition, Reconstruction and Palaeoenvironmental Interpretation.* Geol. Soc. Special Publication 120, London.

Williams, H. and McBirney, A.R. 1964. Petrological and structural contrast of the Quaternary volcanoes of Guatemala. *Bull. Volcanol.*, 27, 61.

Williams, M.A.J., De Decker, P. and Kershaw, A. P. 1991. *The Cainozoic in Australia: a re-appraisal of the evidence.* Spec. Publ. 18, Geol. Soc. Australia. Sydney.

Wilson, J.T. 1963. Evidence from islands on the spreading of ocean floors. *Nature*, 197, 536–538.

Windley, B.F. 1977. *The Evolving Continents*. Wiley, New York.

Wise, D.U. 1963. Owl Creek Uplift. Keystone faulting and gravity sliding driven by basement uplift of Owl Creek Mountains, Wyoming. *Am. Assoc. Petrol Geol. Bull.*, 47, 1586–1598.

Woodward, L.A. 1977. Rate of crustal extension across the Rio Grande Rift. *Geology*, 15, 269–272.

Woodward, L. A., 1983. Potential oil and gas traps along the overhang of the Nacimiento Uplift, northwestern New Mexico. In Lowell, J. D. ed. *Rocky Mountain Foreland Basins and Uplifts*. Rocky Mountain Assoc. Geologists, Denver, 213–218.

Worzel, J.L. 1976. Gravity investigations of the subduction zone. In Sutton, G.H., Manghnani, M.H. and Moberly, R. eds. *The Geophysics of the Pacific Ocean Basin and its Margin*. American Geophysical Union Monograph, 19. Washington, 1–16.

Wu, F.T. and Wang, P. 1988. Tectonics of western Yunnan Province, China. *Geology*, 16, 153–157.

Yano, T. and Kunisue, S. 1993. Folding mechanisms in the central Nigata Oil Field, central Japan and its origin. *Hokuriku Geology Institute Report* 3, 71–94.

Yano, T. and Wu, G. 1995. Middle Jurassic to Early Cretaceous arch tectonics in East Asian continental margin. *Proc. Int. Symp. Kyungpook National University*, 199, 5177–5192

Yano, T. and Wu, G. 1999. *Basement structure around the Lijiang Basin, Northwest Yunnan, China*. Disaster Prevention Research Institute, Kyoto University.

Yin, A., Nie, S., Craig, P. and Harrison, T.M. 1998. Late Cenozoic tectonic evolution of the southern Chinese Tian Shan. *Tectonics*, 17, 1–27.

Young, A. and Saunders, J. 1986. Rates of surface processes and denudation. In Abrahams, A.D. ed. *Hillslope Processes*, Allen and Unwin, Boston, 3–27.

Zeil, W. 1979. *The Andes; A geological review*. Gebruder Borntraeger, Berlin.

Zhang, D.D. 1998. Geomorphological problems of the middle reaches of the Tsangpo River, Tibet. *Earth Surface Processes and Landforms*, 23, 889–903.

Zhu, Z. 1997. A coupled climatic–tectonic system – the Tibetan Plateau and the east Asian continent. In Jablonski, N.G. ed. *The Changing Face of East Asia during the Tertiary and Quaternary*. University of Hong Kong Press, Hong Kong, 303–312.

Index

Acholi Surface 39–40
active margins 14
Adriatic-Po Basin 75–6
Africa, inselberg 39
African Surface 39, 49, 205
Agulhas Slump 150–1
Aldan Plateau 147
alluvial plains 227
Alps 8, 285
Altai Mountains 145–6
altiplanation surfaces 233–4
Amazon 118
Amazon Basin 8
Andean Cordillera fault spreading 160;
  orogeny 6
Andes 112–27; Avenue of Volcanoes
  112; drainage patterns 118; faults
  119–21; Lake Titicaca 112;
  mushroom tectonics 160; planation
  surfaces and uplift 112–18; plate
  tectonic explanations 121–6;
  Pleistocene stratovolcanoes 119;
  puna surface 114–18; Quaternary
  119; summary 127; symmetry
  287–8; volcanic plateau 118–19;
  volcanoes 118–19
andesite 168, 174, 177–8
andesite line 178–9
anhydrite 153
Antarctica 268–9; mountains on
  passive margins 213–14
antecedent drainage 254, 285–6
Apennines 66–82, 229; Adriatic-Po
  Basin 75–6; drainage patterns 73–5;
  emergence and uplift 72–3; Ionian
  Basin 75; lubricant 153; nappes
  153–4; outline geology 66–70;
  planation surface 71–2; plate
  tectonics explanations 77–82; Po
  valley and subduction 76–7;
surrounding seas 75–6; Tyrrhenian
  Basin 75; volcanoes 75
Appalachian Mountains 2; drainage
  247; main features 206–9;
  mountains on passive margins 193,
  206–10; plate tectonic explanations
  209–10
areal volcanism 179; lava cones 179;
  maars 179; scoria cones 179
Argentina-Chile Cordillera 117
Arizona, gneiss mantled domes 191–2
Arkansas River 103
Arun River 132, 285
Ascension Island 172
ash 169
Asian mountains 128–49; Altai
  Mountains 145–6; continental East
  Asia 148–9; Himalayas 130–8;
  Japan 147–8; Karakoram 147;
  Kunlun Mountains 142–4; other
  Asian mountains and tectonic
  features 142–7; plate tectonic
  explanations 138–42; Tarim Basin
  144; Tibetan Plateau (Qinghai-
  Xizang Plateau) 128; Tien Shan
  144; Transbaikal Mountains 146–7;
  Turfan Depression 144–5
asymmetry of passive margins 215–17
Atlantic Ocean 13; Mid-Ocean Ridge
  175–6
Atlas mountains 94–5; Rif Atlas 95;
  Saharan Atlas 95; Tel 95
Australia, fault spreading 160
Auvergne, volcanoes 179–80
Avenue of Volcanoes, Andes 112
Ayers Rock (Uluru) caprock plateau
  28–30; inselberg 39

Baetic Cordillera 86
Baikal rift, fault block mountains 52–3

Bald Rock (Australia), granite dome 40
Bangalore Plain 236
Baram Delta 152
Barcena 178
basalt 168, 199; Continental Basalts 168–9; dating 238; Island Arc Basalts 169; Mid Ocean Ridge Basalts (MORBS) 168–9; Ocean Island Basalts (OIBS) 168; plateaus 172–4
Basin and Range fault block landscape 43–6
Bengal Fan 137–8, 150–1
Benioff zone 16
Bighorn Mountains 100
Bighorn River 105
Black Canyon 103, 105
blocks 169, 271
Blue Mountains (NSW) 8
bombs 169
Bornhardt 39
bourrelets marginaux 193
Brahmaputra River 132, 136, 150
Branden der Deckenstirne 279
Brazil, mountains on passive margins 210–11
Brooks Range 96, 106
Buchan, John 134–6
Burg, J. P. 5

Cainozoic 18–19
caldera 174–5
Caledonian Mountains 2
Californian Great Valley 96
cambers 161
Canadian Rockies 106
Canadian Shield 8
Cantabrica Range (Cordillera Cantabrica) 85
caprock plateaus 21–30; Ayers Rock 28–30; Kimberley and Carr Boyd Ranges 23–7; mesas (tepuis) of Venezuela and Guyana 21–3; Ozark Plateau 27–8
carbonatite 180
Caribbean arc 18; ocean-ocean collision 18
Carpathians 1, 86–7; Eastern Carpathians 89; plate tectonic explanations 90–1; Southern Carpathians 89–90; Western Carpathians 88–9
Carr Boyd Ranges 23–7

Cascade Range 96, 108; volcanoes 108, 178
Castille 85
Caucasus 91–3; denudation chronology 92–3; Mt Elbrus 91–2; planation surfaces 91–2
Cenozoic 18–19
Central America 109–10; volcanoes 109–10
Central Cordilleras 85–6
Chile 121
Chimborazo 170
Chuan-Jin fold belt 154–5
cinders 169
cirques 31–2
classification and distribution of mountains 271–3
climate 262–70
climatic effects: global 265–70; regional 262–5
Clyde River 201
Coast Ranges (USA/Canada) 106–8
Colima 178
Colorado and other plateaus 97–100; Bighorn Mountains 100; Mogollon Rim 99; mushroom tectonics 160; Owl Creek uplift 100; Sierra Madre 97; valley anticlines 162; Yukon plateau 97
Colorado River 103, 105; valley anticlines 162
Columbia plateau 96
Columbia River, basalt plateau 169, 172–3, 178
Compassberg 194
compression, crustal shortening 274–5
continent-continent collisions (Himalayan type) 15
continent-ocean collision 15
Continental Basalts 168–9; ash 169; blocks 169; bombs 169; cinders 169; lapilli 169; pyroclastics 169; scoria 169
continental drift 12–13
continental East Asia 148–9
convection currents 13
coral 172, 286
Coral Sea 199
Cotton, Charles 311–12
Crater Lake 108
crustal shortening 274–5
crustal thickening 18
cuesta, bevelled 24–7, 229

Danube 89
dating of planation surfaces 237–9;
    drainage patterns 239; fission track
    data 239; offshore or basin
    sediments 239; overlying rock 238;
    relative height 237; soils and
    weathering profiles 238
Dayman Dome 189
Dead Sea 46, 49
Death Valley 41–2
Deccan traps, basalt plateau 173–4,
    203, 227
depositional plains 227
Devil's Tower, volcanic intrusions 184
differential load structures 161–6;
    cambers 161; North-west Dolomites
    164–6; valley bulges 161–2; valley
    inclines of Colorado Plateau 162;
    valley inclines of Himalayas 162;
    valley inclines of Zagros Mountains
    162–4
Dimbovita River 89
dipping sedimentary rocks 243–6
Disturbed Zone, Rocky Mountains
    103
divide location changes 251–2
dolerite 184
dolomite 152–3
Dolomites: differential load structures
    164–6; gravity tectonics 164–6;
    Marmolada complex 164–5;
    morphological development 237
downfaulting/downwarping at margin
    224–5
drainage patterns 21–7; dating 239;
    dipping sedimentary rocks 243–6;
    folded sedimentary rocks 246–7;
    horizontal sedimentary rocks 242–3;
    interpretation 241–7; relationships
    262; simple patterns 242; structural
    control 242–7; summary 262
Drakensberg 8; great escarpments 195,
    216; mountains on passive margins
    193, 204–5
Dronning Maud Lane 214–15
Duero 85
dykes 184
Dynamic Equilibrium 28

Eagle River 105
Earth volume 10–11
earthquakes, plate boundaries 13–14
Eastern Carpathians 89–90
Eastern Ghats 217

Eastern Highlands, mountains on
    passive margins 193, 199–202
Ebor Volcano 201
Eocene 18–19
epeirogeny, definition 3, 6–8
Eras 18–19
erosion 241, 265–70; surface 231
erosion of volcanoes 180–3; dating
    180; glaciers 183; stages 182
eruption types 170–1; areal 170;
    fissure 170
etchplains 232–3
Europe, symmetry 288–9
European Alps Gipfelflur 60; Insubric
    Line 64; nappe theory 60; planation
    surface 63; plate tectonic
    explanations 63–6; root zone 62;
    Tauern window 61–2; Tethys Sea
    61; Vienna Basin 62
European mountains 60–95; Apennines
    66–82; Atlas mountains 94–5;
    Caucasus 91–3; European Alps
    60–6; Iberian Peninsula 82–91;
    Sudeten 94; Urals 93–4
exhumed plains 234

fault block mountains 41–59; Baikal
    rift 52–3; Basin and Range province
    43–6; New Zealand mountains
    57–9; Oregon type fault block
    landscapes 41; Owen Stanley
    Mountains, Milne Bay 55; Rhine
    Graben 53; rift valleys 46–52;
    Ruwenzori Mountains 53–5; Shanxi
    Mountains and Shanxi Graben
    system 56–7; Sierra de Famatina 59;
    Sierra Nevada 41–2
fault spreading 160
faulting effects 261–2
faults, Andes 119–21
Fen River 57
feral relief 31
Fergusson Island 190
Finland 8–9
fission track data 239
Fitzroy River, antecedent drainage 254
fluvial modifications 247–9
folded sedimentary rocks 246–7
folding causes: applied force and body
    force 275–6; lateral compression vs
    gravity tectonics 276; thin skinned
    tectonics and wedge 276–8
folding, nappes and decollement
    152–9; Apennine nappes 153–4;

Chuan-Jin fold belt 154–5; Jura 154; lubricants 152–3; Naukluft Mountains 158–9; Papuan fold belt 155–7; Pelvoux Massif 155; Taiwan 157–8
folds/gravity structures 273–81; compression, crustal shortening 274–5; field evidence 273–4; folding causes 275–8
foothill growth by gravity push 167
Ford, M. 5
fossils 18

Ganges River 150, 227
Geikie, J. 9
geological time 18–20
geology, Himalayas 131–2
geomorphologist 1
geosynclines 9
Giant Rhinoceros, Tibetan Plateau (Qinghai-Xizang) 128
Gipfelflur (summit level) 34, 229–30; European Alps 60–1
glaciated mountains 31–4; cirques 31–2; nunataks 31
glacis 231
gneiss mantled domes 257, 261; Arizona 191–2; Papua New Guinea 186–91
Gondwanaland 15, 131, 205; fossils 132
Goodenough Island, gneiss dome 186, 189–91
Gough Island 171
Grand Canyon 99, 105–6, 227; basalt plateau 173
granite 168
granite in fold belts 185
granite mountains 184–6; hornfels 185; Mole Granite 185–6; Mt Duval 185–6; Mt Kinabalu 186
gravity collapse structures 159–60
gravity and isostasy 282–5; Airy's theory 282; Bouguer gravity anomaly 282–5; Pratt's theory 282
gravity spreading 160, 278–81; fault spreading 160; mushroom tectonics 160; post-uplift gravity tectonics criteria 280–1; pre-planation gravity sliding criteria 279–80; recognition criteria 279; symmetry, double subduction 279; valley bulges and valley anticlines 279

gravity structures 150–67; folding, nappes and decollement 152–9; post-uplift gravity structures 159–67; underwater folding and faulting 150–2; *see also* folds/gravity structures
Great Bight 217
great escarpments 193–9
Green River 103, 105
greenhouse gases 267, 269
Greenland, mountains on passive margins 213–14
Gunnison River 103, 105
Guyana, mesas (tepuis) 21–3
guyots 172, 286
gypsum 153

Hawaiian Islands 176
Heart Mountain 103
Henry Mountains 180; volcanic intrusions 184
high plains 229
High Plains (Australia) 1
High Tatra 88
Highveld (southern Africa) 1
Himalayas 130–8; age of uplift 137–8; geology 131–2; High Himalayas 130; Indus-Tsangpo suture 130; Lower Himalayas 130; Main Boundary Fault 130; Main Central Thrust 130; Namcha Barwa 130; Nanga Parbat 130; plate tectonic explanations 138–42; rivers and drainage patterns 132–6; valley anticlines 162; vertical uplift 136
horizontal sedimentary rocks 242–3
Hornad River 87
hornfels 185
horst 41
hotspot volcanoes 176–7
Hsü, K. 4
Hungary, volcanoes 180

Iberian Peninsula 82–91; Baetic Cordillera 86; Cantabrica Range (Cordillera Cantabrica) 85; Carpathians 86–7; Central Cordilleras 85–6; Eastern Carpathians 89; plate tectonic explanations 90–1; Pyrenees 83–5; Southern Carpathians 89–90; Western Carpathians 88–9
Ice Age 269
ice caps 268–9

Iceland: ocean ridge 176; spreading sites 15
igneous rocks, magma and lava 168–9; andesite 168; basalt 168; granite 168; ignimbrite 168; rhyolite 168
ignimbrite 118–19, 168; andesite 174; dating 238; nuées ardentes 174; plains and plateaus 228; plateaus 174; rhyolite 174; welded tuffs 174
Indian Shield 131
Indonesian arc 178; island arc subduction 18
Indus River 132, 138; sediment 227
inselbergs and rock domes 34–40
Insubric Line, European Alps 64
intraplate volcanoes 180
introduction 1–20; book outline 11–12; geological time 18–20; plate tectonics 12–18; study of mountains 11; terminology 4–9; theories 9–11
Ionian Basin 75
Iron Gate Gorge 89
island arcs 178; subduction 16
island chains/hotspot volcanoes 176–7
isostasyand mountain building 285–6; and volcanoes 286; see also gravity and isotasy

Jackson, J. A. 5
Japan 147–8, 152; ignimbrite plateaus 174; island arc 178; Kitakami Mountains 147–8
Jarrahwood Axis 202
Jones, O. T. 314
Jorullo 178
Jura Mountains 65–6; folds 154, 276; salt (halite) 152

Kalahari Basin 204, 216
Karakoram 147
Karroo 205; basalts 169, 174
Kathmandu Nappe 131
Kilimanjaro 180, 183
Kimberley and Carr Boyd Ranges caprock plateau 23–7; caprock plateaus 23–7
King, P. B. 5
Kitakami Mountains 147–8
Klamath Mountains 96, 107
Koolau dome 264
Krakatoa 268
Kunlun Mountains 142–4, 149; Pulu 142

laccoliths 184
Lake Aiding, Turfan Depression 144–5
Lake Baikal 149; Transbaikal Mountains 146–7; see also Baikal rift
Lake District 261
Lake Eyre 216
Lake Titicaca 112
Lake Toba 174
lapilli 169
Laramide orogeny 96–7
Laramie River 105
Lassen Peak 108, 178
lateral tectonic extrusion 166–7
lava cones 179
lava plains 227–8
Lewis Thrust, Rocky Mountains 103
Limpopo River 204
Long Plain Fault 160
lubricants anhydrite 153; dolomite 152–3; gypsum 153; salt (halite) 152; shales 153

maars 179
Mackenzie Mountains 96
magma 168–9, 174–5
marginal bulges 193
marginal swells 193–4; bourrelets marginaux 193; causes 225–6; marginal bulges 193; Randschwellen 193–4; rim highlands 193
marine erosion 286; see also plains of marine erosion
Matterhorn (Switzerland), glaciated mountains 32
Mauna Kea 169
Maybole Mountain 180–1
Meckering Line, great escarpments 195, 202
median plateaus and outfacing ranges 287
Meiji Seamount 176
mesa 229
mesas (tepuis) of Venezuela and Guyana 21–3
Mesozoic 18–19
Midway 176
Milankovich cycles 265, 269
Mio-Pliocene 20
Miocene 18–19; Mio-Pliocene 20; Upper Miocene 20
Mississippi, sediment 227
mobile belts 9–10
Mogollon Rim 99
molasse deposits, Himalayas 137–8

Mole Granite 185–6
MORBS *see* basalt
morphotectonic features, mountains on passive margins 217–18
Mount Conner, caprock plateau 21
Mount Cook, glaciated mountains 33
Mount Elie de Beaumont, glaciated mountains 33
Mount Fuji 178
mountain building: definition 5; mechanisms 307–10; plate tectonics 297
mountain building and climate change 262–70; global climatic effects 265–70; regional climatic effects 262–5
mountain tectonics 271–310; classification and distribution of mountains 271–3; folds, fold belts and gravity structures 273–81; gravity and isostasy 282–5; isostasy and mountain building 285–6; isostasy and volcanoes 286; mountain building mechanisms 307–10; mountain building ultimate mechanism 310; plate tectonics and mountains 296–301; Plio-Pleistocene uplift 303–7; still-stands and planation speed 301–3; summary of planation and uplift in mountain areas 304–6; symmetry of mountains 286–96; twenty causes of tectonic uplift 308
Mountains of the Moon *see* Ruwenzori Mountains
mountains on passive margins 193–226; absence of morphotectonic features 217–18; asymmetry of passive margins 215–17; cause of marginal swells 225–6; causes of uplift 219–24; downfaulting/downwarping at margin 224–5; examples 199–202; great escarpments 194–9; marginal swells 193–4; time of uplift 218–19
Mt Duval 185–7
Mt Elbrus 91–2, 169
Mt Elgon 180–1
Mt Etna 170, 184
Mt Hood 178
Mt Katmai 175
Mt Kinabalu 186, 188
Mt Kosciusko 193; glaciated mountains 34

Mt Mazama 174, 178
Mt Monadnock, granite dome 40
Mt Napak 180
Mt Rainier 178, 183
Mt Robson 106
Mt St Helens 108
Mt Shasta 108
multiple planation surfaces 235–7; multiple plains of same age 237
Murray Basin 8, 227
mushroom tectonics 160

Namcha Barwa 130
nappes 11–12; Himalayas 131–2; *see also* folding, nappes and decollement
Naukluft Mountains 158–9; dolomite 153
Nawakot Nappe 131
Neogene 18–20
Nepal 135–6
New Zealand mountains, fault block mountains 57–9
Niger Delta 151–2
North America *see* Western North America
North European Plain 88
North Platte River 105
North-west Dolomites 164–6
Northamptonshire, differential load structures 161
Novaya Zemlya 93
nuées ardentes 174
nunataks 31

obduction 16–18
ocean ridges 175–6
ocean surveys 12–18
oceanic volcanoes 171–2; Hawaiian Islands 172; pillow lava 172; seamounts 171–2
offshore/basin sediments dating 239
OIBS *see* basalt
Oldoinyo Lengai 180
Olgas 28
Oligocene 18–19
Olt River 87
Oregon type fault block landscapes 41
Orizaba 110
orogeny, definition 3–6
Oslo Graben, shales 153
Ouachita Mountains 109
overlying rock dating 238
Owen Stanley Mountains 155; fault block mountains 55

Owl Creek uplift 99–100
Ozark Plateau 27–8; Dynamic
　Equilibrium 28

Pacific, symmetry 289–94
Pacific Border mountains 96
Pacific margin 177–8; island arcs 178
Pacific Ridge 175
Pakaraima Mountains: Angel Falls 22;
　Kaitur Falls 22; mesa 21
Pakistan 147, 164
Palaeogene 18–19
palaeoplains 234
Palaeozoic 18–19
Paleocene 18–19
Pannonian Basin 87
Papua New Guinea: fold belt 155–7;
　obduction 16–18; Ultramafic Belt
　17; volcanoes 178
Paracutin volcano 267
Parana 217; basalts 169, 174
Paricutin 178
passive margin volcanoes 179–80
passive margins 14, 273; asymmetry
　215–17; see also mountains on
　passive margins
Patagonia 117
pediplains 231–2
Pelvoux Massif 155
peneplains 231
Periods 18–19
Perm 93
Phanerozoic 18–19
Pico da Bandiera 195
plains 227–8; depositional 227;
　ignimbrite plains and plateaus 228;
　lava 227–8; structural 228
plains of marine erosion 234–5
plains/planation surfaces,
　drainage/climate 227–70; dating of
　planation surfaces 237–9; drainage
　pattern relationships 262; drainage
　patterns 241–7; exhumed plains
　234; general surface lowering 241;
　mountain building and climate
　change 262–70; multiple planation
　surfaces 235–7; palaeoplains 234;
　plains 227–8; plains of marine
　erosion 234–5; planation surfaces
　228–34; river valley landscapes
　247–62; significance 240–1; tectonic
　deformation 239–40
planation 1–4
planation surfaces 228–34;

altiplanation surfaces 233–4;
　etchplains 232–3; evidence 228–30;
　pediplains 231–2; peneplains 231;
　significance 240–1; varieties 230–1
planation and uplift, summary 304–6
planation and uplift in mountain
　areas, summary 304–6
plate tectonics 1–2, 9, 175, 271;
　outline 12–18; spreading sites 15;
　subduction sites 15–18; Wegener,
　A. 12; see also mountain tectonics
plate tectonics and mountains
　296–301; difficulties 298–300;
　mountain building 297; orthodoxy
　300–1; Wilson Cycle 297–8
plateaus and erosional mountains 1–4,
　21–40; ancient plateaus 31; caprock
　plateaus 21–30; glaciated mountains
　31–4; inselbergs and rock domes
　34–40; rugged mountain ranges 31
Playfair's Law 242
Pleistocene 18–19
Plio-Pleistocene 20; uplift 303–7
Pliocene 18–19
Po valley and subduction 76–7
polyorifice volcanism 179
Popacatapetl 109–10, 178
post-uplift gravity structures 159–67;
　differential load structures 161–6;
　foothill growth by gravity push
　167; gravity collapse structures
　159–60; gravity spreading 160;
　lateral tectonic extrusion 166–7
Precambrian 18–19
Prince Charles Mountains 215
puna surface, Andes 114–18
Pyrenees 83–5
pyroclastics 169

Qinghai-Xizang Plateau 128
Quaternary 18–19
Quinzang movement 129–30

radial drainage 257–61
Randschwellen 193–4
Recent 18–19
relationships of plains/mountains to
　folding/non-folding areas 8–9
relative height dating 237
reversed drainage 254–7
Rhine Graben, fault block mountains
　53
rhyolite 168, 174
rift valleys 46–52, 180

rim highlands 193
Rincon Mountain 191–2
river capture 249–51
river valley landscapes 247–62;
  antecedent drainage 254; divide
  location changes 251–2; faulting
  effects 261–2; fluvial modifications
  247–9; radial drainage 257–61;
  reversed drainage 254–7; river
  capture 249–51; superimposed
  drainage 252–4
rock domes *see* inselbergs and rock
  domes
Rocky Mountains 96, 100–3;
  Disturbed Zone 103; fault spreading
  160; Heart Mountain 103; Lewis
  Thrust 103; mushroom tectonics
  160
root zone, European Alps 62
Roraima Formation 22
rugged mountain ranges 31; feral
  relief 31
Ruwenzori Mountains, fault block
  mountains 53–5

St Helena 171–2
salt (halite) 152
Santiago River 118
Sayan Mountains 146
Scandinavia, mountains on passive
  margins 211–13
science and origin of mountains
  311–15; causes 313–14; modelling
  and ground truth 315; orthodoxy
  314–15; physics vs landscape 313;
  theories and bandwagons 312
scoria 169, 171, 179
Scottish Highlands 9
seamounts 170–2, 176–7, 272, 286;
  guyots 172
Serro do Mar 210; great escarpments
  195
shales 153
Shanxi Mountains and Shanxi Graben
  system, fault block mountains 56–7
Sicily 68
Sierra de Famatina, fault block
  mountains 59
Sierra Madre Occidental 96–7
Sierra Nevada 9, 86, 262, 264; fault
  block mountains 41–2
Sierra Nevada (USA) 96
simple subduction (Andes type)
  15–16

Snake River 9; basalt plateau 172–3;
  lava 227
Snowdon, glaciated mountains 31
Snowy Mountains, great escarpments
  195
soils and weathering profiles 238
Solomon Islands, island arc subduction
  16
South Sandwich Islands: island arc
  subduction 16; ocean-ocean collision
  18
South-west Australia, mountains on
  passive margins 202–3
Southern Africa, mountains on passive
  margins 204–6
Southern Carpathians 89–90
spreading sites 13–15
Stanvoy Ranges 146
Steens Mountain, fault block 41
still-stands and planation speed 301–3
Stille, H. 8
Stone Mountain, granite dome 40
structural plains 228
study of mountains 11
subduction 1, 13
subduction sites 15–18; continent-
  continent collisions 15; continent-
  ocean collision 15; crustal
  thickening 18; island arc subduction
  16; obduction 16–18; ocean-ocean
  collision 18; simple subduction
  15–16
subsidence and caldera collapse 174–5
Sudeten 94; Walbrzych Upland 94
Sugarloaf Mountain, inselberg 39
superimposed drainage 252–4
surface lowering 241
Surtsey 172
symmetry of mountains 286–96; Andes
  287–8; conclusion 295–6; Europe
  288–9; lack of symmetry 294–5;
  median plateaus and outfacing ranges
  287; Pacific 289–94

Table Mountain, caprock plateau 21
tableland 229
Tablelands (southern Africa) 1
Taiwan 157–8
Tajo 85
Taklimakan Desert 144
Tamboro 268
Tamilnad Plain 235
Tarim Basin 144; Taklimakan Desert
  144; Tien Shan 144

Tasman Sea 199, 201
Tasmania, glaciated mountains 33–4
Tauern window, European Alps 61–2,
  167
Tawonga Fault 160
tectonic deformation of planation
  surfaces 239–40; *see also* mountain
  tectonics
tectonic uplift, twenty causes 308
terminology 4–9; epeirogeny 6–8;
  orogeny 4–6; plains/mountains and
  folding/non-folding areas 8–9;
  uplift, orogeny and mountain
  building 4
terrestrial volcanoes 171; cones 171;
  scoria cones 171; shield volcanoes
  171; stratovolcanoes 171
Tertiary 18–19
Tethys Sea, European Alps 61
theories 9–11
Tibetan Plateau (Qinghai-Xizang) 1,
  128; climatic effects 262–70; Giant
  Rhinoceros 128; Lunpola basin 128;
  Main Surface 128; Peak Surface 128;
  planation surfaces 128–30; plate
  tectonic explanations 138–42,
  271–3; uplift (Quinzang movement)
  129–30
Tien Shan 144, 149
Tierra del Fuego 117
Transantarctic Mountains 214–15;
  great escarpments 195
Transbaikal Mountains 146–7; Lake
  Baikal 146–7; Sayan Mountains
  146; Stanvoy Ranges 146
Transverse Ranges 96, 107
Tristan da Cunha 172
Troodos Mountains, obduction 18
Tsangpo River 136
Turfan Depression 144–5; Lake Aiding
  144–5
Tweed Volcano 201
Tyrrhenian Basin 75

Uinta Mountains 99, 105; Uplift 103
Uinta Plateau, mushroom tectonics 160
Uluru *see* Ayers Rock
underwater folding and faulting
  150–2; Agulhas Slump 150; Bengal
  Fan 150–1; Niger Delta 151–2
uplift, orogeny and mountain
  building, definition 4
Upper Miocene 20
Urals 93–4

Vah River 87
valley anticlines 279; Colorado Plateau
  162; Himalayas 162; Zagros
  Mountains 162–4
valley bulges 161–2, 279
Venezuela Gran Sabana 22–3; mesas
  (tepuis) 21–3
Vesuvius 171
Vienna Basin 62, 184
volcanic cones 171–2, 257; complex
  171; monogenetic 171; oceanic
  volcanoes 171–2; terrestrial
  volcanoes 171
volcanic intrusions 184; dolerite 184;
  dykes 184; laccoliths 184; volcanic
  necks 184
volcanic landforms 171–4; volcanic
  cones 171–2; volcanic plateaus
  172–4
volcanic necks 184
volcanic plateaus 172–4; basalt
  plateaus 172–4; ignimbrite plateaus
  174
volcanoes 4, 13–14, 271; Andes
  118–19; Apennines 75; ignimbrites
  118–19; obduction 18;
  stratovolcanoes 118–19; *see also*
  volcanoes and granite mountains
volcanoes distribution 175–80;
  andesite line 178–9; areal volcanism
  179; intraplate volcanoes 180; island
  arcs 178; island chains/hotspot
  volcanoes 176–7; ocean ridges
  175–6; Pacific margin 177–8;
  passive margin volcanoes 179–80;
  rift valleys 180
volcanoes and granite mountains
  168–92; distribution of volcanoes
  175–80; drainage disruption 184;
  erosion of volcanoes 180–3; eruption
  types 170–1; examples of granite
  mountains 185–6; gneiss mantled
  domes 186–92; granite in fold belts
  185; granite mountains 184–5;
  igneous rocks, magma and lava
  168–9; major volcanic landforms
  171–4; subsidence and caldera
  collapse 174–5; volcanic intrusions
  184; volcanic mountains 169–70

Waianae dome 264
Walbrzych Upland, Sudeten 94
weathering 241, 265–70
Wegener, A. 12

welded tuffs 174
Western Australia 9
Western Carpathians 88–9; High
  Tatra 88; North European Plain 88;
  Pannonian Plain 88
Western Ghats 211, 217; great
  escarpments 195; mountains on
  passive margins 193, 203–4
Western North America 96–111;
  Canadian Rockies 106; Cascade
  Range 108; Central America
  109–10; Coast Ranges 106–8;
  Colorado and other plateaus 97–100;
  drainage anomalies (Rockies-plateau
  belt) 103–6; Laramide orogeny
  96–7; Ouachita Mountains 109;

plate tectonic explanations 110–11;
  Rocky Mountains 100–3
Western Victoria Plains 9
Wilson Cycle 297–8
Wilson, J. Tuzo 297

Yellow River 57, 129
Yukon plateau 97
Yunnan Province 148

Zagros Mountains: salt (halite) 152;
  valley anticlines 162–4; Zagros
  Crush Zone 163; Zagros Fold Belt
  163
Zangpo Basin 137
Zoige Basin 130